2024 Surface Preparation and Cleaning Conference

Chandler, Arizona, USA
24 - 25 April 2024

ISBN: 979-8-3313-1298-5

Printed from e-media with permission by:

Curran Associates, Inc.
57 Morehouse Lane
Red Hook, NY 12571

Some format issues inherent in the e-media version may also appear in this print version.

Printed by Curran Associates, Inc. (2025)

Surface Preparation and Cleaning Conference (SPCC)
c/o Mark Thirsk
Linx Consulting-PO Box 384
Mendon, MA 01756-0384
USA

Phone: 617-273-8837

mthirsk@linx-consulting.com

Additional copies of this publication are available from:

Curran Associates, Inc.
57 Morehouse Lane
Red Hook, NY 12571 USA
Phone: 845-758-0400
Fax: 845-758-2633
Email: curran@proceedings.com
Web: www.proceedings.com

TABLE OF CONTENTS

PRESENTATIONS

Performance Characterization of Surface Enhanced Particle Sizing for Reliable Detection of Sub-10nm Particles on 300mm Wafer 1
A.O. Altun, M.K. Tripp, K. Mehrabi, R. Coulthard, M. Litchey, G. Van Schooneveld, L. Zazzera

Cu Pillars: Challenges and Solutions to Strip Thick Photoresist 11
A.D. De Zeria, I. Saucedo, J. Taddei, P. Tyler, E. Altamirano-Sanchez, N. Reddy

Real Time Analysis of PPT Levels of Organics in UPW 22
B. Dufek, T. Herek, S. Ketkar, A. Schultz, D.R. Wiederin

Molybdenum Wet Etchant with Dual Dissolver System for 3D NAND Fabrication 33
C.W. Wang, S.H. Yeh, G. Cleveland, R. Siegelman, J. Yang, J. Kim

Suppression of Lateral Etching of Work Function Metal for PN Boundary Scaling 41
D. Ueda, H. Kawarazaki, T. Nakano, T. Tanaka, A. Harumoto, J.G. Lai, E.A. Sanchez

Extreme Silicon Thinning for Back Side Power Delivery Network Application: Need for a Highly Selective Alkaline Silicon Etch Stopping on SiGe Etch Stop Layer 51
F. Sebaai, R. Loo, A. Jourdain, E.A. Sanchez, E. Beyne, H. Kawarazaki, T. Nakano

High Selective Wet Etching of Different SiGe Layers for CFET Structures with Selectipur E-2504 Solution 62
F.J.L. Villanueva, A. Klipp, F. Sebaai

Wet Etching Mechanism of Ruthenium: Effect of Oxidants and Inhibitors 73
H. Park, H. Ha, T. Jung, Y. Ko, Y. Cho, W. Cho, A. Klipp, H. Chae

Modeling and Simulation of Boron Removal from Water Using N-Methylglucamine 81
I. Basu, N. Kumar, N. Kulkarni, A. Bhabhe

Highly Uniform PolySi Channel Etch 90
J. Ke, O. Liu, J. Yang, K.S. Chin, J. Kim

Investigation of Surface Characterization of 4H-SiC for Better Cleaning Performance 98
Y.J. Ra, J. Lee, J. Yang, T.U. Kim, T.G. Kim, J.G. Park

Use of Hybrid Metrology to Develop Holistic Filtration Solutions in Hydrogen Peroxide 113
K. Maharjan, N. Williams, S. Huang, S. Sampath, B. Dufek, S. Ketkar, A. Schultz

Quantitation of Potential Critical Organic Contaminants from Polymer Components 121
L. Zazzera, A. Chan, G. Van Schooneveld, A.O. Altun, B. Mcintosh

Effect of SDS Surfactant in H2-DIW on Cavitation Bubble Dynamics in Ultrasonic and Megasonic Fields 127
M.J. Khan, J. Hazarika, Y.H. Cho, W. Cho, A. Klipp, J.G. Park

Plasma-Based Dry Clean Process Monitoring by Exhaust Gas Analysis with ToF-MS 141
M. Kim

Backside Metallic Contamination Removal Focused on Noble Metals Using an Halogen-Based Solution 147
M.C. Pignie, L. Gabette, V. Loup, L. Magnin, J. Daviot, M. Audouin

Dummy Poly Gate Etch Clean Challenges in Advanced Nanosheet Devices 154
M. Sankarapandian, W.S. Lo, A.V. Ramirez, S. Rimal, S. Reynoso, M. Medikonda, C.J. Yang, N. Flores, X. Liu, T. Hinode, H. Zhang

Superiority of Bevel Nozzle Dispense Spin Over Backside Center Dispense Spin for Wafer Bevel Cleaning: Insights from Computational Fluid Dynamics 162
R. Dong, A. Fujita, J. Nogami, D. Bassett, I. Simms, A. Rotondaro, J. Barnett

Investigation of the Water Scatter in Single Cleaning Process by Deriving a Splash Parameter 173
S. Baek

A Study on the Causes of Carbonyl Sulfide Residue by Etching and Cleaning Process Interaction 179
Y. Kim

Etching of Tungsten Via a Combination of Thermal Oxide Formation and Wet Chemical Oxide Dissolution 186
A. Pacco, T. Nakano, J. Loyo-Prado, K. Lai, H. Kawarazaki, E.A. Sanchez

Enhanced Surface Activation Through Advanced Hydrogen Peroxide Delivery 198
A. Alvarez

The Intersection of Semiconductors and Governments: Increased Funding, Support…and Regulations 203
K.J. McLaughlin

Is Ultrahigh Vacuum Technology Potential Approach to Prepare Industrial Si Surfaces? 213
Z.J. Rad

Insights into Hydrophobicity Or Hygroscopicity of Non-Volatile Residue 222
D. Oberreit, M. Walker, J. Thomas

Contamination Monitoring on Photomask Substrates with TXRF 233
S. Mezhenny, M. Slaton, B. Buford, M. Beebe, H. Kono, M. Nakanishi, S. Kikuta, M. Kuhn

Evolution and Characterization of PVA Brush in Post CMP Cleaning for 10nm Below Metal CMP Process 246
J.G. Park

POSTERS

Sustainable Pyrrolidinone-Based Solvents Refinement for Wet Cleaning Process 272
C.H. Ko, J. Powell, K. Hampton

Study of Quaternary Ammonium Hydroxide (QAH) for Photoresist Applications 273
H. Zhou, V. White, K. Zhang

A Copper (Cu) Post-CMP Cleaning Formulation for Both Wafer and Pad Clean 274
J. Cheng, P. Sun, W. Lin, W.E. Huang, P. Hsu

Application of TMAH in Semiconductor Wet Etching 275
K. Liu, Z. Ma, Y. Wu, X. Zhang, J. Wang, S.A. Henry

Development of a Wafer Heating Stage for Thermal Desorption Spectroscopy 276
X. Zou, M. Fisher, H. Gotts

Conjointly Address Stringent Device Specifications, Long Batch Life Considerations and HSE/ESG Standards for Specific Copper Seed Wet Etching Applications 277
J. Daviot, A. Passelande

Reduced Chemical Consumption and Metal Loss for Polymer Clean by Batch Spray 278
B. Hammerl

Author Index

Performance characterization of surface enhanced particle sizing for reliable detection of sub-10nm particles on 300mm wafer

Ali O. Altun, Marie K. Tripp, Kamyar Mehrabi, Riley Coulthard
Mark Litchey, Gary Van Schooneveld, Larry Zazzera

UNISERS UNIVERSAL CONTAMINATION ANALYZER

CT Associates, Inc.

Outline

- General technology introduction
  - Surface enhanced particle sizing (SEPS)
  - Surface enhanced Raman speciation (SERS)
- UNISERS lab suite
- Split-lot experiment: Correlation with liquid nanosizer (LNS)
- UNISERS contamination classification roadmap

2

Why do we need surface-enhanced optical scattering?

We need surface-enhanced optical scattering?

- Enhanced **Rayleigh** Scattering for Surface Enhanced Particle Sizing (SEPS) can size particles on the wafer below 10nm
- Enhanced **Raman** scattering (SERS) enables identification of the composition of the particle to aid in the identification of the source(s) and understand the risk to yield.

3

Typical process flow of SEPS&SERS analysis

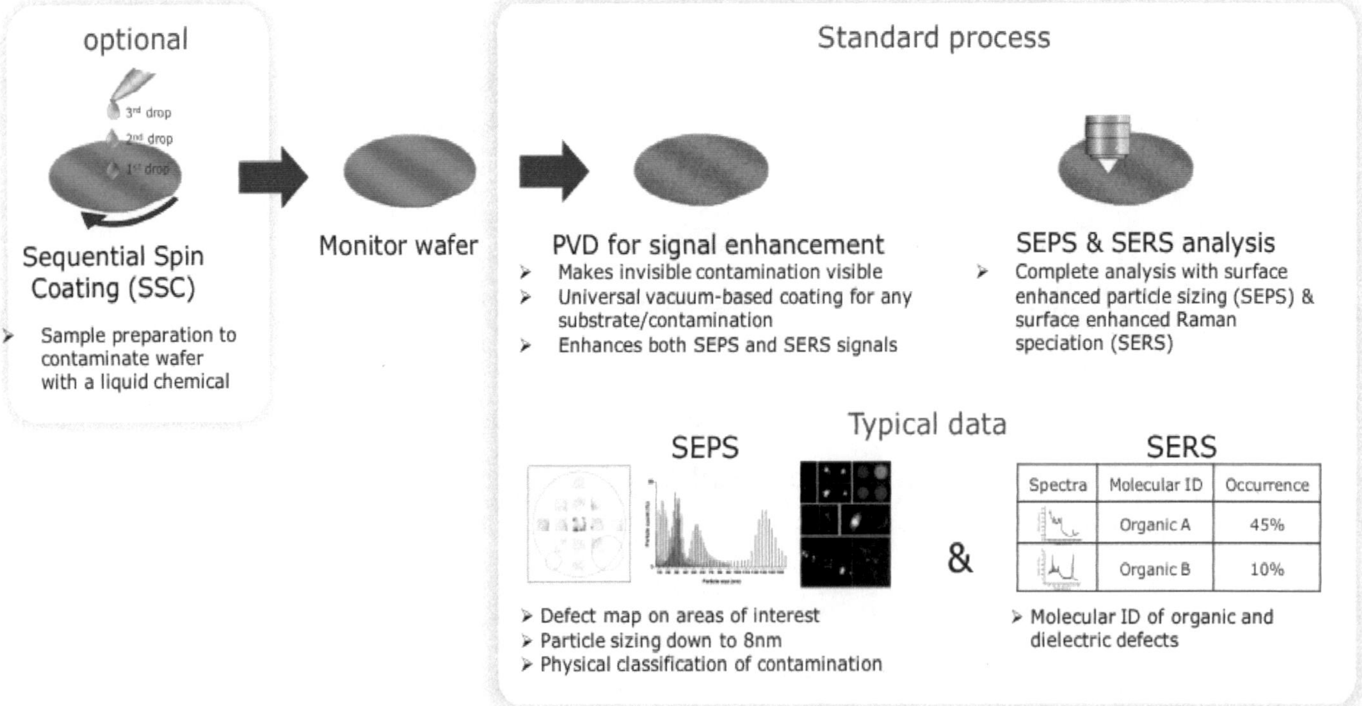

4

Surface Enhanced Particle Sizing (SEPS) for defect classification

Examples of SEPS Images of defects

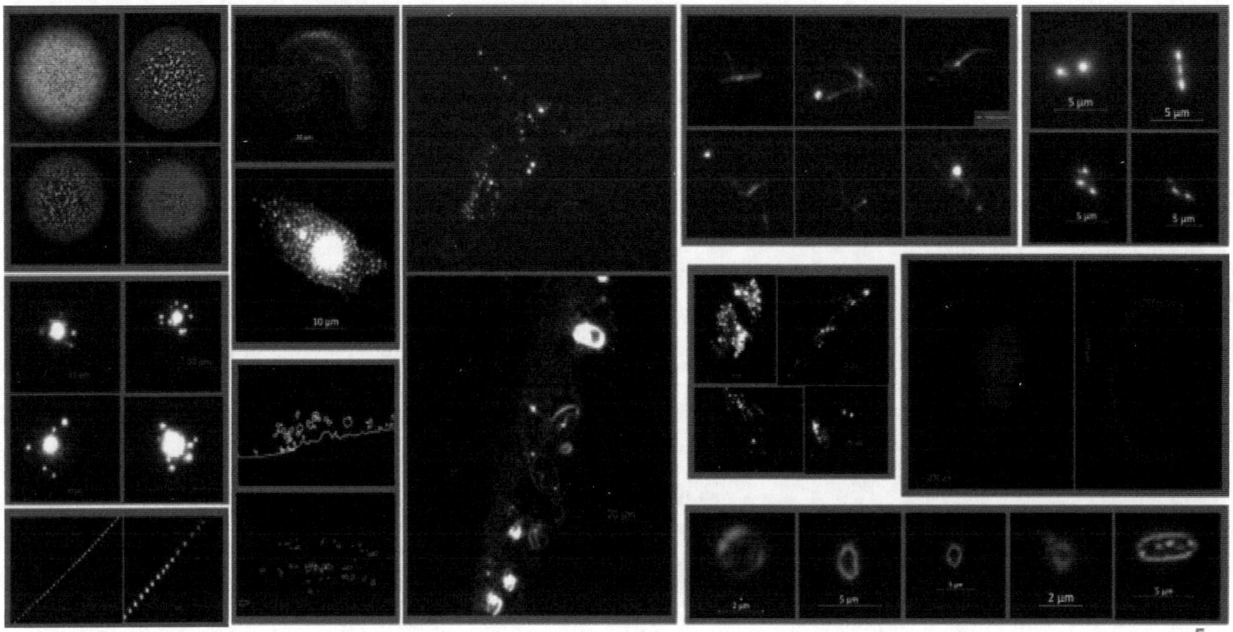

5

Surface Enhanced Particle Sizing (SEPS)

- SEPS images are used to assign size to each individual particle

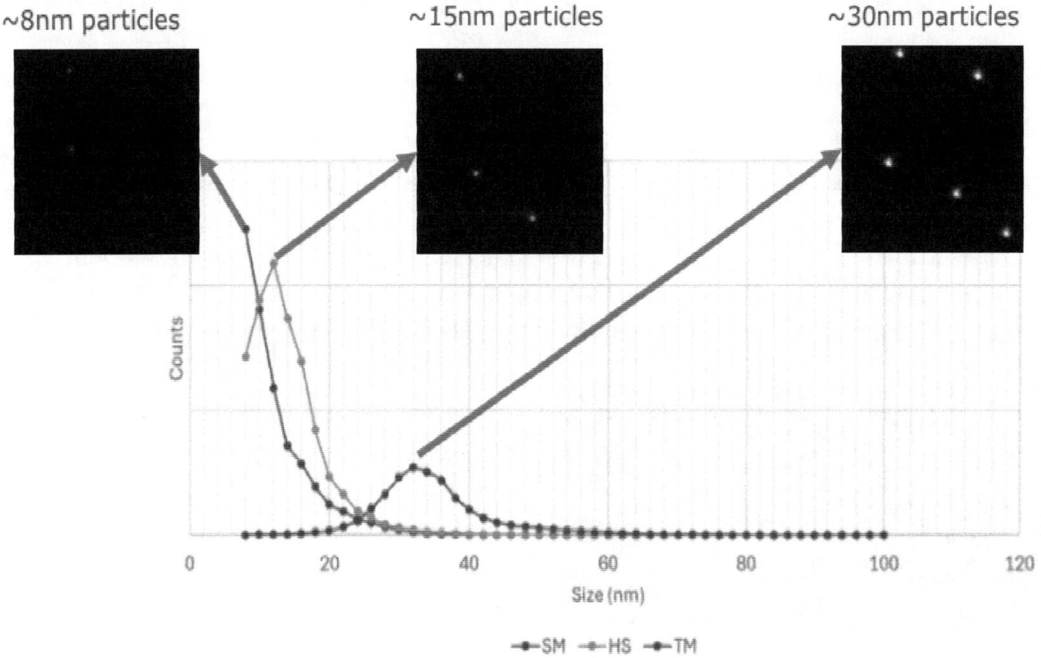

6

SEPS: Particle detection & sizing reliability

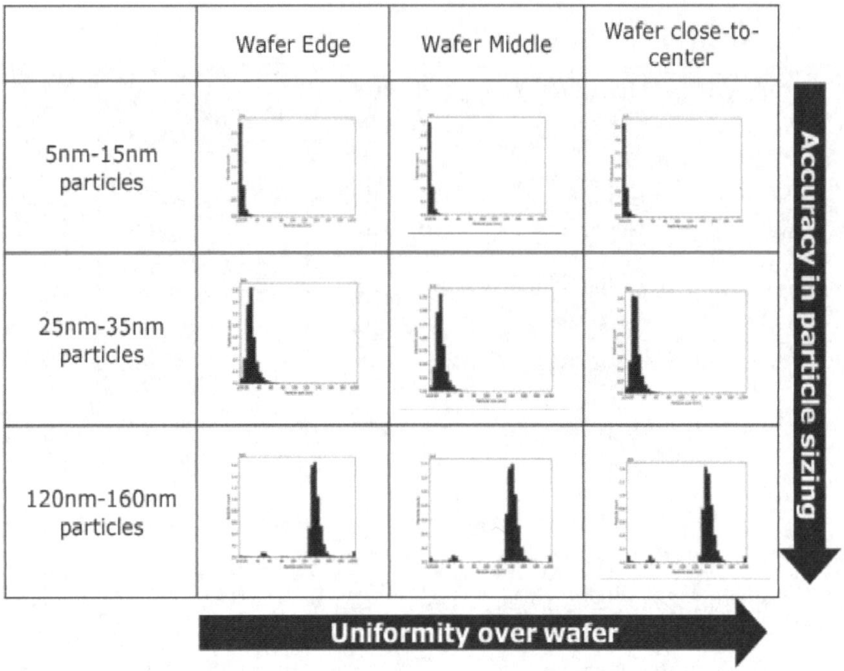

7

Correlation studies with other particle detection tools

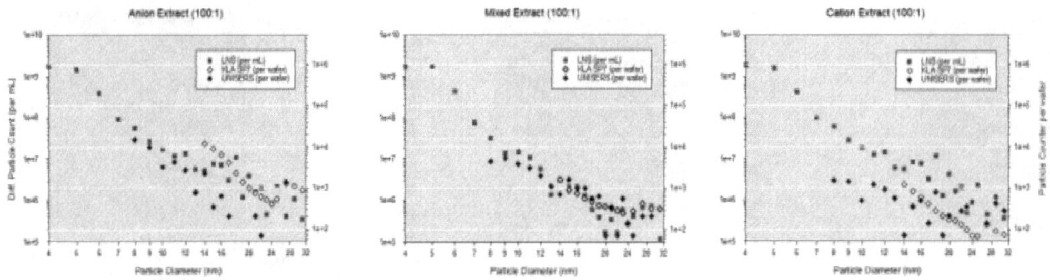

Source: IRDS, UPW task-force test

- Early UNISERS data from 2022
- Proven ability to detect particles below 8nm
- Out-performed classical on-wafer techniques

Reused with Permission: Presented by Gary Van Schooneveld in the UPM Community Event June 2023

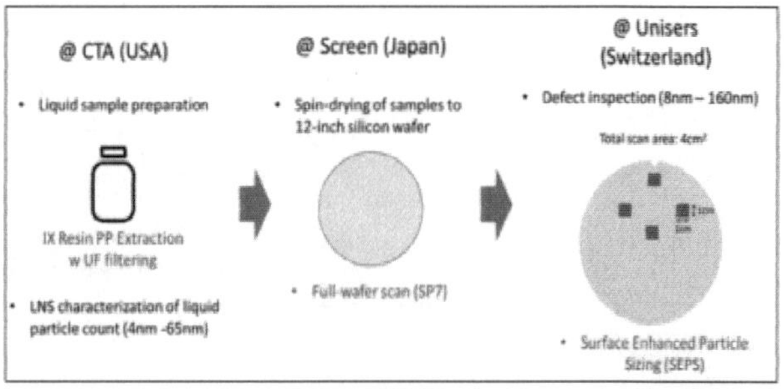

8

Advantage of UNISERS for defect review

➢ More information than SEM review for faster yield improvement

Various wafers with Surfscan defect map

Source: IRDS, UPW task-force test

SEM defect review

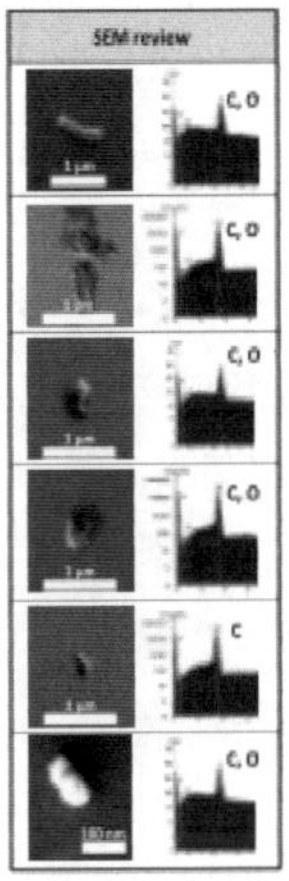

173 defects reviewed.
120 invisible defect.
53 defect images.

Learnings

SEM images:
- Particles and agglomerations

EDX analysis:
- Carbon and oxygen

UNISERS defect review

1513 defect images.

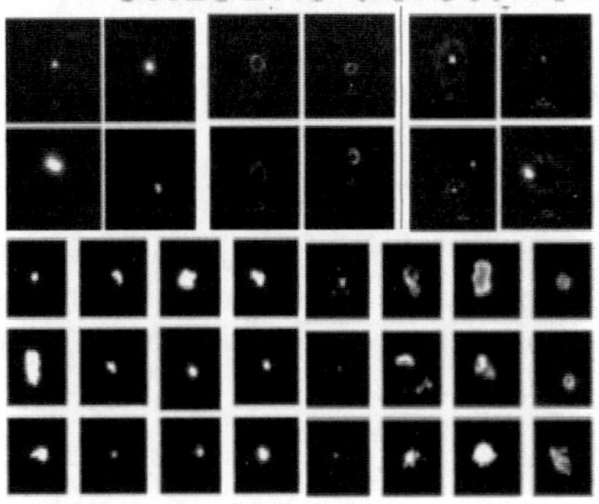

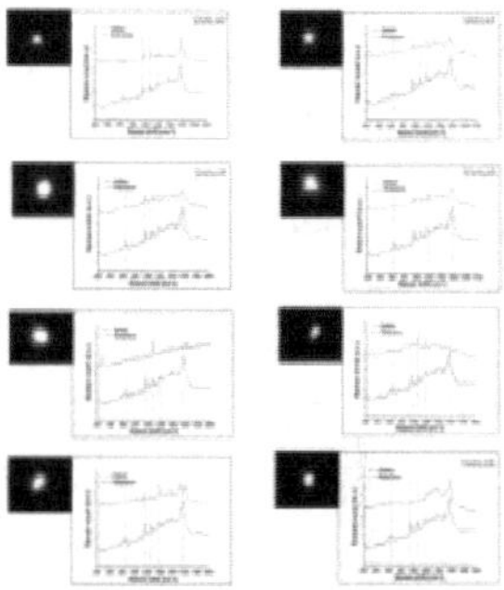

Learnings

SEPS images:
- 96% single particles
- 1% evaporation marks
- 3% organic films

SERS analysis:
- 60% polystyrene sulfonate (ion exchange resin extract)
- 27% unspecified organics
- 13% silicates

9

UNISERS laboratory suite

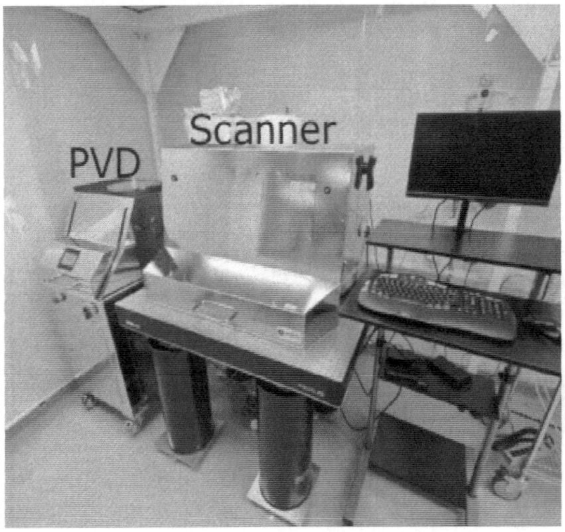

- Wafer size: Up to 300mm
- Typical coating time (including pumping, venting): 15 mins
- SEPS particle size lower limit: 8nm
- SEPS speed (8nm mode): 22 mins to 30 mins /cm2
- SERS speciation lower limit: 50nm - 80nm
- Typical SERS acquisition time per particle: 5s - 60s

30nm deposited wafer

Before enhancement

After enhancement

10

Preparation of validation wafers for SEPS using sequential-spin-coater (SSC)

Sequential spin coating

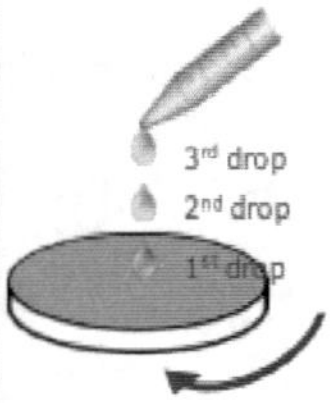

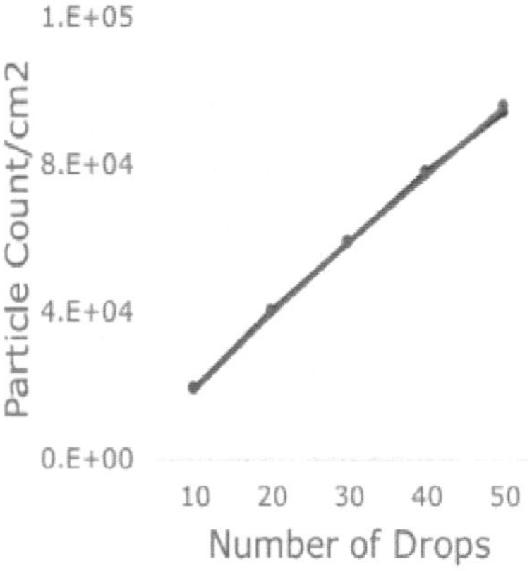

- Wafer size: up to 300mm
- Up to now: UPW, IPA, 1% dilute acids
- Automated droplet generator

UNISERS validation wafers

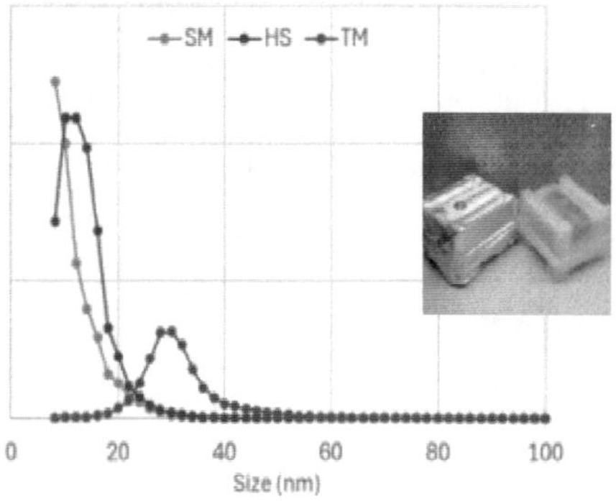

- SM-Ludox: 9nm silica
- Ludox HS: 15nm silica
- Ludox TM: 30nm silica
- Si150: 145nm silica

11

Tool Matching: Particle Sizing of validation wafers

Particle Size (nm)		HS	TM	Silica_150
Alpha1	Avg.	14nm	32nm	144nm
	Sigma	0.5	0.9	4.2
	3 Sigma	2	3	13
	CV (%)	3.6	2.8	2.9
Alpha2	Avg.	14nm	32nm	147nm
	Sigma	0.4	1.0	6.6
	3 Sigma	1	3	20
	CV (%)	2.9	3.1	4.5
Matching criteria (mean ± 3-sigma)				
Overall	Avg.	14 ± 2	32 ± 3	145 ± 17
Alpha 3	Avg.	13nm	30nm	154.3nm
		Pass	Pass	Pass

- Test setup: 3 size standards (HS, TM, and Silica_150), 5 wafers for each particle type, 2 systems at different sites
- Results: CV less than 4.5% across all 3 particle sizes
- **Data from the first two systems are used to set tool calibration matching success criteria**

12

Split-lot test for particle detection and sizing

Carrier Position	Description	Order	Concentration (#/mL)
1	Pre-rinse blank - 2 minutes	3	0
2	18nm/28nm DR1	6	7,2E+03
3	18nm/28nm DR1	7	7,2E+03
4	18nm/28nm DR1	12	7,2E+03
5	Pre-rinse blank - 2 minutes	15	0
6	18nm/28nm DR2	18	1,6E+05
7	18nm/28nm DR2	19	1,6E+05
8	18nm/28nm DR2	24	1,6E+05
9	Pre-rinse blank - 2 minutes	27	0
10	18nm/28nm DR3	50	3,2E+06
11	18nm/28nm DR3	51	3,2E+06
12	18nm/28nm DR3	56	3,2E+06
13	18nm/28nm DR4	39	7,0E+07
14	18nm/28nm DR4	40	7,0E+07
15	18nm/28nm DR4	46	7,0E+07
16	Pre-rinse blank - 2 minutes	59	0

Sample preparation recipe:

- Particle size distribution will be measured during SSC by liquid nanosizer (LNS).
- Wafer size: 2-inch. 2xTriplicates of each condition.
- Number of sequential spin coating (SSC) cycles is 20.
- Pre-rinse of wafers with UPW for 2 minutes at 500 rpm, dry at 5,000 rpm before SSC.
- SSC recipe:
  - 5000rpm
  - 20 cycles (20 secs delay between cycles)
  - 0.5ml sample volume per cycle
  - 200 secs of final drying time (@5000rpm)

Scan recipe:

- Two UNISERS systems are used. One @CTA, one @UNISERS
- Scan area: 12 areas of 5mmx5mm (total 3 cm2)

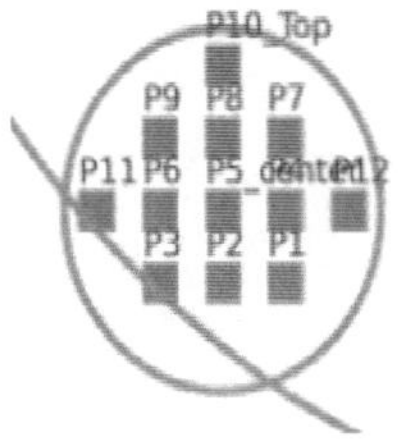

13

Tool-to-tool comparison

UNISERS Lab Suite @ CTA
Average of 3 wafers (subset) background subtracted

UNISERS Lab Suite @ UNISERS
Average of 3 wafers (subset) background subtracted

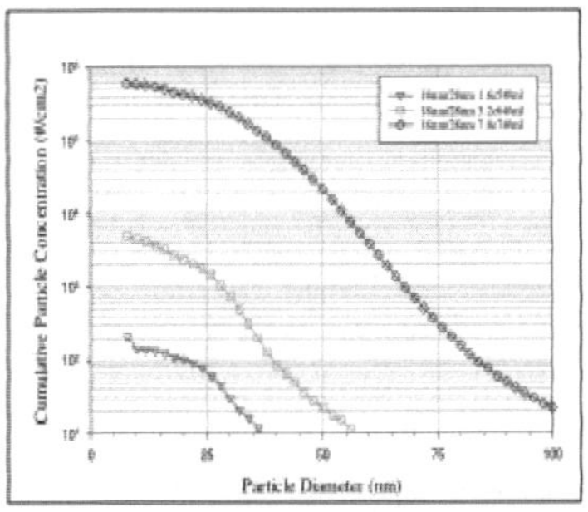

Cumulative Particle Concentration

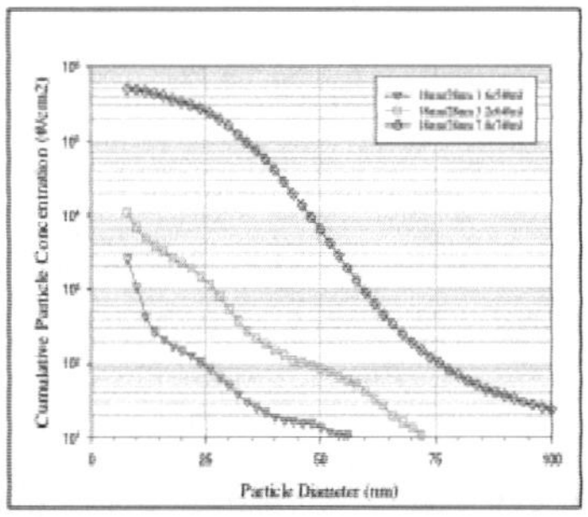

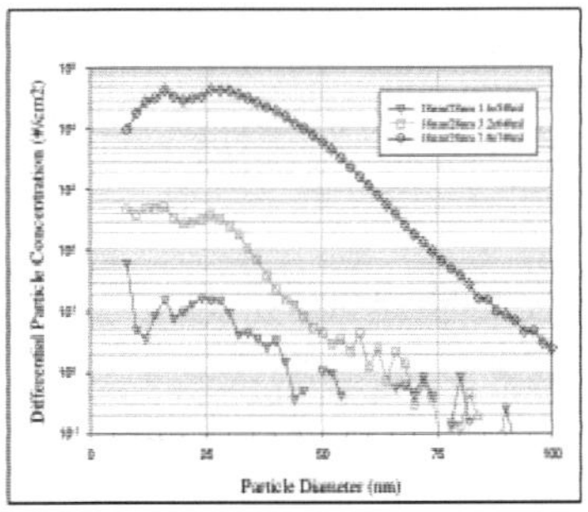

Differential Particle Concentration

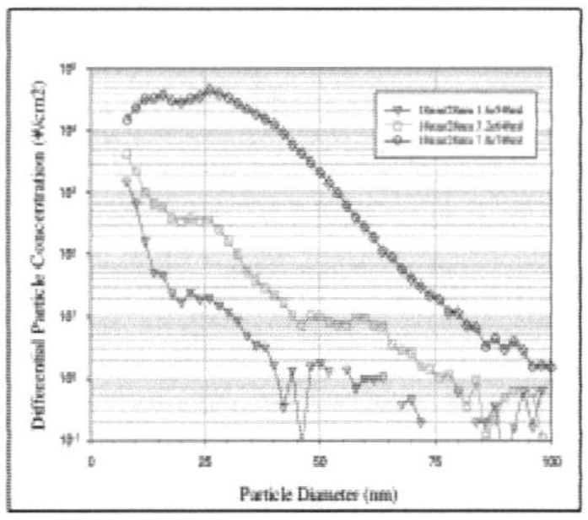

14

Particle deposition uniformity over wafers

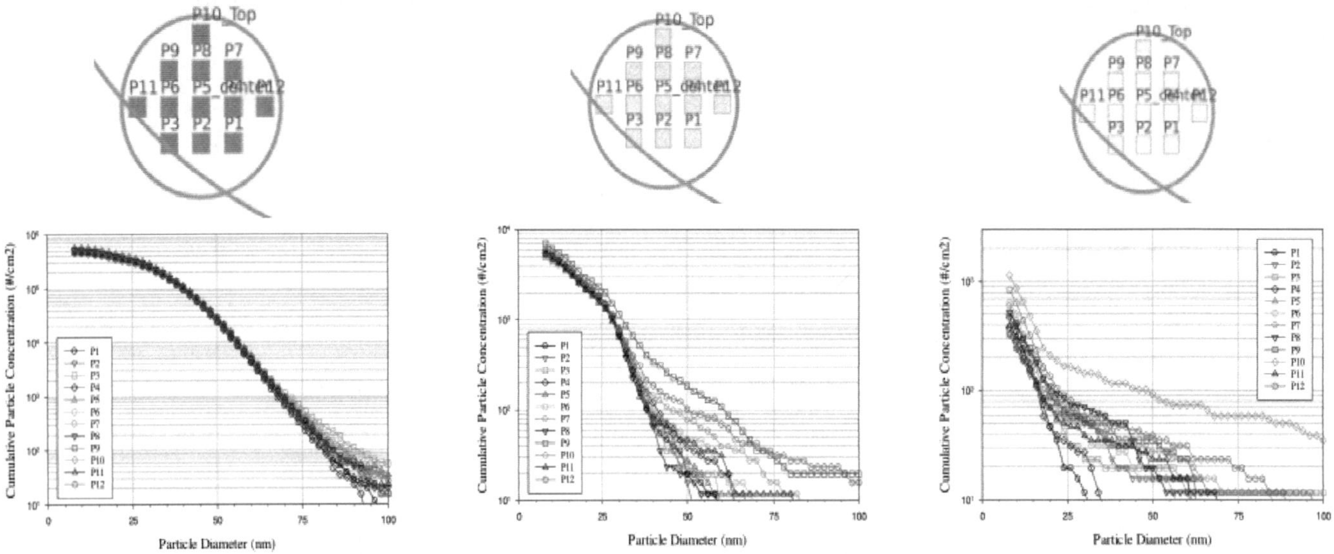

Decreasing particle concentration in dispensed liquid

Particle sizing comparison: UNISERS & LNS

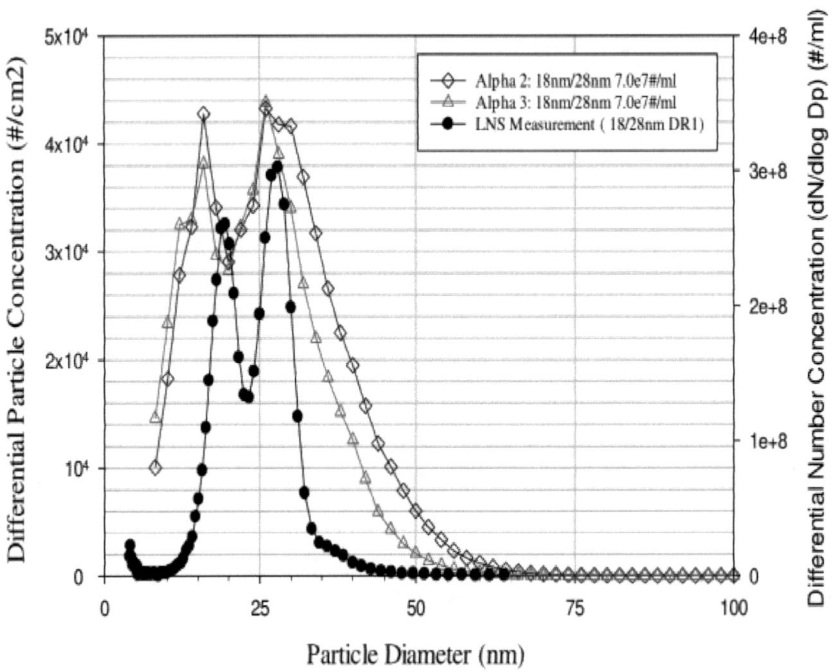

Contamination classification
Roadmap

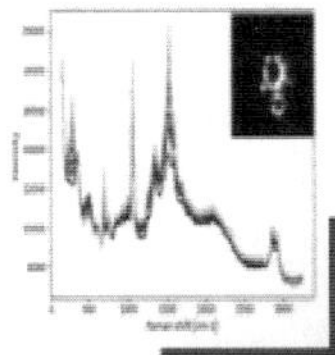

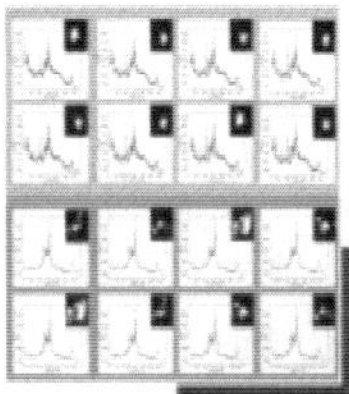

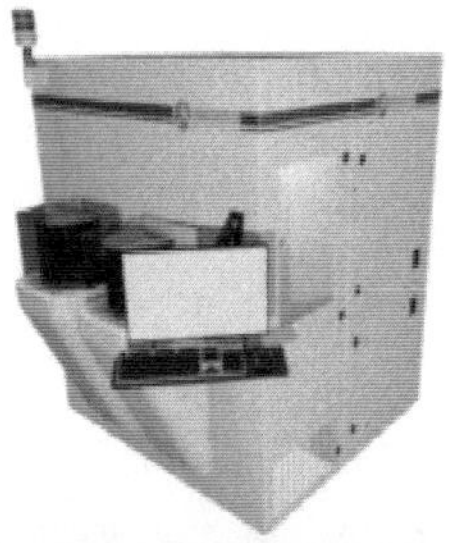

2024

- Automatic SERS Acquisition
- SERS Spectra Clustering
- SEPS Image Clustering
- Smart SERS

2025+

- Database Comparison
- Advanced SERS
- Advanced SEPS Classification
- KLARF support

17

Conclusions

- UNISERS laboratory suite has been developed and key performance specifications are validated across multiple systems.
  - Sub-10nm particle detection is demonstrated by detecting different particle size distribution of 9nm & 15nm silica particles.
  - Tool-to-tool consistency in particle sizing is demonstrated.
  - SEPS particle sizing is correlated with LNS
- Classification of contamination is done with both SEPS and SERS data supported by a dedicated database.

18

Thank You!

ali.altun@unisers.ch

19

Cu Pillars: Challenges and Solutions to Strip Thick Photoresist

3D

Amaia Diaz De Zeria, Ivonne Saucedo, John Taddei, Phillip Tyler, Efrain Altamirano-Sanchez and Naveen Reddy

SPCC 2024

Outline

- Problem statement and challenges
- Solution hypothesis
- Lab results
- Tools results
- Conclusion
- Acknowledgements

Problem statement

Strip thick photo resist (PR)

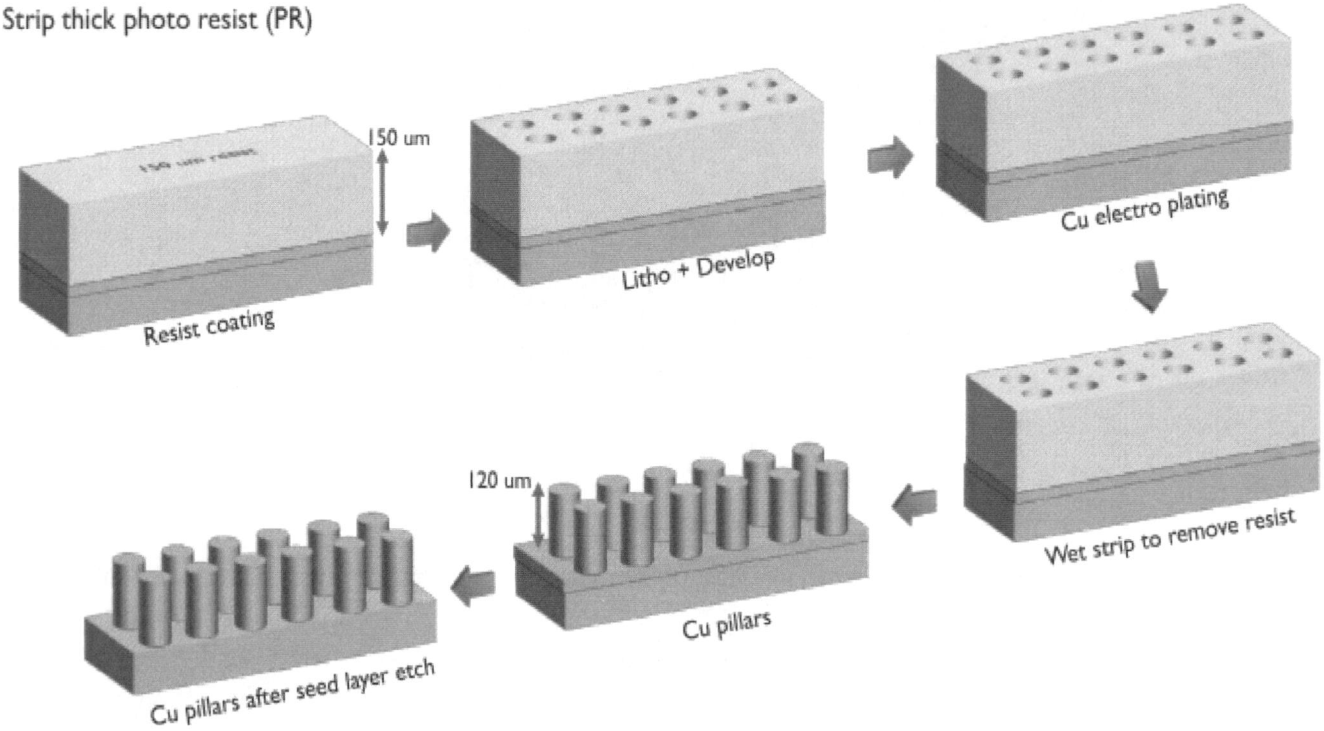

Current POR-1: Challenges

Solvent @ 60 C: 12 min solvent strip + BKM DIW rinse + IPA rinse + Dry

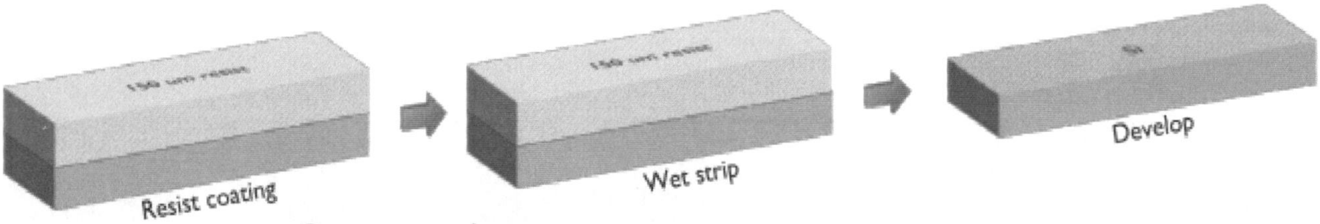

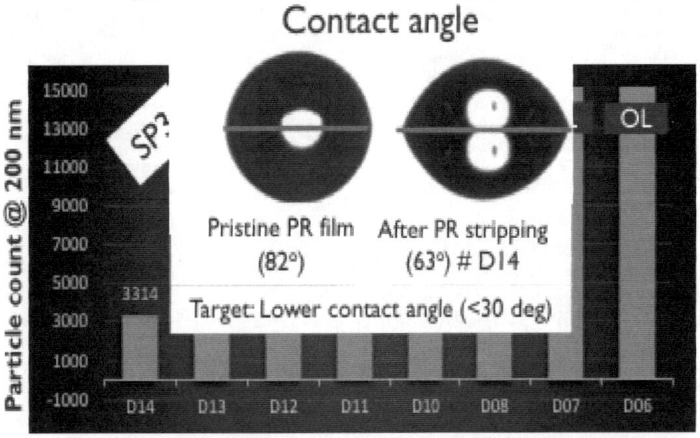

Stripping thick layer of thick PR is hard and consumes 15-20L solvent per wafer

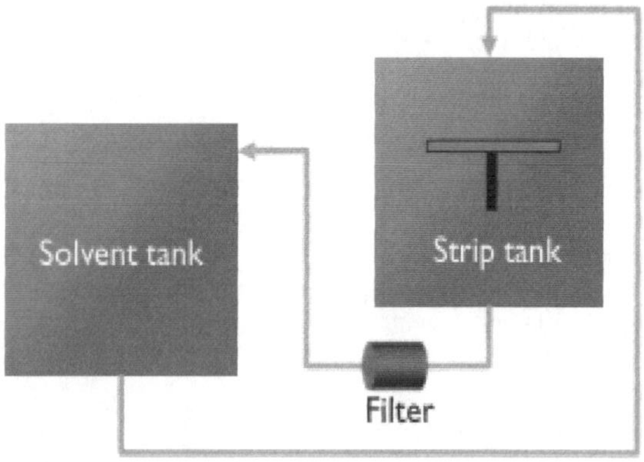

Solvent under recirculation

Current POR-2: Challenges

Solvent @ 60 C: PR strip + BKM DIW rinse+ IPA rinse + Dry

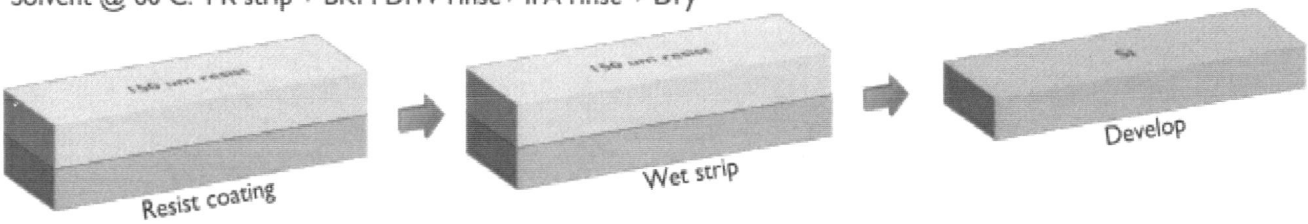

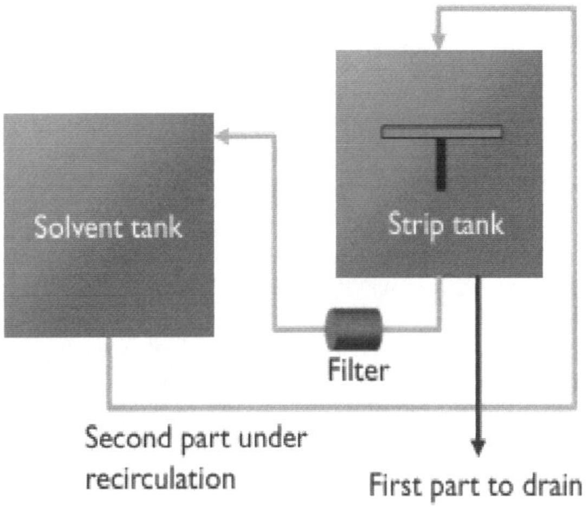

10 min to drain
5 min recirculation

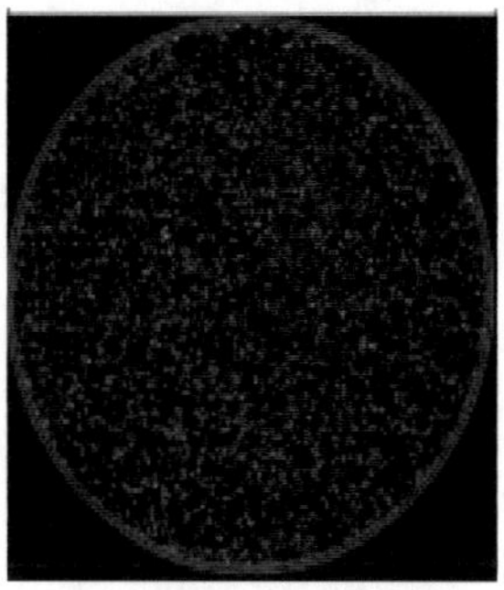

Still lot of particles

Thick Photoresist

	Thick PR
Coating Thickness	140-150 um
Composition	• Novolak Resin • Acrylic type resin (unknown) • Polyhydroxy styrene resin (unknown) • 3-methoxybutyl acetate • Propylene glycol monomethyl ether acetate • Cresol • Photoacid Generator
Type of PR	Positive
Annealing steps	• Dehydration bake • Pre-anneal bake • Post-exposure bake
Developer	TMAH based

Photoresist workings

Novolac synthesis

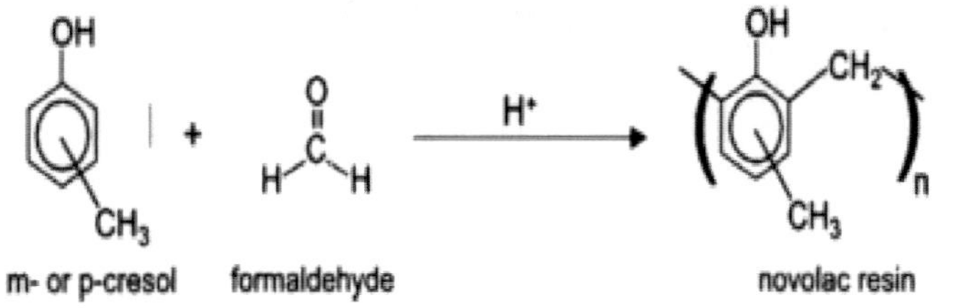

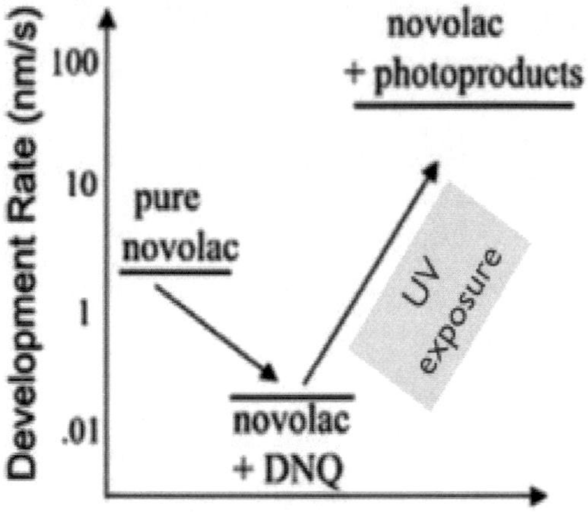

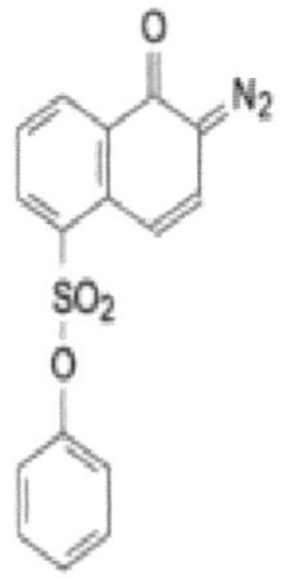

DNQ reaction sequence (Photo acid generator)

Problem statement

Strip thick photo resist (PR)

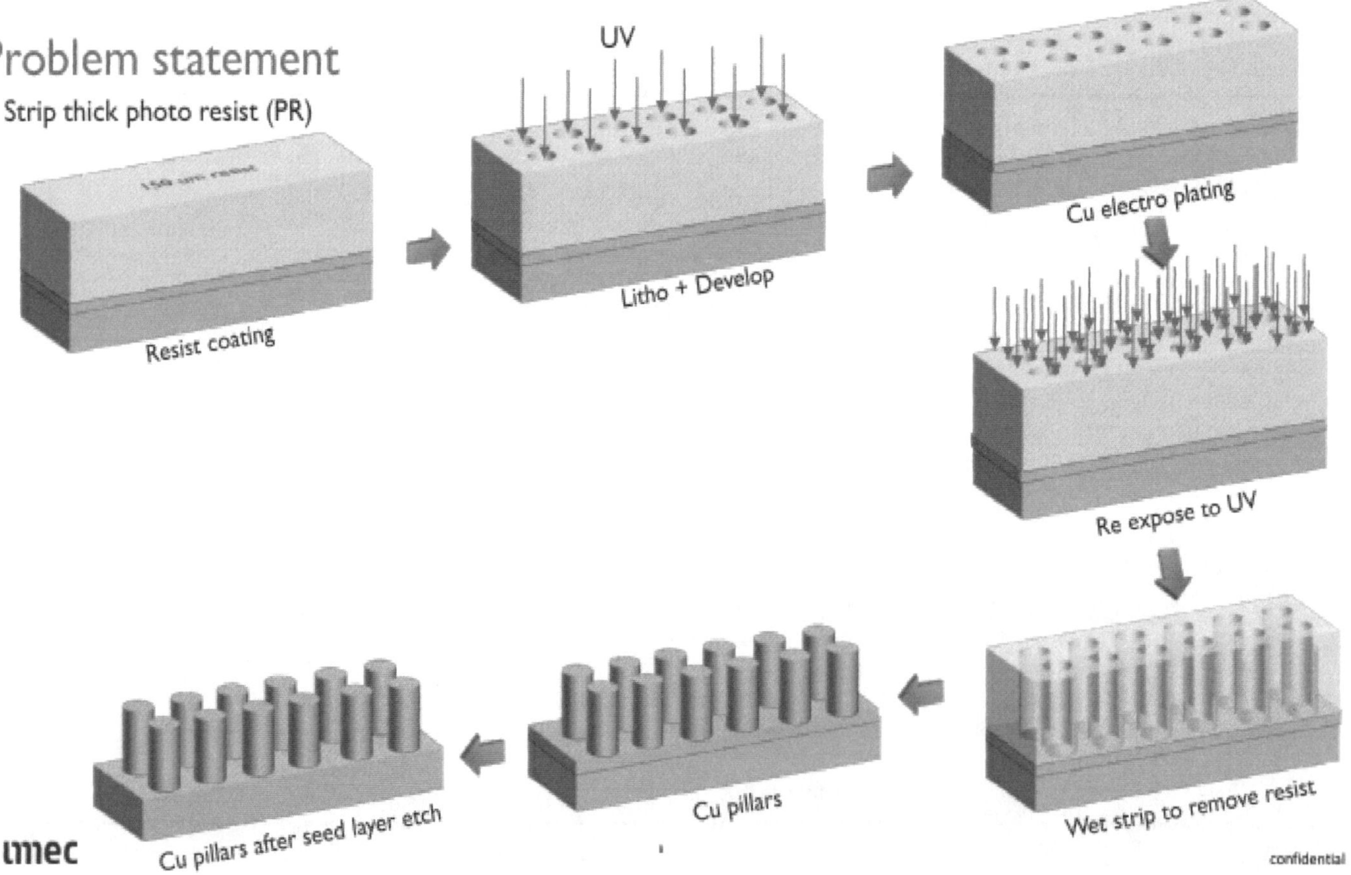

Methodology and characterization

Methodology

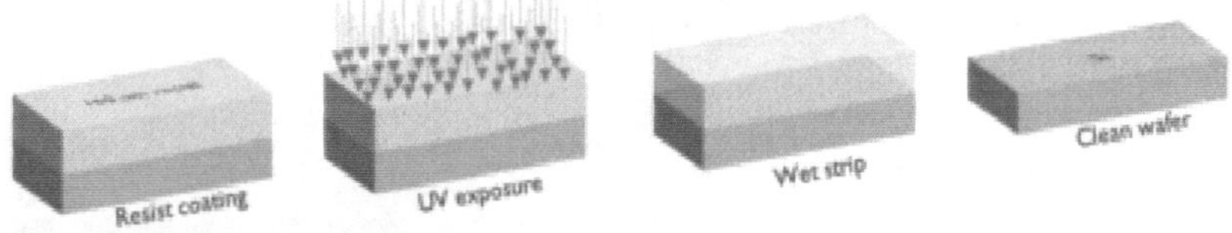

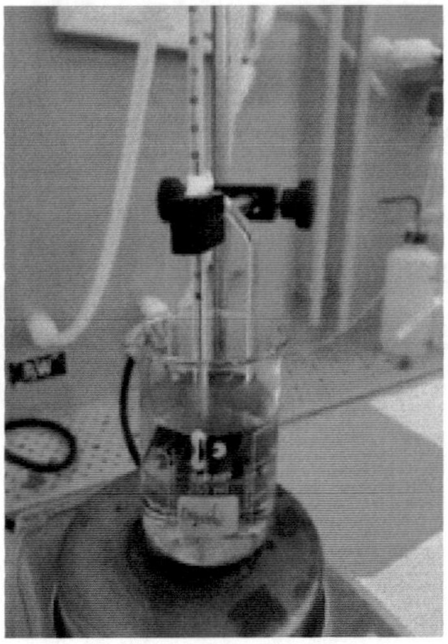

Setup

- Samples: 0, 600, 1200, 1800 mJ/cm^2
- Solvent: 200 mL
- Stripping speed: 300 rpm
- 6 samples on a glass holder
- Temperature: RT, 40 C and 60 C
- Time: 5, 10, 15, 25, 45, 90 min
- Rinse
  - DIW: 1 min spray
  - IPA: 5-10 sec
  - Dry: N2

Characterization

- Visual
- Contact angle
- FTIR
- SP3
- Solvent lifetime study

Results: Lab

- Visual
- Contact angle

Q&A

- FTIR
- Solvent lifetime

Solvent at RT, 40 and 60 °C - visual

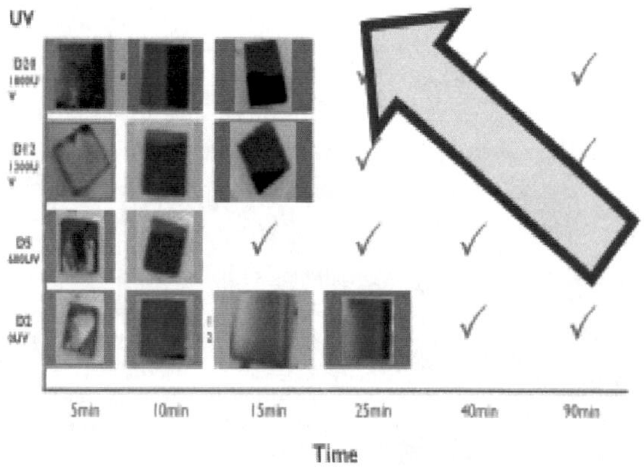

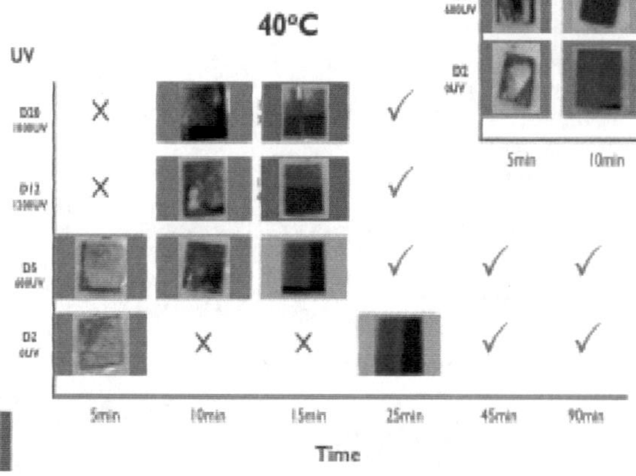

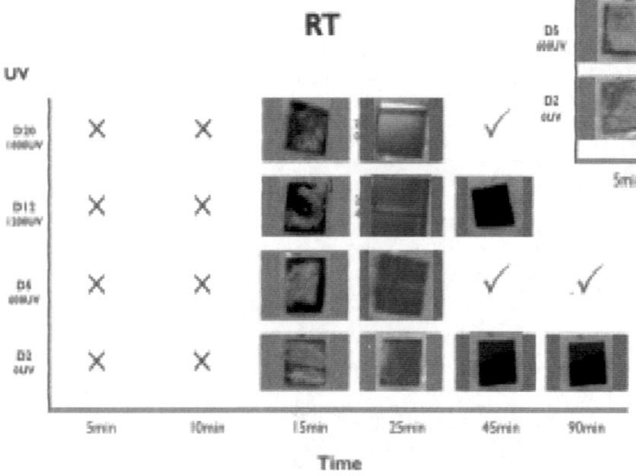

12 confidential

Results: Lab

- Visual
- Contact angle

Solvent at RT, 40 and 60 °C - CA

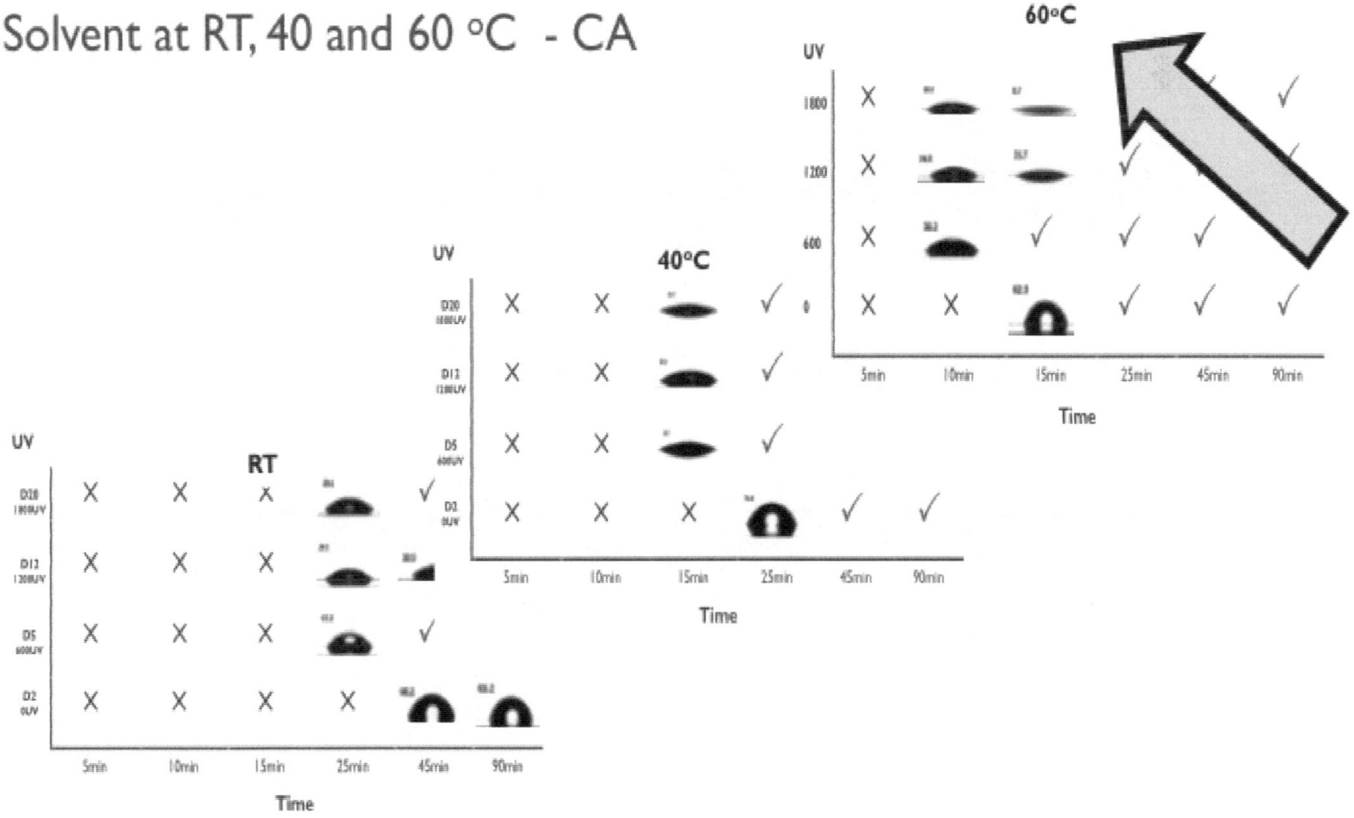

imec

14 confidential

Results: Tool

- SP3
- Contact angle
- Solvent lifetime

	Solvents	Conditions	Comments
PR Strip	• TMAH based commercial solvent 60 C	• Pre wet < 1 min • *1-10 min* solvent spray	The solvent spray time has been varied in the test.
Rinse	• DIW • IPA • Dry	• 3 min DIW • 1 min IPA • 1 min Dry	

Results: SP3 particle comparison (Day1)

Wafer #	UV mJ/cm²	Spray time (min)	100 nm	200 nm	comments
D05	600	10	522	209	
D06	600	5	736	399	One point not clean
D07	600	5	190	69	
D08	600	3	350	92	
D09	600	1	11668	2958	

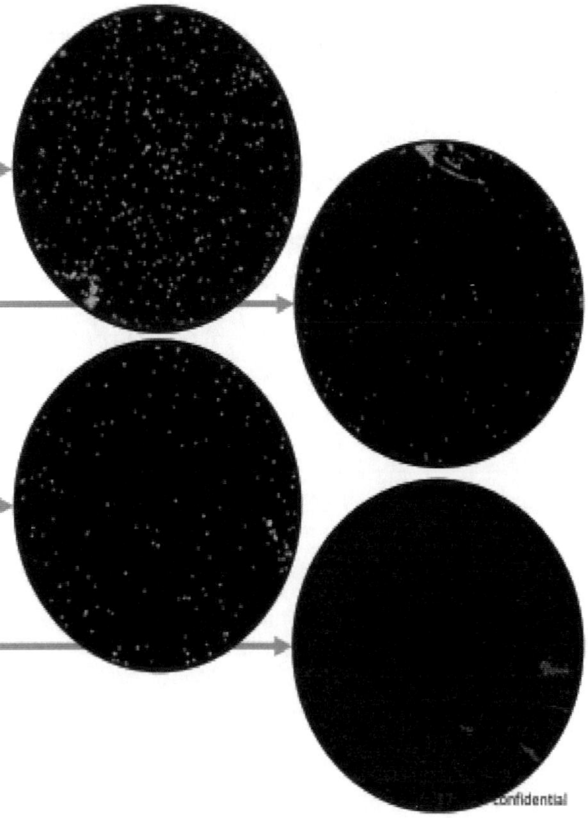

Results: SP3 particle comparison (Day2)

Wafer #	UV mJ/cm²	Spray time (min)	100 nm	200 nm	comments
SPC		3	107	28	
SPC		3	122	32	
D20	1800	3	99	42	
D12	1200	3	1025	694	
D10	600	3	1259	300	
SPC		3	88	28	
D13	1200	5	735	531	2 comet
D14	1200	7	667	460	3 comet
D15	1200	10	499	328	
D16	1200	10	781	533	3 comet

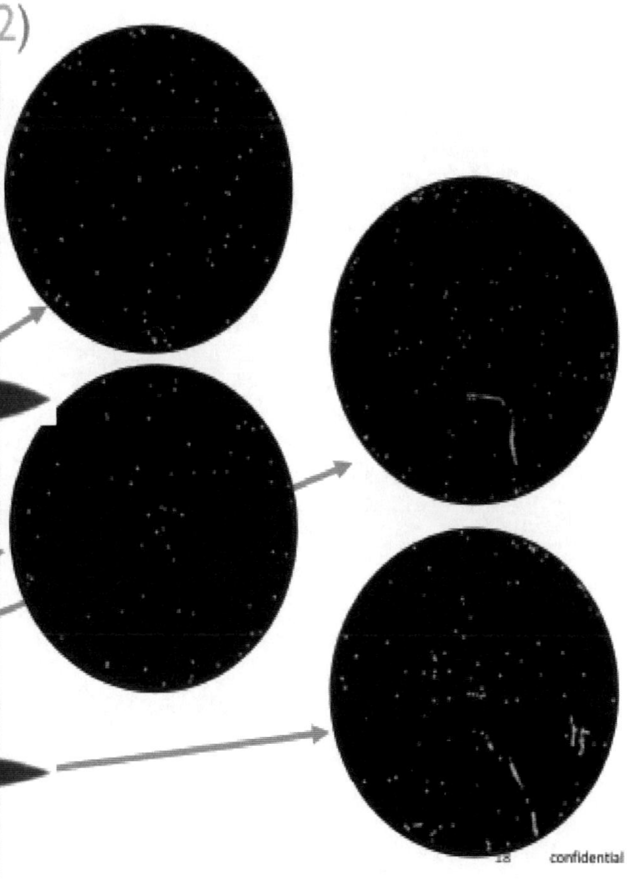

Results: SP3 particle comparison (Day2)

Wafer #	UV mJ/cm²	Spray time (min)	100 nm	200 nm	comments
D21	1800	10	240	162	
D23	1800	10	247	160	
D24	1800	10	364	221	
SPC		3	411	121	

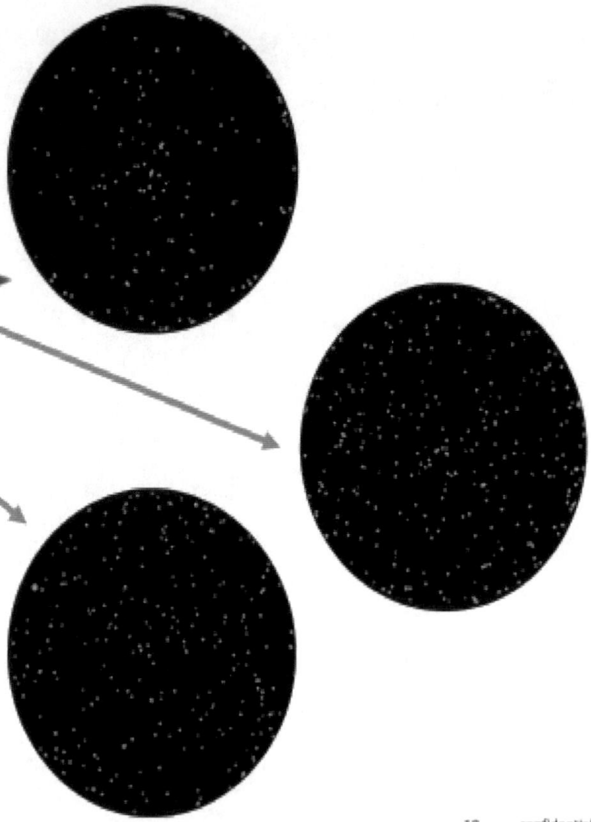

imec

Conclusions

	POR FAB	Lab (UV coupons)	FAB (UV wafers)
Stripping time	>10 min	<10 min	3 min
Stripping solvent	15-20 L/wafer	4.5 wafer/L	<2 L /wafer Still good SPC
Particles (@200nm)	>3000	-	~500
Contact angle	> 60 deg	< 25 deg	< 30 deg

- ❑ Photo resist stripping timing is shorter ↓
- ❑ Photo resist volume removed is higher per unit solvent ↑
- ❑ Post strip CA is lower → Surface is cleaner ↑

imec

Acknowledgements

PR coating – Paulina, Angel and Nelson

Wafer dicing – Geert Schoofs and Akito

Lab – Ivonne

Discussion – Quoc Toan Le

Program Support
3D Team, Gerald Beyer, Andy Miller and Eric Beyne

imec 21 confidential

Elemental Scientific
ICP | ICPMS | AA

Real Time Analysis of PPT Levels of Organics in UPW

Brianna Dufek, Tyler Herek, Suhas Ketkar, Austin Schultz, Daniel R. Wiederin

From IRDS Yield Enhancement Tables

From Table YE3 2023 IRDS Roadmap
Technology Requirements for Surface Contamination Control

Ultrapure Water	2023	2024	2025	2026	2027
Total Organic Carbon (ppb) for Immersion Litho*	< 1	< 1	< 1	< 1	< 1
Critical Organics as C (ppb)*	< 1	< 1	< 1	< 1	< 1
Non-Critical Organics as C (ppb) *	< 3	< 3	< 3	< 3	< 3

*TOC online measurement remains the reference detection method for total organics in UPW.
Users of this roadmap are encouraged to speciate organics in their UPW

Organic compounds are categorized as critical with respect to their ability to form a hydrogen bond with any oxide surfaces including gate, tunnel, or native oxide
Compounds with BP >200°C are considered critical organics

From IRDS Yield Enhancement Tables

From Table YE2 2023 IRDS Roadmap
Yield Enhancement Difficult Challenges

Near Term Challenges: 2022-2028	Description
Challenge # 3 Critical Organics in UPW	Insufficient metrology for online speciation of organics

- Online measurements are currently possible only for TOC
  - No speciation available

Elemental Scientific

UPW

- UPW is one of the largest volume chemical used in semiconductor fabrication
- >2000 gallons of UPW required per wafer
  - Typical Fab uses 1 -3 million gallons/day
- UPW used in
  - Wet Etch
  - Photoresist Stripping
  - Wafer Cleaning
- Onsite purification of surface/ground/recycled water
  - Urea concentration can vary in the feedwater
    - Urea difficult to completely remove
  - High Molecular weight organics leach from IX resins, tubing, piping, fluid handling components
    - Could be left as residue on wafer surface
      - Impact further processing

© Elemental Scientific | www.icpms.com | sales@icpms.com

Current Status of Organic Analysis in UPW

- Online analysis for TOC
- Other organic analysis performed in the lab
  - Analytical Techniques
    - GC-MS
      - Semi Volatile contaminants like Trihalomethanes, phthalates, siloxanes
    - LC-OCD
      - High Molecular Weight Organics
      - Slow Analysis

Meeting the IRDS Challenge

- Developed an online system
  - Trace organic analysis in UPW
    - ppt detection limits
    - Up to 40 sampling points
    - Remote samplers with central analyzer
    - Sampling point can be up to 500 m from analyzer
  - System uses
    - Electrospray Ionization for its high sensitivity
    - TOF MS for speciation with high mass accuracy

Online UPW Organic Analysis System

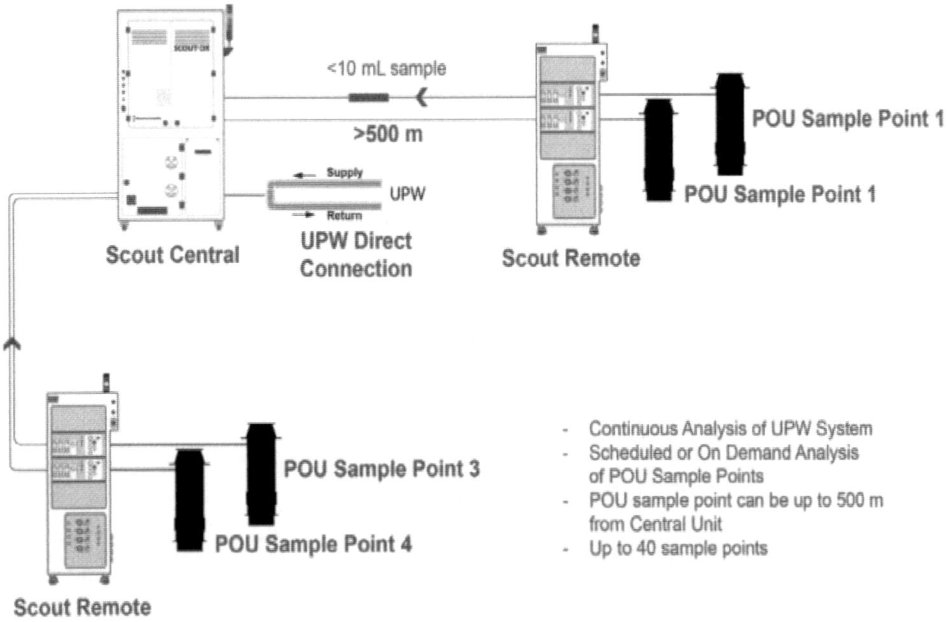

- Continuous Analysis of UPW System
- Scheduled or On Demand Analysis of POU Sample Points
- POU sample point can be up to 500 m from Central Unit
- Up to 40 sample points

Elemental Scientific

scout CARBON/ prepFast CARBON

- Uses Electrospray Ionization (ESI) coupled to a Time of Flight(TOF) Mass Spectrometer(MS)
  - Ionization occurs at atmospheric pressure
  - Liquid sample is nebulized through a charged capillary
    - Fine mist of charged droplets are sprayed from the tip of the capillary
    - Spraying process assisted by a coaxial gas flow
    - Droplets reduce in size due to solvent evaporation
    - Droplets further reduce in size, to a few nm, due to coulomb repulsion
  - Ions detected by TOFMS
    - A Quadrupole MS can be added to help with identification of unknown contaminants using MS/MS

Electrospray Ionization

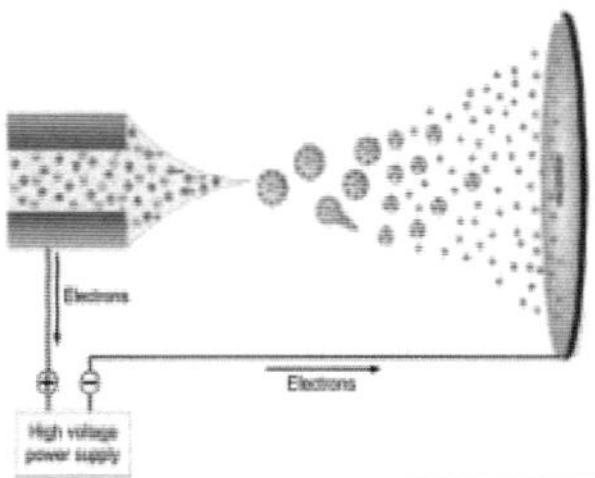

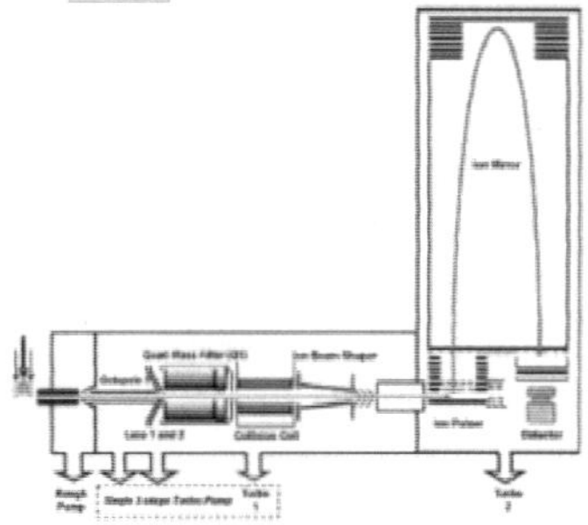

Separation using Quadrupole –Time of Flight Mass Spectrometer

System Calibration

- For Targeted contaminants
  - Urea, TMA, Triazole,.....
    - Preform MSA calibration to obtain quantitative information
- For other contaminants
  - Perform MSA calibration using surrogates by Class
    - Diethylaminoethanol for -N class of contaminants
    - Triethylphosphate for -P class of contaminants
    - Diethylphthalate for -CHO class of contaminants
  - Obtain Near Quantitative information for these contaminants

Calibration for Targeted Contaminants

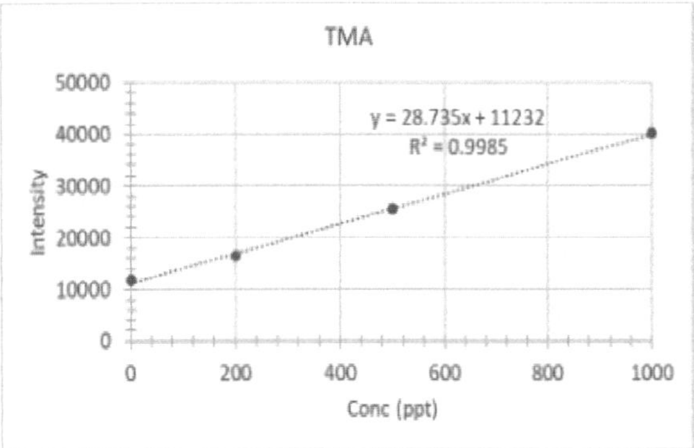

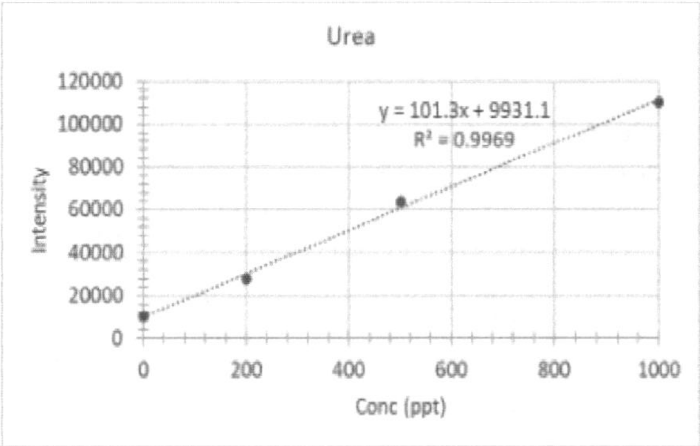

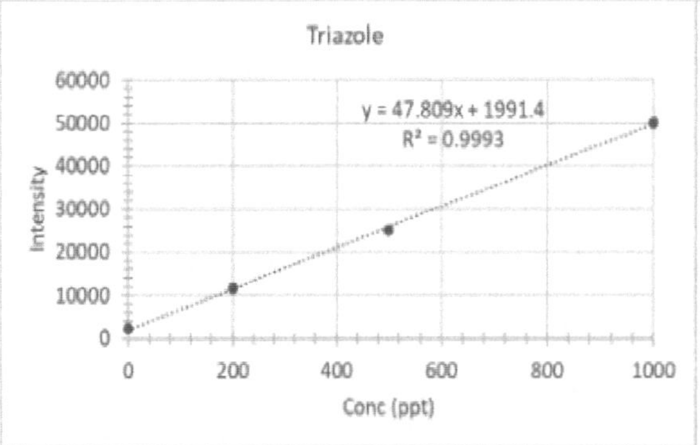

Calibration by Class of Contaminants

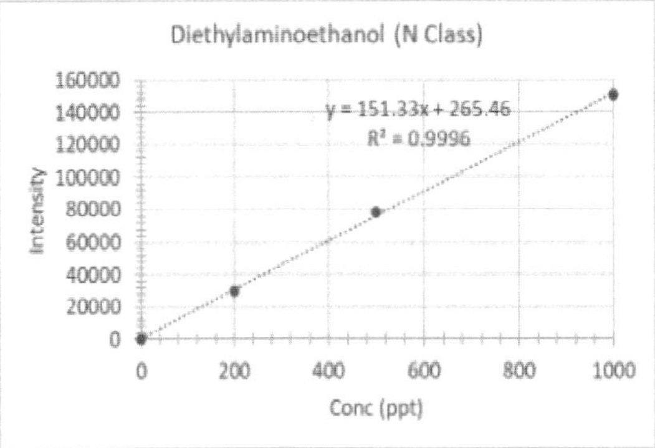

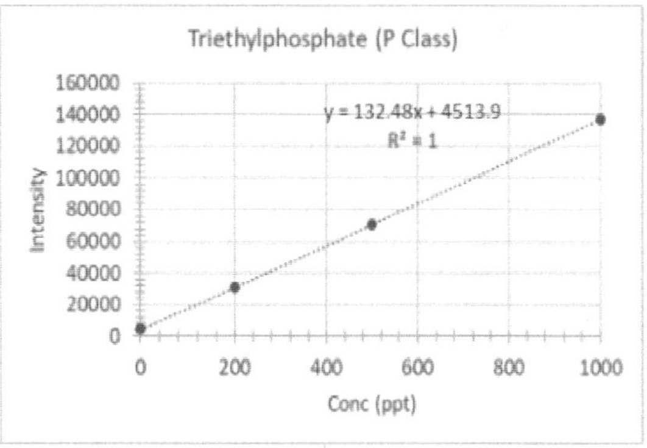

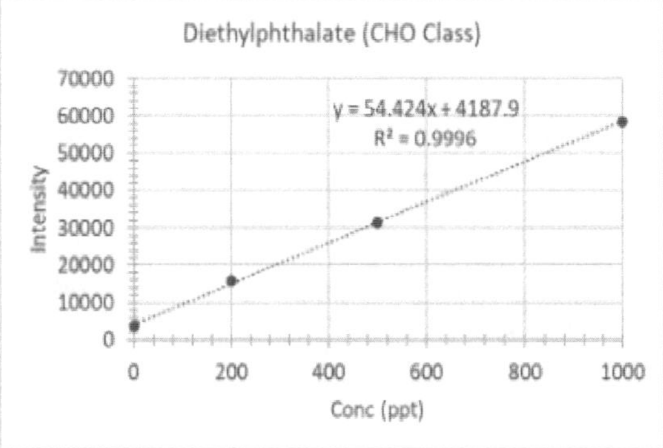

Mass Spectrum for UPW Samples

Positive Ions

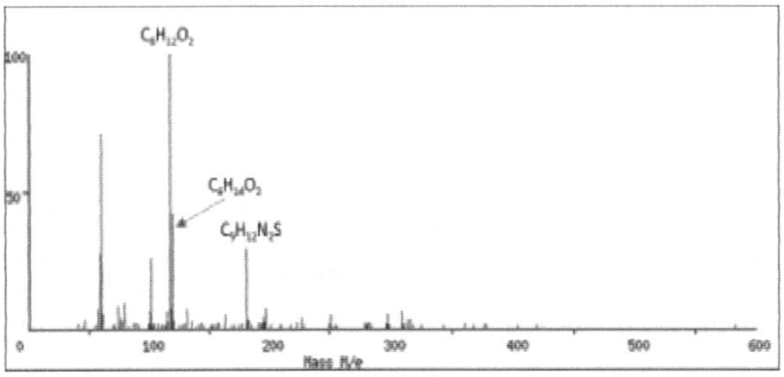

UPW Sample 1

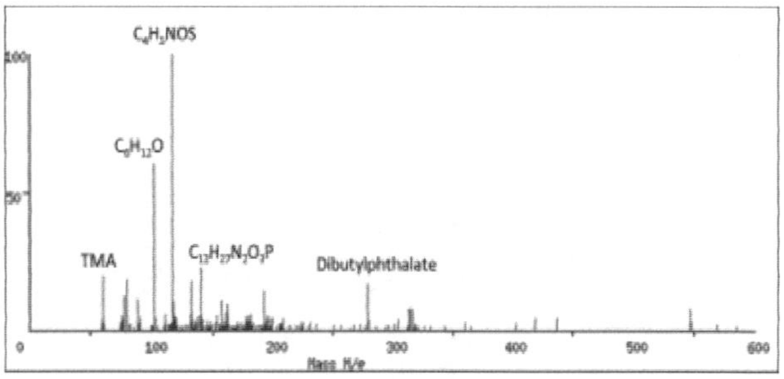

UPW Sample 2

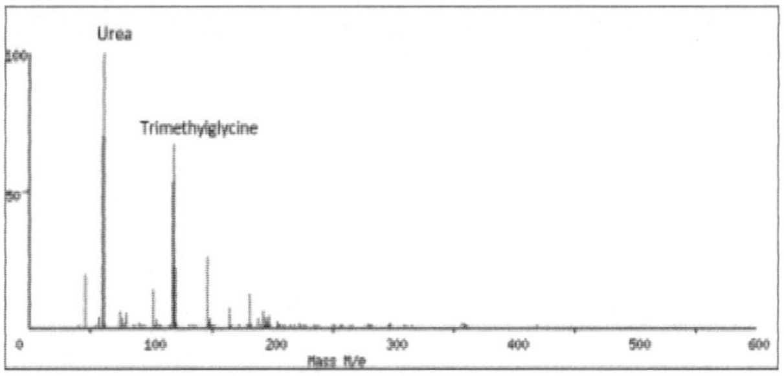

In house system is purifying city water which has large amount of Urea

Mass Spectrum for UPW Samples

Negative Ions

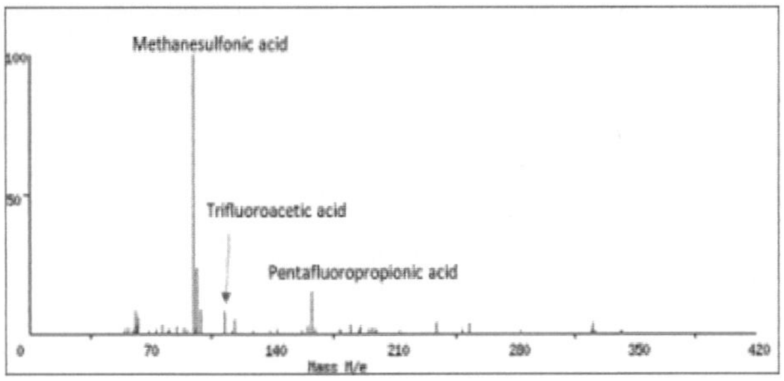

UPW System 1

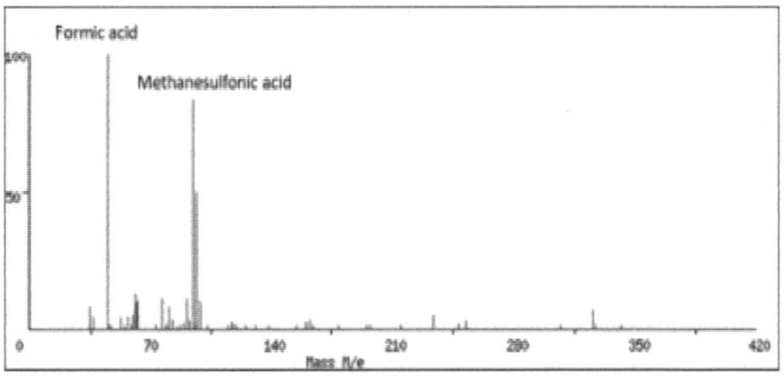

UPW System 2

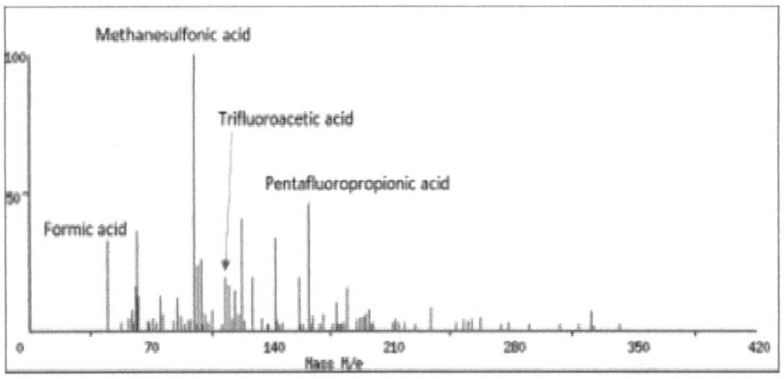

In house system is purifying city water which has large amount of Urea

Elemental Scientific

Organic Contamination in UPW Samples

Sample	TMA (ppt)	Urea (ppt)	Triazole (ppt)	-N Class (ppb)	-P Class (ppb)	-CHO Class (ppb)
UPW System 1	420	109	24	2.2	0.8	1.6
UPW System 2	359	56	ND	12.4	5.3	8.0
In House UPW	744	3698	ND	2.7	0.8	8.5

Monitoring of TMA in Different UPW Sources

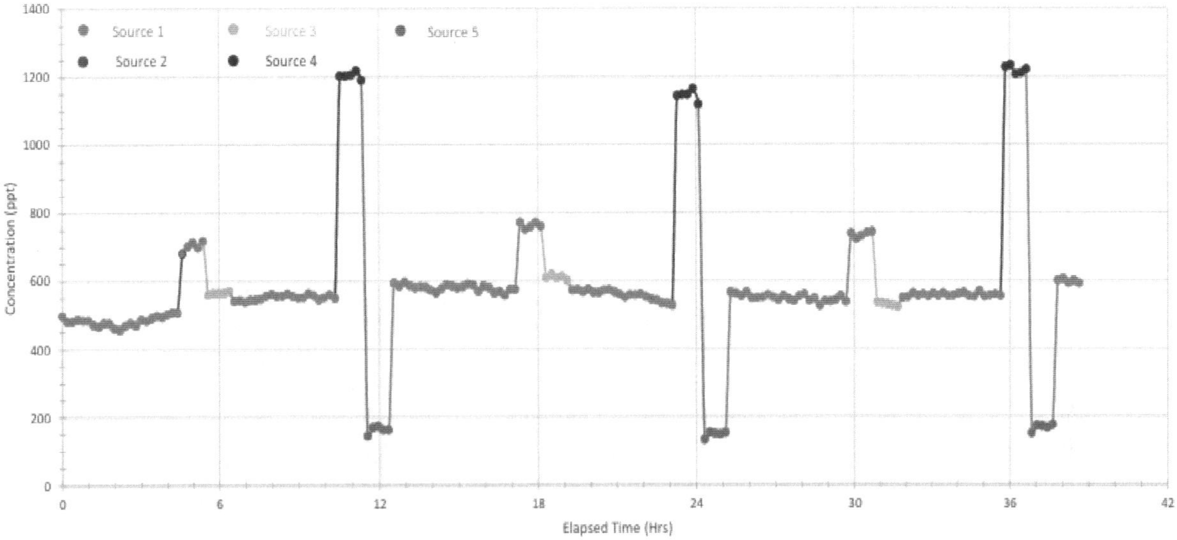

Monitoring of Urea in Different UPW Sources

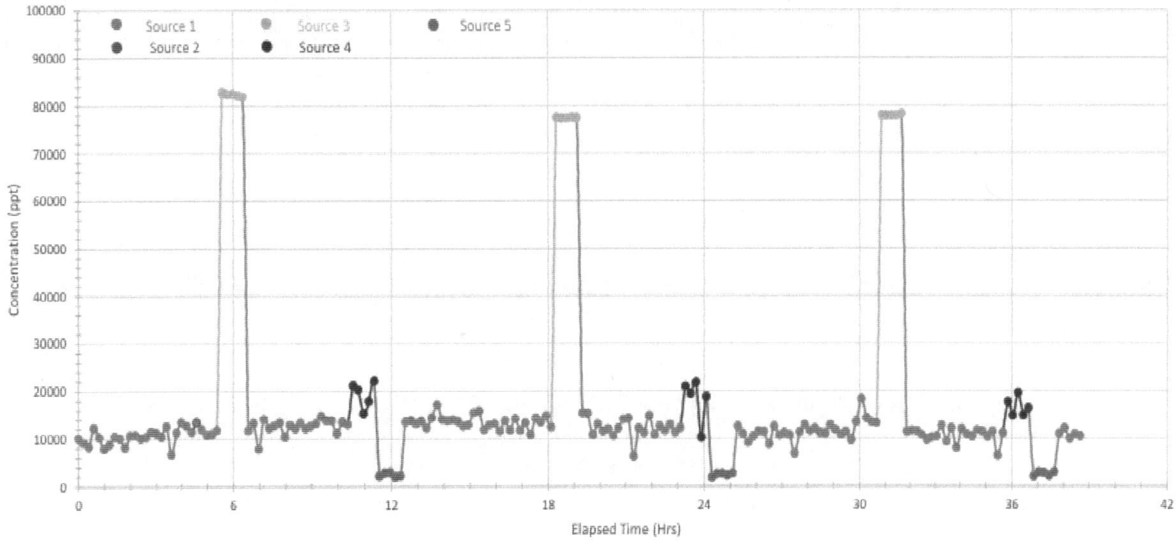

Elemental Scientific

Monitoring of Diethylphthalate in Different UPW Sources

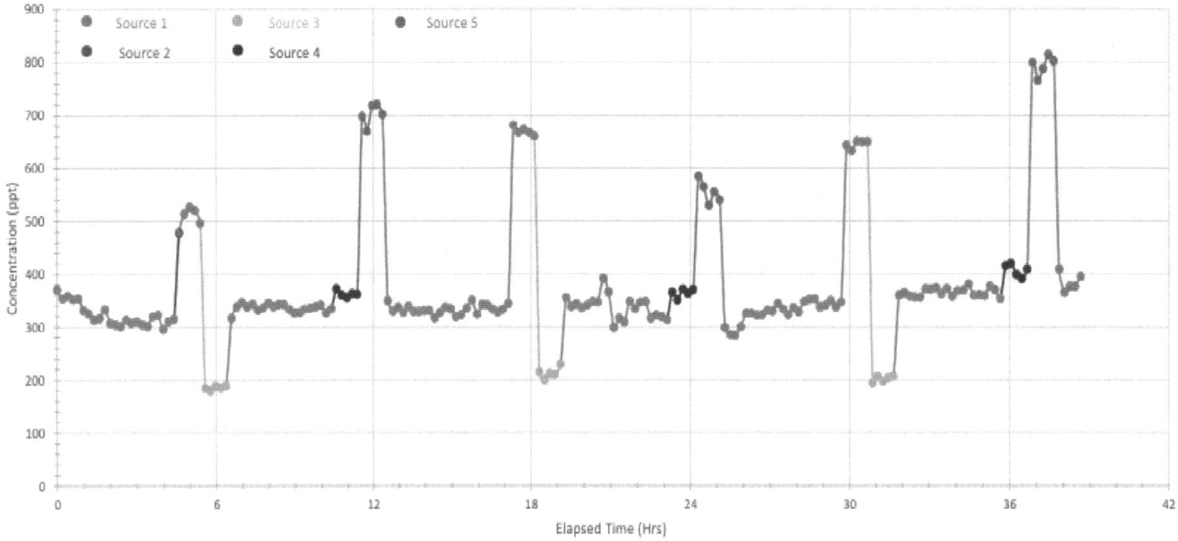

Elemental Scientific

Comparison of UPW Sources

Source	TMA (ppt)		Urea (ppt)		TEP (ppt)		DEP (ppt)	
	Average	SD	Average	SD	Average	SD	Average	SD
Source 1	554	47	11715	1911	34	4	338	22
Source 2	734	28	13158	1933	388	63	607	75
Source 3	568	34	79262	2265	26	1	200	14
Source 4	1190	36	18104	3302	33	2	378	23
Source 5	158	12	2428	374	98	9	675	102

	TMA	Urea	TEP	DEP
LOD	15 ppt	8 ppt	1 ppt	1 ppt

Elemental Scientific

Monitoring of Methanesulfonic Acid

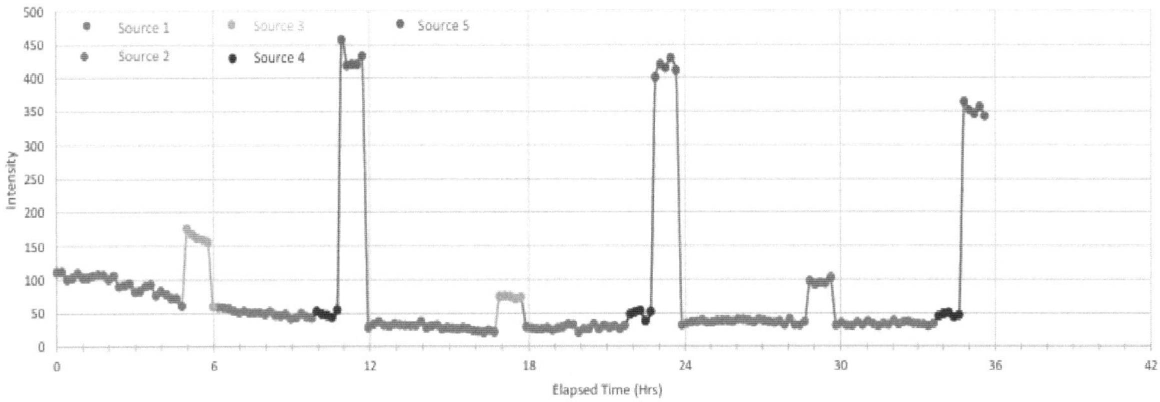

Elemental Scientific

Monitoring of Trifluoroacetic Acid

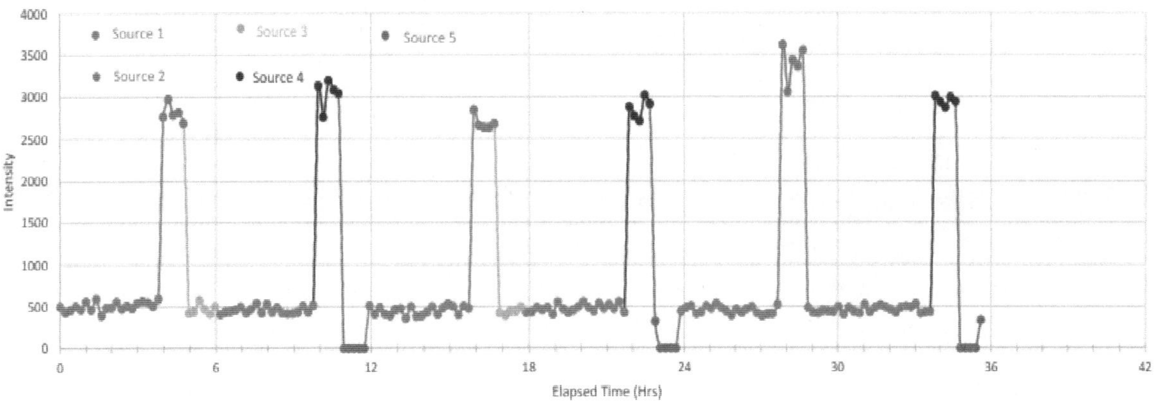

Monitoring of Pentafluoropropionic Acid

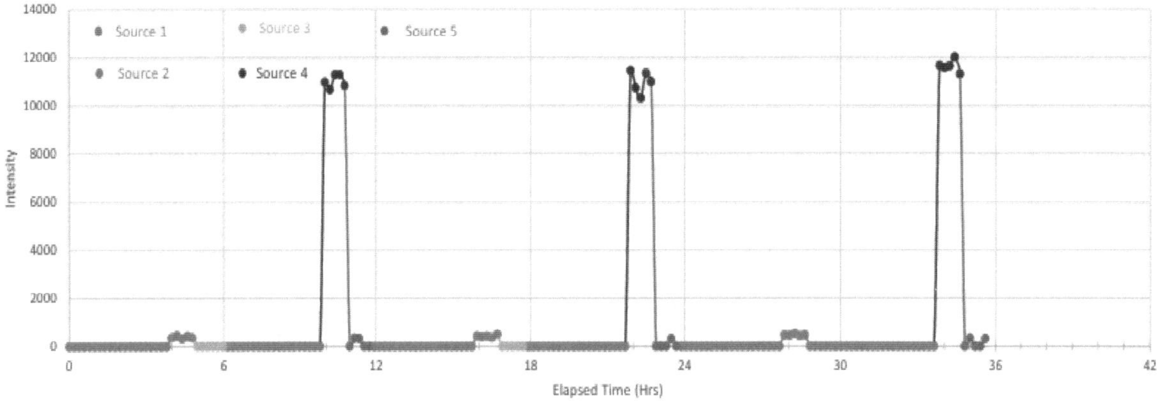

Instrumentation Used for Analysis

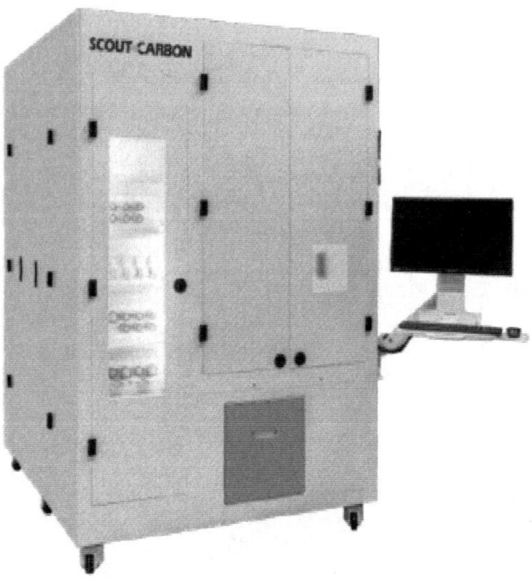

SCOUT CARBON

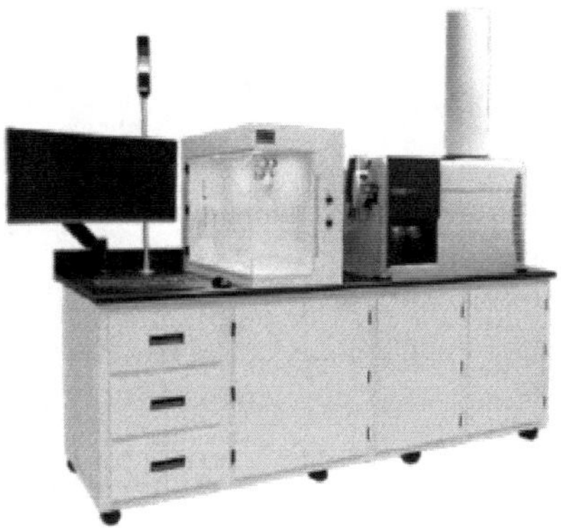

prepFAST CARBON

Molybdenum Wet Etchant with Dual Dissolver System for 3D NAND Fabrication

Chia-Wei Wang, Shu-Hao Yeh, Gregory Cleveland, Rebecca Siegelman, Jean Yang, Justin Kim

Phone Number: +886-912252982
E-mail: chiawei.wang@dupont.com
Address: No.2 Li-Hsin 4th Road, Hsinchu Science Park, Hsinchu, 30078, Taiwan

Presentation at the Surface Preparation and Cleaning Conference (SPCC) 2024
Apr. 24, 11:50 – 12:10 MST

DuPont E&I ST
Advanced Cleans & Slurry Technologies

Agenda

- **Introduction**
  - Introduction of 3D NAND
  - Word line material and recess process
  - Molybdenum etching mechanism
- **Molybdenum wet etchant for 3D NAND**
  - Design concept
  - Dual MoOx dissolver system
  - MoOx passivator
- **Summary**

Evolution from 2D NAND to 3D NAND

- 3D NAND was developed to address the limitation of conventional planar 2D NAND.

Structure of NAND memory

2D NAND

- Increase density by shrinking

3D NAND

- Increase density by stacking

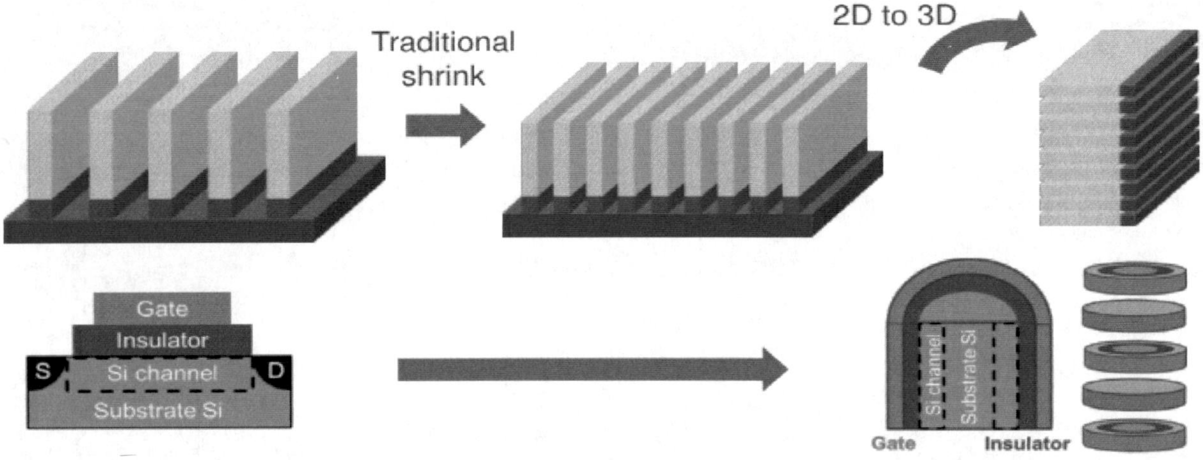

Ref: *"Equipment and Materials for 3D NAND Manufacturing"* Report by Yole (2020)

Word line material selection for advanced 3D NAND

- New word line material is needed to replace W for advanced 3D NAND and currently Mo is the most promising candidate.

Stacking scheme

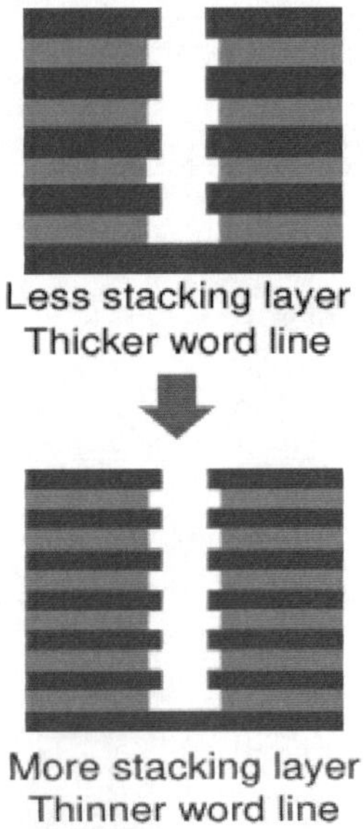

Physical properties of word line materials

Material	Bulk resistivity (µohm-cm)	Electron mean free path (nm)	Bulk resistivity x mean free path
W	5.3	15.5	81.8
Mo	5.3	11.2	59.8
Ru	7.8	6.6/4.9	51.4/38.1

Ref: *J. App. Phys. 119, 085101 (2016)*

Comparisons between Mo and Ru as word line material

Resistivity	Sub-threshold swing	Retention	Cost
Ru > Mo	Mo > Ru	Ru ≈ Mo	Mo << Ru

Ref: *IMEC in 2021 Symposium on VLSI technology*

Word line recess process in 3D NAND fabrication

- Novel Mo wet etchant is required to meet all the criteria for word line recess process.

3D NAND fabrication flow

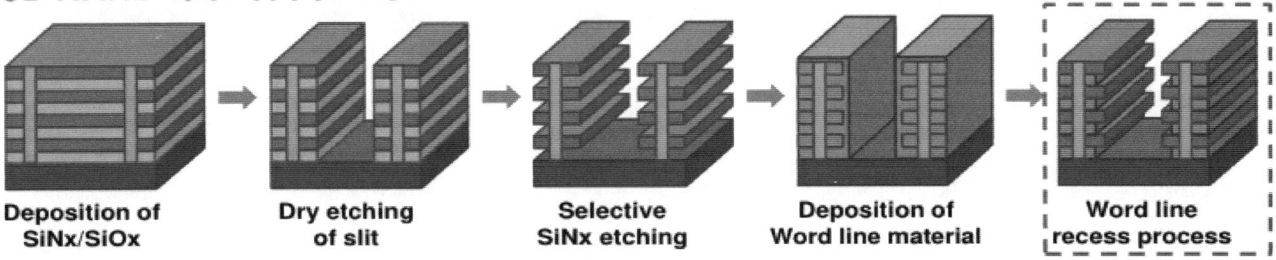

Typical defect type of word line recess

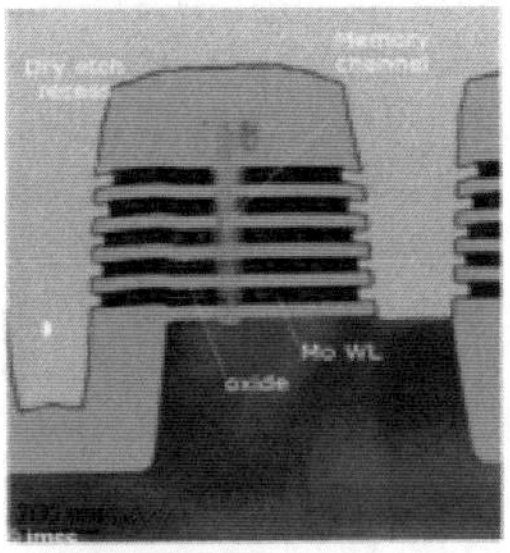

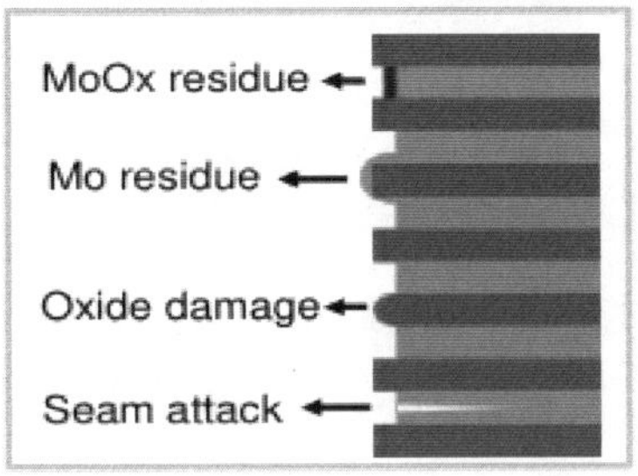

Ref: *IMEC in 2021 Symposium on VLSI technology*

Key performance requirement

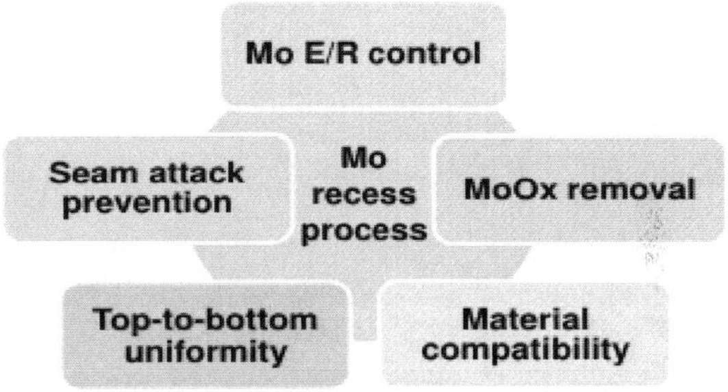

Mechanism of Mo etching under different pH

- In order to perform Mo etching, it is necessary to first oxidize Mo into MoOx and subsequently dissolver it.
- We developed our Mo etchant in strong acidic range (pH < 1) due to controllable MoOx dissolution.

Mo etching mechanism

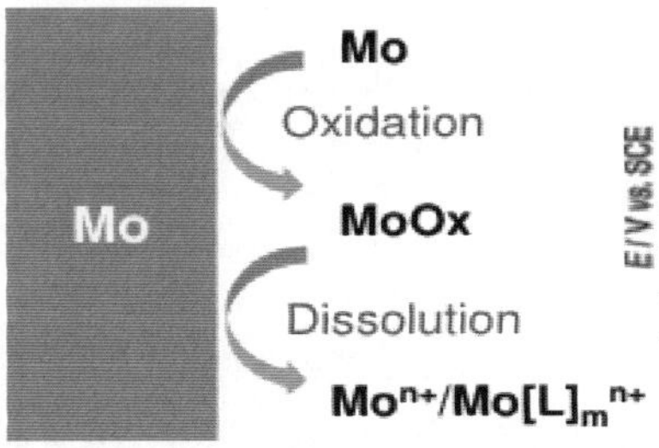

Pourbaix diagram of Mo

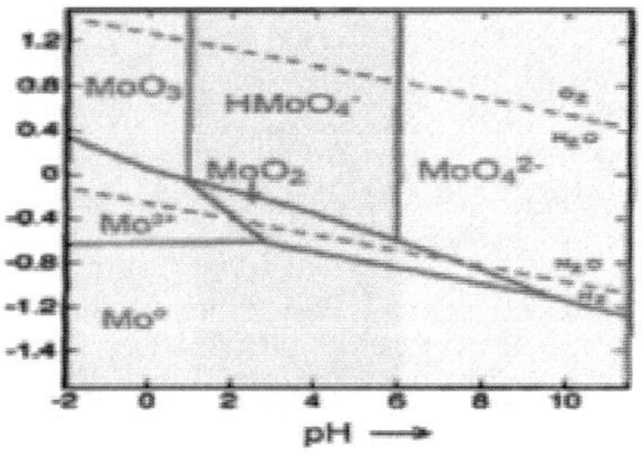

Ref: *Adv. Mater.* **2017**, *29*, 1701619.

Mo etching behavior under different pH range

pH range	pH < 1	1 < pH < 6	pH > 6
Mo oxidation	$Mo \rightarrow Mo^{3+} \rightarrow MoO_3$	$Mo \rightarrow MoO_2 \rightarrow HMoO_4^-$	$Mo \rightarrow MoO_2 \rightarrow MoO_4^{2-}$
MoOx dissolution	Slow ($MoO_3 + H_2O \rightarrow HMoO_4^- + H^+$)	Fast ($HMoO_4^-$ directly dissolve in water)	Fast (MoO_4^{2-} directly dissolve in water)
Remark	Need MoOx dissolver for MoOx dissolution control	Difficult to control MoOx dissolution since the dissolution of $HMoO_4^-$ and MoO_4^{2-} are too fast	

Relationships between MoO_3 dissolution and performance

- Etching performance is strongly correlated with MoO_3 dissolution rate.

Fast MoO_3 dissolution rate

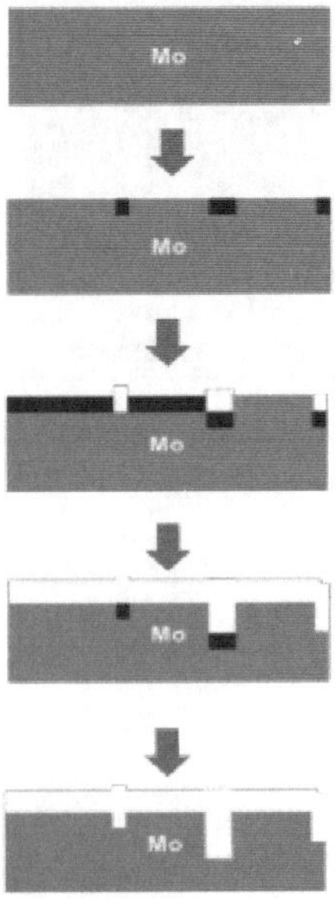

Nonuniform oxidation to MoO_3

MoO_3 quickly dissolved while some of the Mo surface was not oxidized into MoO_3

Non uniform etching and seam attack

Optimal MoO_3 dissolution rate

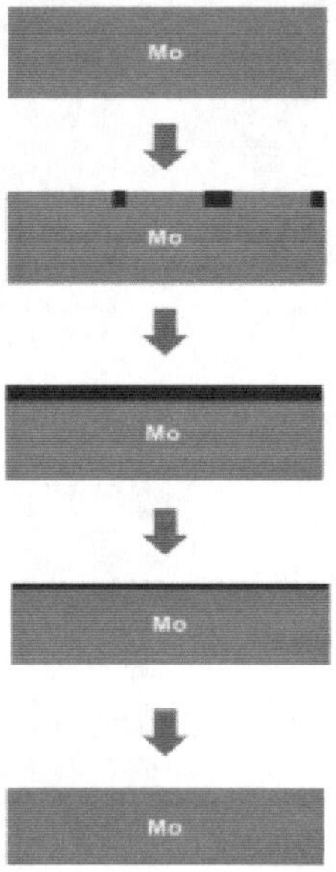

Nonuniform oxidation to MoO_3

MoO_3 was slowly dissolved due to weak solubility while Mo was further oxidized into MoO_3.

Uniform etch without seam attack issue

Design concept of Molybdenum etchant

Molybdenum wet etchant

Mo Oxidant A
- Oxidize Mo into MoO_3
- Control Mo oxidation rate

\+

MoOx dissolvers B & C
- Dissolve MoO_3
- Dual dissolver system to control dissolution.
- Control E/R and etch profile

\+

MoOx passivator P (optional)
- Passivate MoO_3
- Slow down MoO_3 dissolution.
- Control E/R and uniformity

Etching behavior of dual MoOx dissolver system

- Mo E/R can be controlled by dual MoOx dissolvers and slow and stable Mo E/R could be achieved under high MoOx dissolver B and low co-dissolver C.

MoOx dissolution mechanism

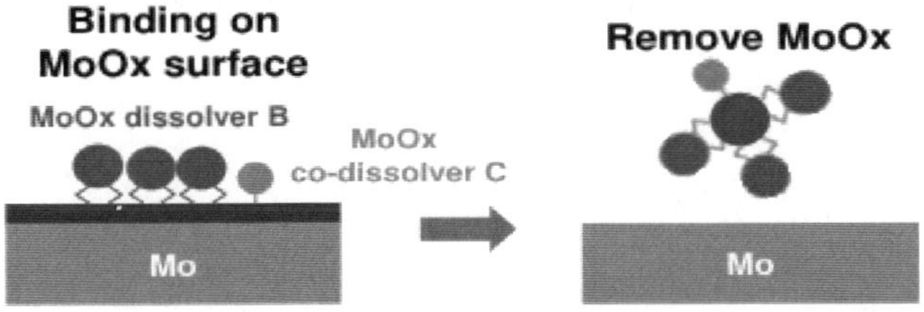

Mo E/R by dual MoOx dissolvers

$$E/R \propto \frac{[B][C]}{(1 + K_B[B] + K_c[C])^n}$$

Mo E/R under different dissolver loading

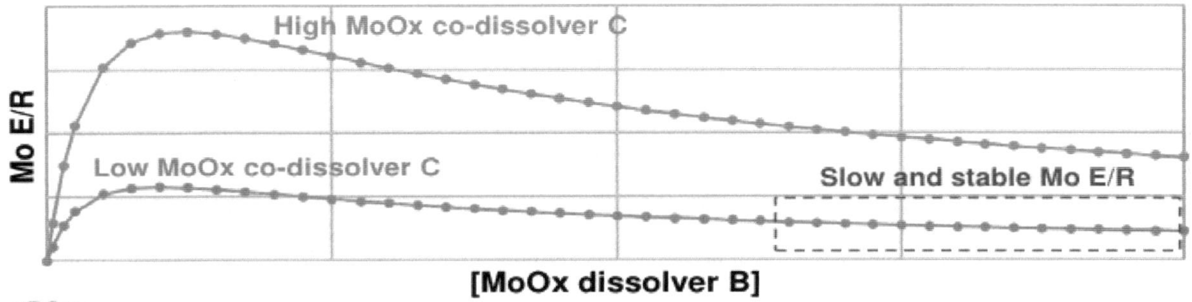

Comparisons between single and dual MoOx dissolver

- Dual MoOx dissolvers showed better Mo E/R control than single MoOx dissolver, which is advantageous for uniformity control.

System	Single MoOx dissolver	Dual MoOx dissolvers
Mechanism		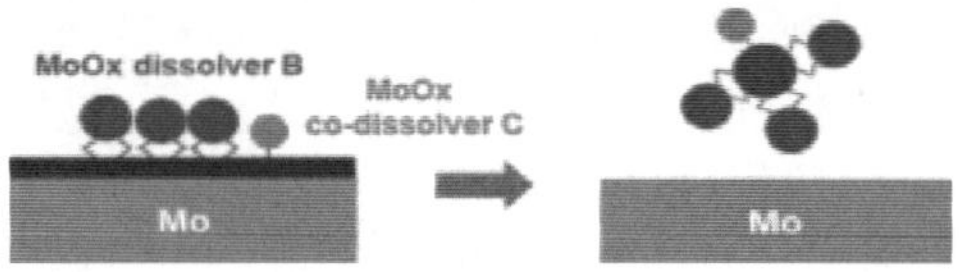
E/R behavior	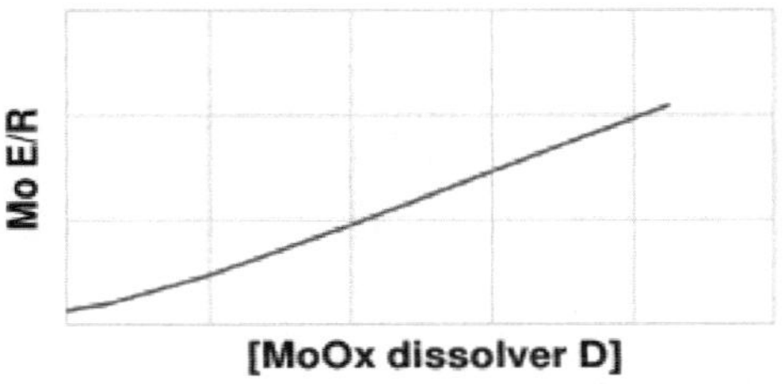	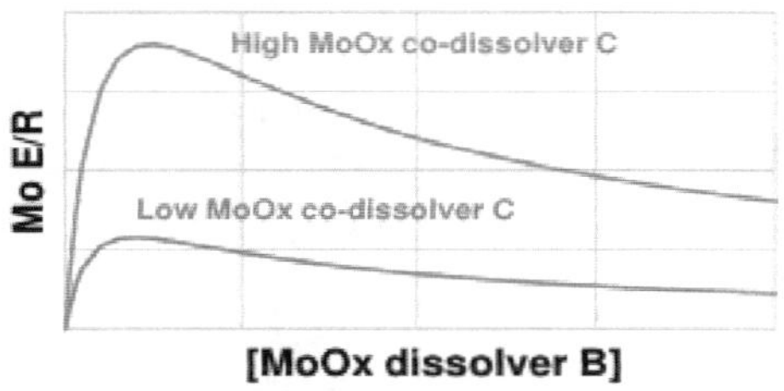
Remark	Mo E/R was proportional to MoOx dissolver loading. → Difficult to control Mo E/R	Mo E/R could be controlled by both dissolver and co-dissolver → Better E/R and uniformity control

Uniformity control by dual MoOx dissolver system

- Uniform etch (E/R_{top} / E/R_{bottom} =1) could be achieved after optimizing dual MoOx dissolver systems (high dissolver B + low co-dissolver C).

Mo E/R in structure (different depth)

$$E/R_x \propto \frac{[B]_x[C]_x}{(1 + K_B[B]_x + K_c[C]_x)^n}$$

$$[B]_x \propto [B]_{bulk} \cdot e^{-x/l_B} \quad [C]_x \propto [C]_{bulk} \cdot e^{-x/l_C}$$

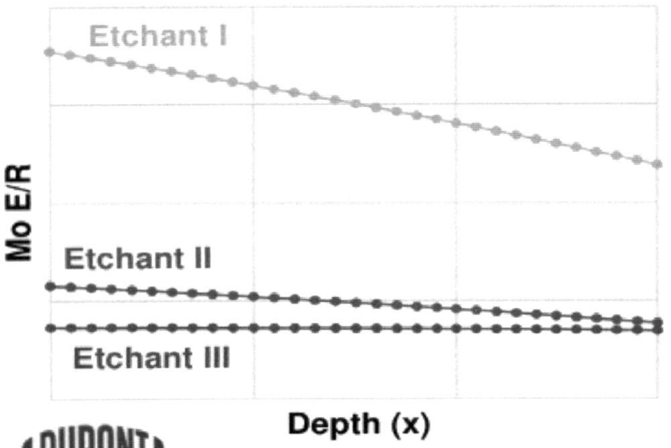

Etchant	I	II	III
MoOx Dissolver B	1X	1X ➡	5X
MoOx co-dissolver C	1X ➡	0.3X	0.3X
E/R_{top}	4.2X	1.5X	1X
E/R_{bottom}	3.1X	1.3X	1X
$\frac{E/R_{top}}{E/R_{bottom}}$	1.4 ➡	1.1 ➡	1.0

MoOx passivator as another uniformity tuning knob

- We further identified MoOx passivator could be another tuning knob to modulate uniformity performance by stronger inhibition on top E/R.

Introduce MoOx passivator to modulate uniformity performance		
MoOx Dual Dissolvers B & C	✓	✓
MoOx passivator P	×	✓
Performance in structure	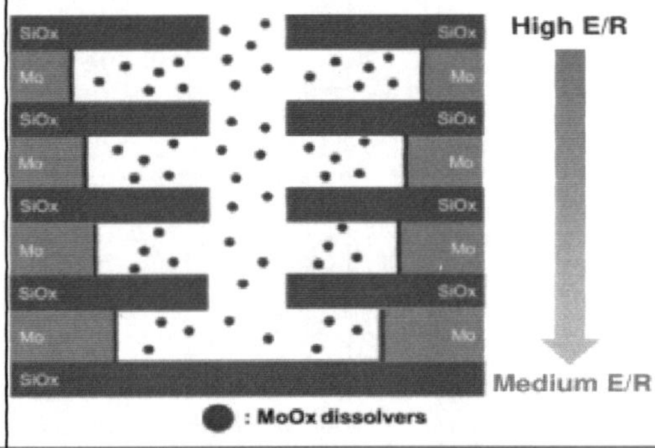	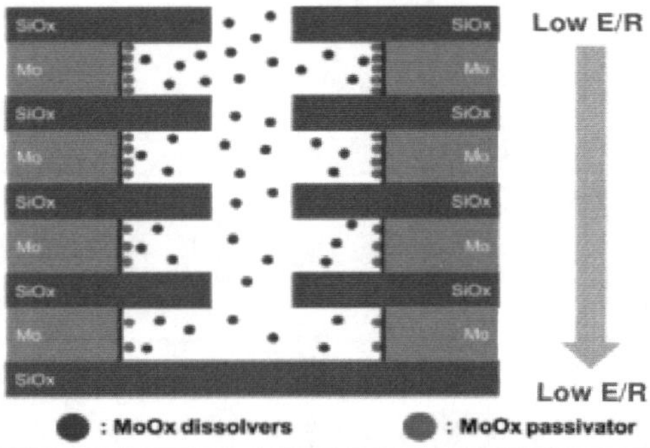
Remark	Higher E/R at via top	Uniform etch by inhibiting top E/R

MoOx passivator as another uniformity tuning knob

- MoOx passivators could block MoOx surface to inhibit MoOx dissolution efficiency and further modulate uniformity performance.

Mo E/R in structure (different depth)

$$E/R_x \propto \frac{[B]_x[C]_x}{(1 + K_B[B]_x + K_c[C]_x + K_P[P]_x)^n}$$

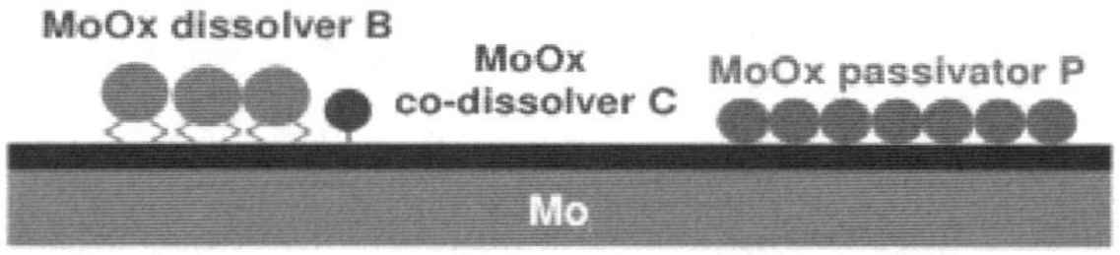

$[B]_x \propto [B]_{bulk} \cdot e^{-x/l_B}$ $\quad [C]_x \propto [C]_{bulk} \cdot e^{-x/l_C}$ $\quad [P]_x \propto [P]_{bulk} \cdot e^{-x/l_P}$

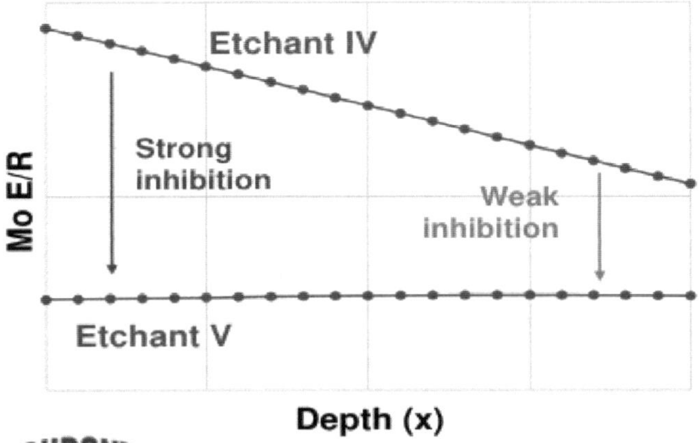

Etchant	IV		V
MoOx Dissolvers	✓		✓
MoOx passivator	×	→	✓
E/R_{top}	3.9X	→	1X
E/R_{bottom}	2.2X	→	1X
$\frac{E/R_{top}}{E/R_{bottom}}$	1.8		1.0

Summary

- Since Molybdenum is the most promising word line material for advanced 3D NAND structure, novel molybdenum wet etchant is necessary to meet the criteria for word line recess process.
- MoOx dissolution efficiency is very critical to etching performance and we developed our Molybdenum etchant in strong acidic range (pH < 1) due to controllable MoOx dissolution efficiency.
- Dual MoOx dissolver systems showed superior E/R control than single MoOx dissolver, and uniform etch could be achieved after optimizing the loadings of MoOx dissolver B and co-dissolver C.
- We also identified MoOx passivators could be another tuning knob to modulate uniformity performance by strong inhibition on Molybdenum E/R at via top.
- Our molybdenum wet etchant not only can achieve uniform etch profile, it can also prevent seam attack and MoOx residue issue after optimizing MoOx dissolution efficiency.

Acknowledgement

❖ **I would like to express my appreciation to all the group members involved in this research.**

- *Shu-Hao Yeh*
- *Jean Yang*
- *Kyong Sik Chin*
- *Carol Cheng*
- *Lester Li*
- *Owen Liu*
- *Amy Chen*
- *Annabelle Wang*
- *Vic Lin*
- *Chin Feng Yeh*
- *Jerry Chang*
- *Jenna Huang*
- *Ivy Xie*
- *Daniel Yi*
- *Tina Liu*
- *Vicky Chang*
- *Justin Kim*
- *Gregory Cleveland*
- *Rebecca Siegelman*
- *Sanahan Vijayakumar*
- *Hugh Mcadams III*
- *Nora Radu*
- *Jessica Klinkenberg*

DuPont Official Use Only

SCREEN imec

Suppression of lateral etching of Work Function Metal for PN boundary scaling

Dai Ueda [1], Hikaru Kawarazaki [1], Teppei Nakano [1], Takayoshi Tanaka [1], Akiko Harumoto [2], Ju-Geng Lai [3], and Efrain Altamirano Sanchez [3]

SCREEN Semiconductor Solutions Co., Ltd. [1]
SCREEN Holdings Co., Ltd. [2]
Imec vzw [3]

Email: D.Ueda@screen-spe.eu, Tel: +49 173 4683601

SCREEN

SCREEN Semiconductor Solutions Co., Ltd.

Outline

- ◆ **Introduction**
  - Scaling challenges for Work Function Metal patterning
  - Previous study of narrow space etching
  - Approach of TiN etching rate suppression
- ◆ **Experimental**
- ◆ **Results & Discussion**
- ◆ **Summary**

Outline

- ◆ **Introduction**
  - Scaling challenges for Work Function Metal patterning
  - Previous study of narrow space etching
  - Approach of TiN etching rate suppression
- ◆ Experimental
- ◆ Results & Discussion
- ◆ Summary

Scaling challenges for Work Function Metal (WFM) patterning

[2] Y. Oniki, *et al.*, *Solid State Phenomena*, 314, (2021) 119-126.

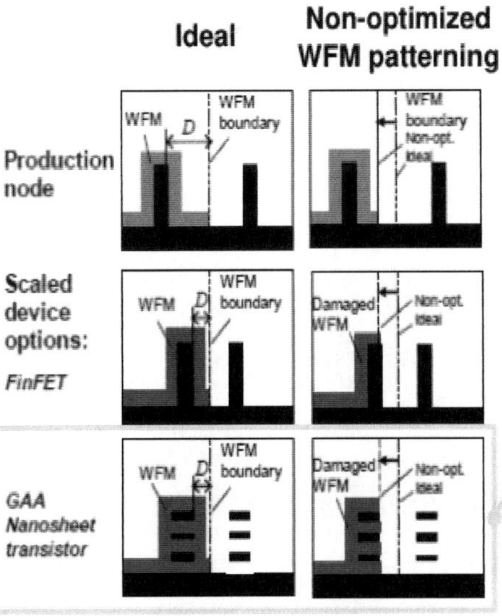

[1] V. Chang, *et al.*, VLSI 2020 (2020).

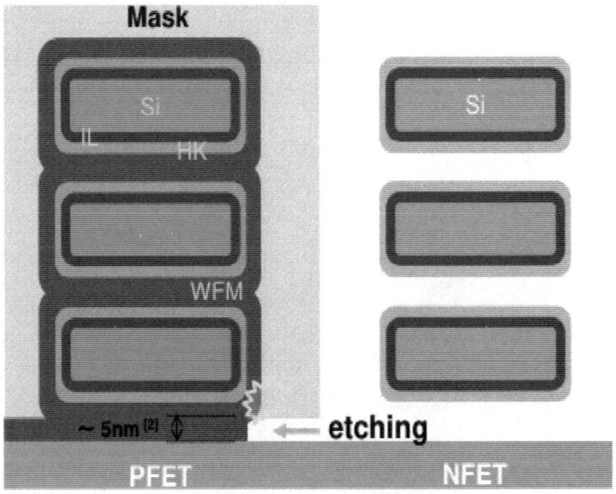

Undercut of WFM in narrow space limits scaling of PN boundary.

Suppression of lateral etching in narrow space (< 5nm) is needed.

Etching rate reduction in narrow space [3]

[3] D. Ueda, *et al.*, *Solid State Phenomena*, 314, (2021) 155-160.

Sample

- Coupon wafer
- SiO_2 thickness: 3, 5, and 10 nm

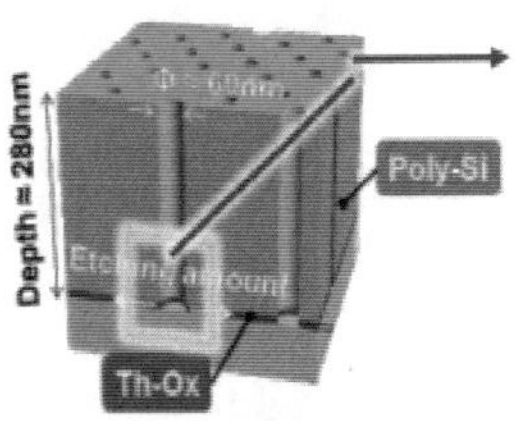

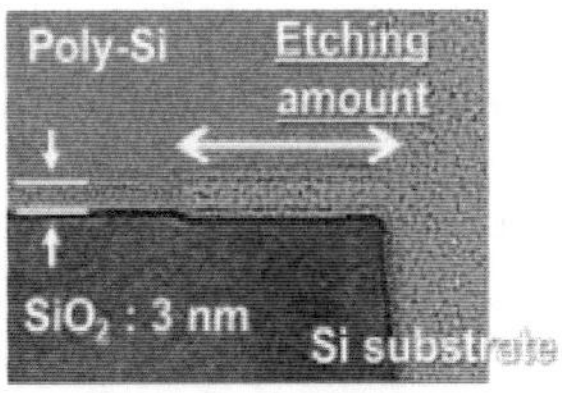

Chemical / Process

dHF (1 : 5) RT, 1 min ⇨ DIW rinse ⇨ Etching amount measurement (SEM, TEM)

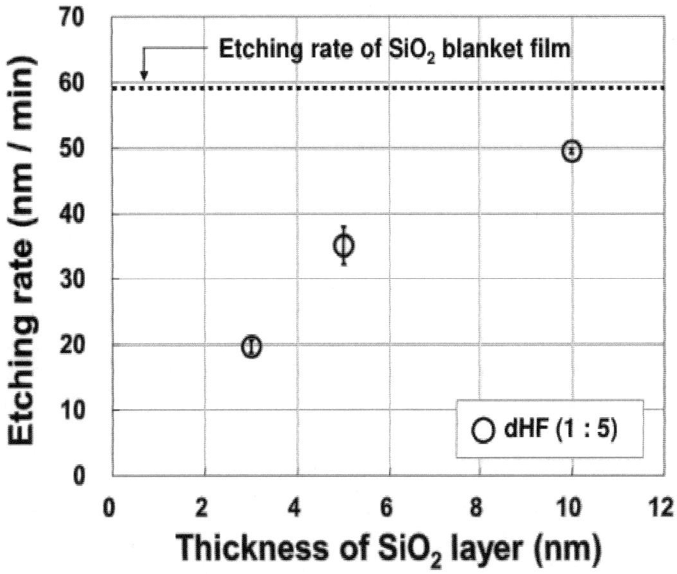

Different etching mechanism from bulk is suggested.

Hypothesis of etching rate reduction

- **Decrease in ion concentration caused by Electric Double Layer (EDL) overlap [4].**

[4] A. Okuyama, *et al.*, *Solid State Phenomena*, **219** (2015), 115-118.

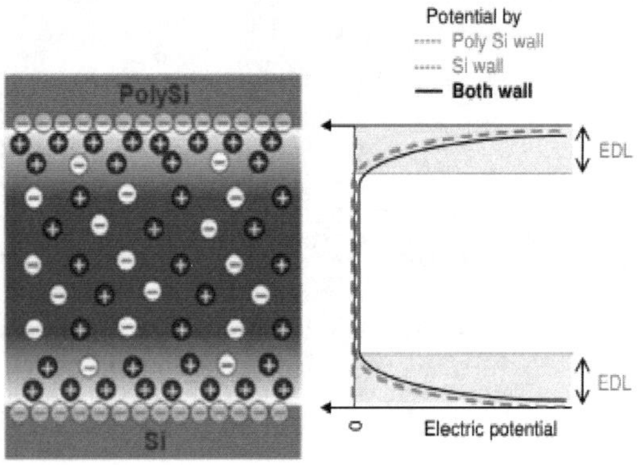

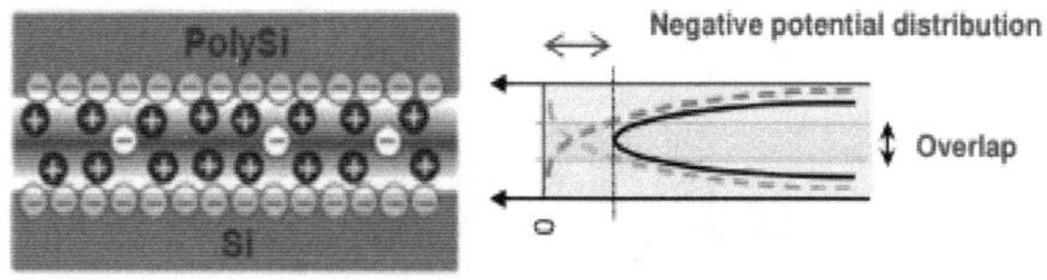

Penetration of HF_2^- ion is prevented.

Further suppression by charged additives was considered.

Change of etching rate by surfactant [3]

[3] D. Ueda, *et al.*, *Solid State Phenomena*, 314, (2021) 155-160.

(1) electrostatic interaction with surfactant and etchant ions, (2) adsorption on surface.

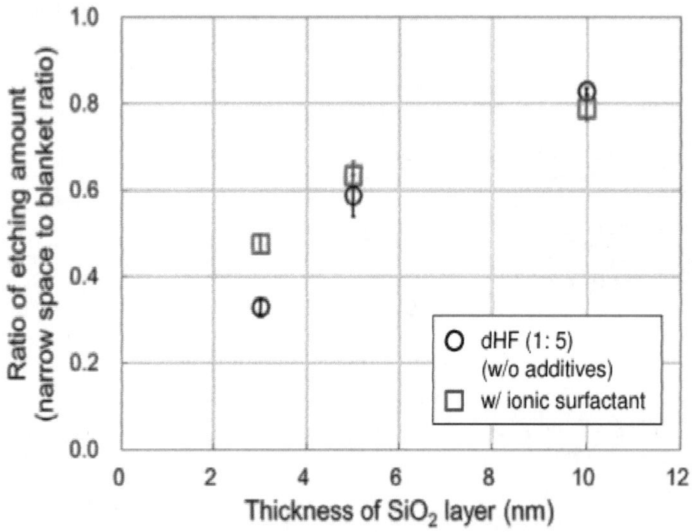

This suggests that the concentration of etchant ions in narrow space can be changed by surfactant.

Approach of TiN etching rate suppression

[5] S. Verhaverbeke, *et al.*, *Materials Research Society*, 477, (1997) 447

- **Etchant ion on H_2O_2 based solution**

  HO_2^- ion is main species for TiN etching [5].

  - TiN is etched by following HO_2^- ion,

$$H_2O_2 \leftrightarrow HO_2^- + H^+$$

oxidation process:

$$TiN + HO_2^- + 4H_2O \leftrightarrow TiO^{2+} + 4OH^- + NH_4OH$$

dissolution process by complex formation:

$$TiO^{2+} + HO_2^- \leftrightarrow TiO^{2+} \cdot HO_2^-$$

Reduction of HO_2^- ion concentration is key.

Approach of TiN etching rate suppression

- **Use anionic surfactant to reduce etchant ion in narrow space**
  - Electrostatic interaction (repulsion) to pushout etchant ion

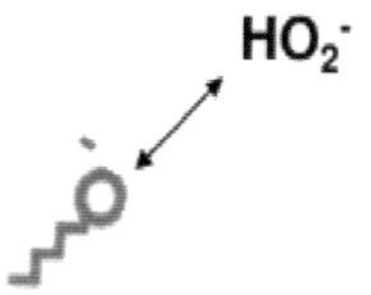

Repulsing due to same charge

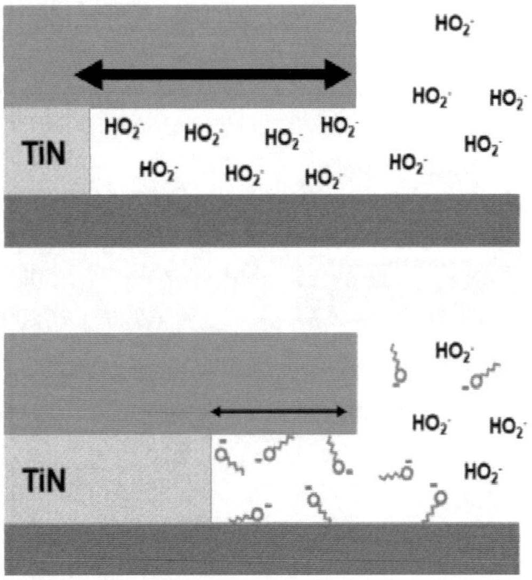

Outline

- **Introduction**
  - Scaling challenges for Work Function Metal patterning
  - Previous study of narrow space etching
  - Approach of TiN etching rate suppression
- **Experimental**
- **Results & Discussion**
- **Summary**

Experimental setup

Sample
- Coupon wafer
- TiN thickness: 4, 5, and 10 nm

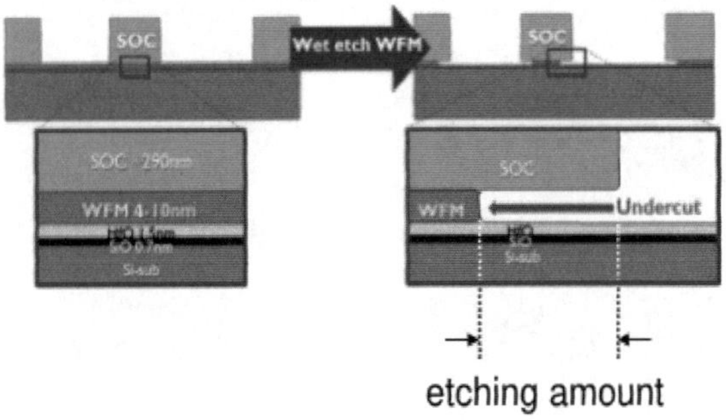

Chemical / Process

HPM (1 : 1 : 8)
50°C, 5, 10, 20 min
w/ or w/o 1 mM surfactant
➩ DIW rinse ➩ Etching amount measurement (SEM, TEM)

Surfactant
- charge impact:
  nonionic, cationic, anionic
- molecular size dependence:
  anionic with three molecular size

Etching rate
- estimated from slops of etching amount vs etching time.

Outline

SCREEN

- ◆ Introduction
  - Scaling challenges for Work Function Metal patterning
  - Previous study of narrow space etching
  - Approach of TiN etching rate suppression
- ◆ Experimental
- ◆ **Results & Discussion**
- ◆ Summary

Charge impact

imec SCREEN

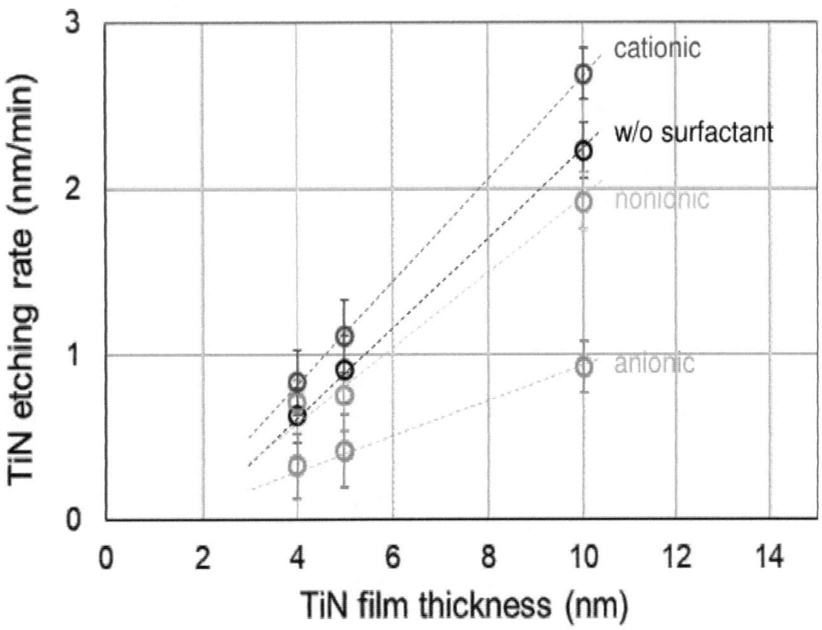

Charge has large impact on narrow space etch.

Molecular size dependence (anionic surfactant)

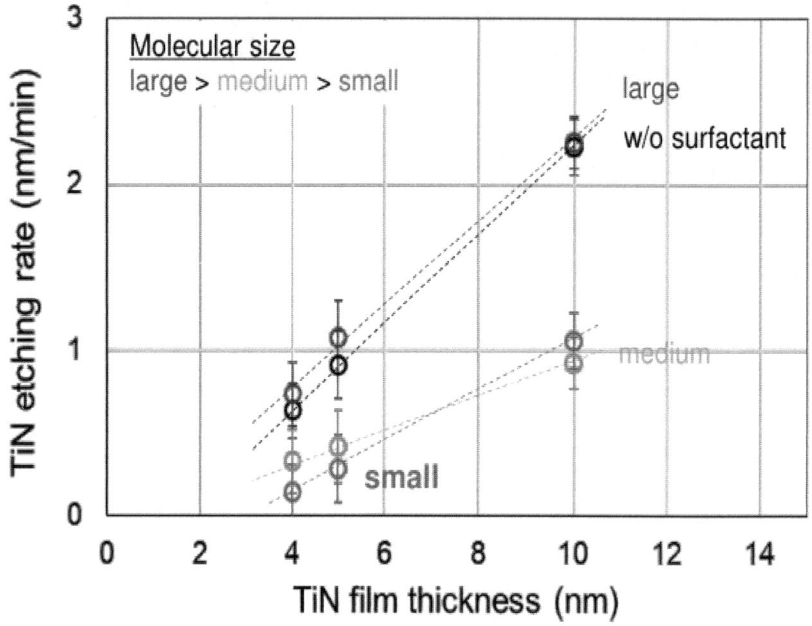

Molecular size affects etching rage in narrow space.

Outline

- Introduction
  - Scaling challenges for Work Function Metal patterning
  - Previous study of narrow space etching
  - Approach of TiN etching rate suppression
- Experimental
- Results & Discussion
- Summary

Summary

imec SCREEN

- Charge of surfactant has large impact on narrow space etch.

- Anionic surfactant with same charge to the etchant ion suppresses the etching rate in narrow areas.

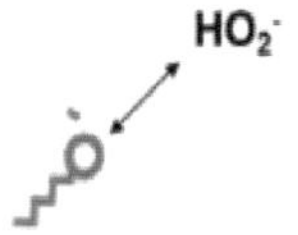

Repulsing due to same charge

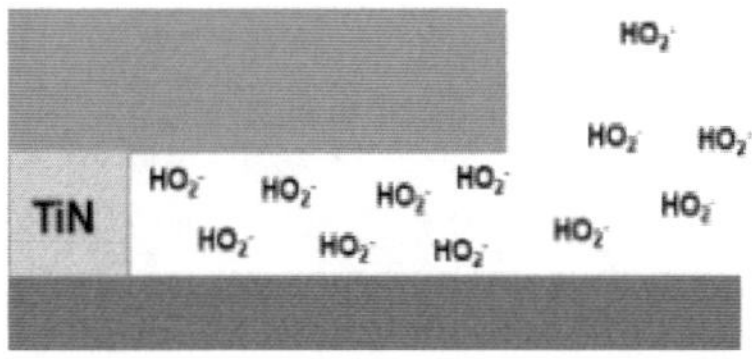

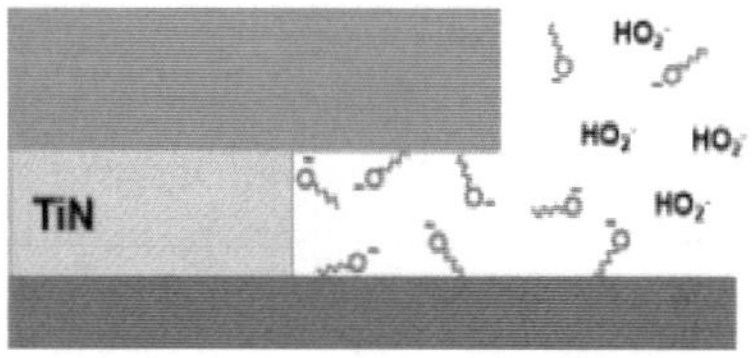

SCREEN

ADVANCING
TECH & INNOVATION

SCREEN Semiconductor Solutions Co.,Ltd.

Innovation for a Sustainable World

SCREEN

Extreme Silicon Thinning For Back Side Power Delivery Network Application: Need For a Highly Selective Alkaline Silicon Etch Stopping On SiGe Etch Stop Layer

Farid Sebaai [1], Roger Loo[1], Anne Jourdain[1], Efrain Altamirano Sanchez[1], Eric Beyne[1], Hikaru Kawarazaki[2], Teppei Nakano[2].

[1] IMEC vzw, Kapeldreef 75, B-3001 Heverlee, Belgium
[2] SCREEN Semiconductor Solutions Co., Ltd., 480-1, Takamiya-cho, Hikone, Shiga 522-0292, Japan

Overview

- **Introduction**
- **Challenge for the current back side silicon thinning**
- **Back side silicon thinning with $Si_{075}Ge_{0.25}$:B ESL**
- **Back side silicon thinning with $Si_{075}Ge_{0.25}$ ESL**
- **Summary**

- **Introduction**
- Challenge for the current back side silicon thinning
- Back side silicon thinning with $Si_{075}Ge_{0.25}$:B ESL
- Back side silicon thinning with $Si_{075}Ge_{0.25}$ ESL
- Summary

imec

Introduction (1)

Back side power delivery network

Conventional approach: Connection through μTSV interconnect

A. Jourdain et al, ETCC, 2020

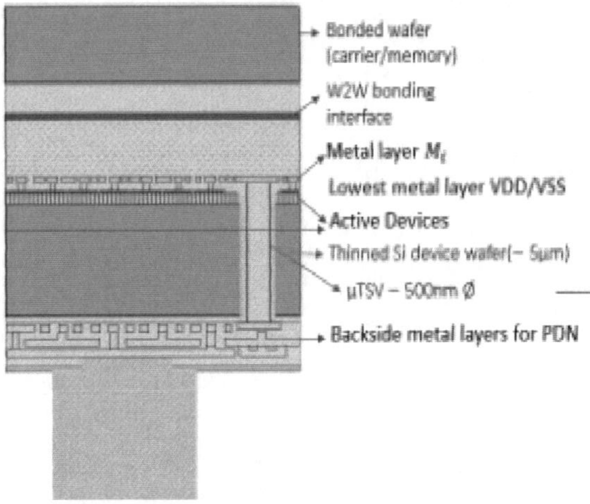

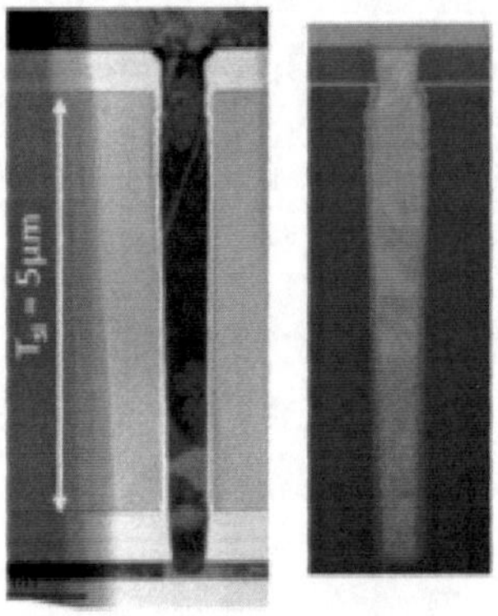

- Conventional approach where power delivery metal tracks are processed in the BEOL of the device wafer.
- Limited area overhead μTSVs : e.g. 500 nm Ø TSV at 5 μm pitch → 1% TSV area.

imec

Introduction (2)

Back side power delivery network

Connection through nTSV

A. Jourdain et al, ETCC, 2020

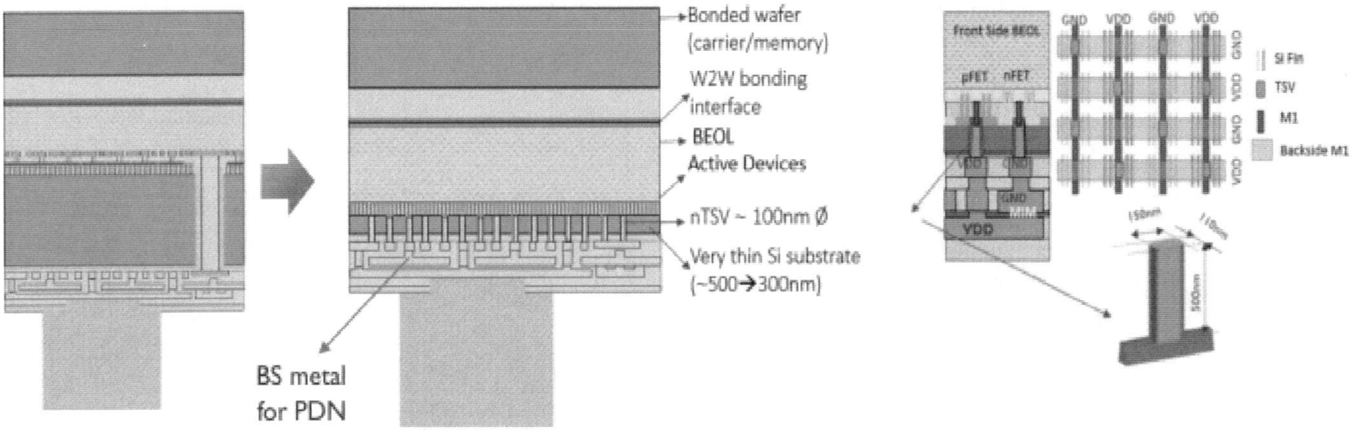

When pushing the limits a bit further and increasing the interconnect density:

- TSV becomes smaller.
- Power delivery lines are moved to the BS: Si becomes very thin
- BS becomes the new FS.

imec

Introduction (3)

Back side wafer thinning description

nTSV connected to buried power rail

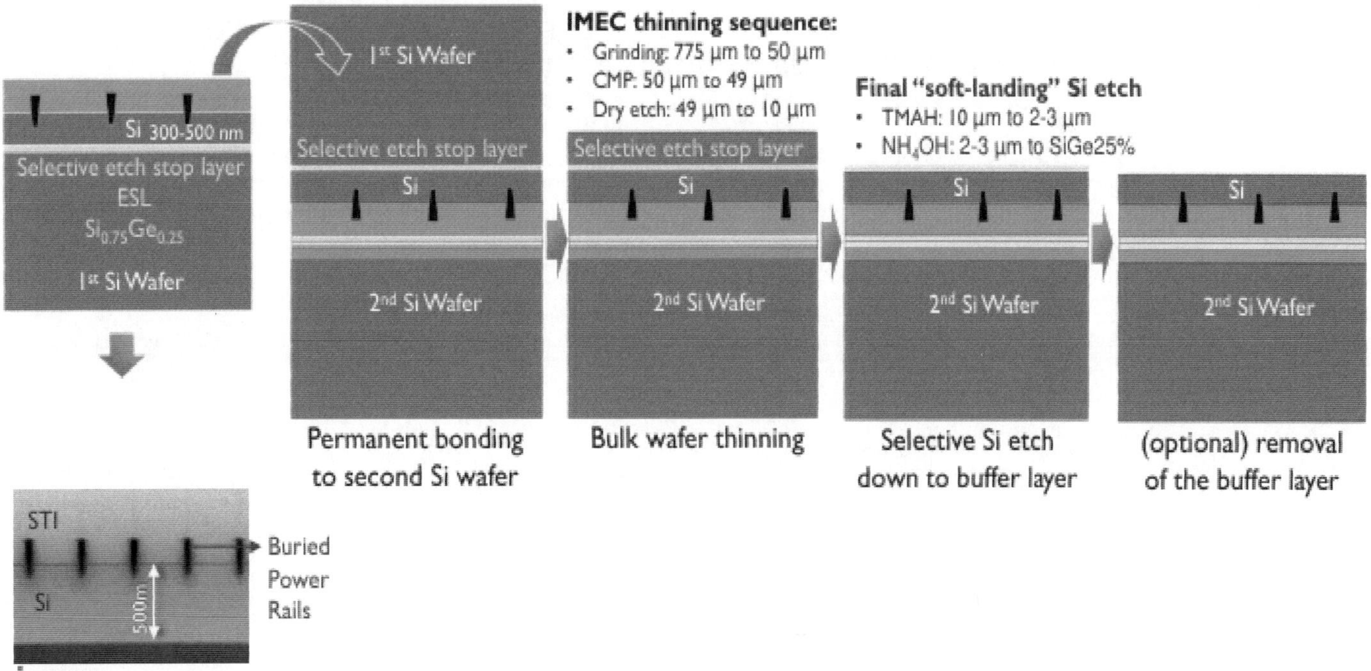

imec

Introduction (4)

Alkaline etching

→ Soft landing Si etching with TMAH + dil NH_4OH stopping on $Si_{0.75}Ge_{0.25}$.

▪ Si etching from 10 µm to 3 µm

- TMAH 5%, 80 °C → for a fast-etching rate.
- TMAH not kept down to ESL because of lack of selectivity to $Si_{0.75}Ge_{0.25}$.

▪ Si etching from 3 µm to the ESL

- Diluted NH_4OH 1_60 at 70 °C. Temperature chosen as a compromise between Si etching rate and selectivity to $Si_{0.75}Ge_{0.25}$ ESL.
- Higher intrinsic selectivity to $Si_{0.75}Ge_{0.25}$ compared to TMAH at the same conditions.

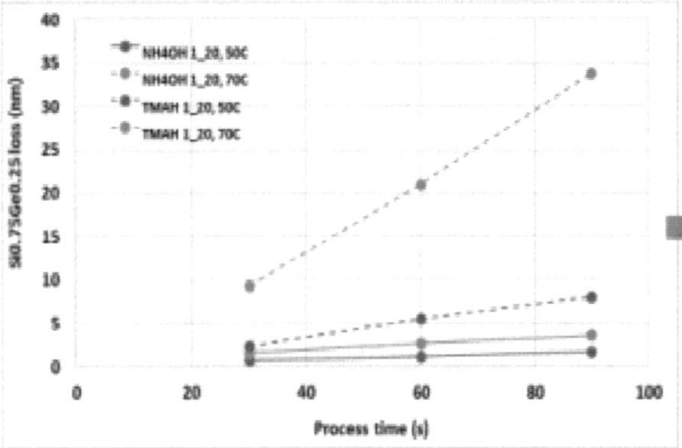

$$Si + 2OH^- + 2H_2O \rightarrow Si(OH)_2^{2+} + 2H_2.$$

- NH_4OH 1_20 → weak base, pH 9-10 → lower $[OH^-]$ →less Si etching rate for the same initial concentration.
- TMAH 1_20 → strong base, pH 14 → higher $[OH^-]$ → More Si etching for the same initial concentration.

- Introduction
- **Challenge for the current back side silicon thinning**
- Back side silicon thinning with $Si_{075}Ge_{0.25}$:B ESL
- Back side silicon thinning with $Si_{075}Ge_{0.25}$ ESL
- Summary

Challenge for the back side silicon thinning (1)

Need to reduce the SiGe ESL thickness

- For 50 nm $Si_{0.75}Ge_{0.25}$ etch stop layer (= $Si_{0.75}Ge_{0.25}$ thickness used in the current BSPDN integration) , the epitaxial growth occurs in the metastable region. → risk of SiGe relaxation is possible.

→ There is a need to growth the $Si_{0.75}Ge_{0.25}$ etch stop layer in the stable region
→ Chosen target at **<u>10 nm $Si_{0.75}Ge_{0.25}$</u>**

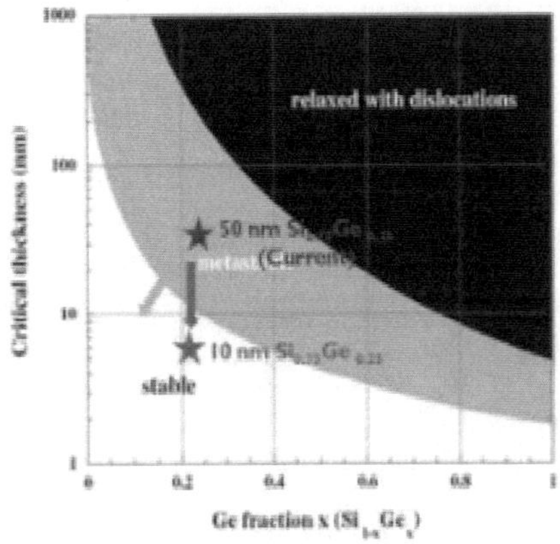

D. J. Paul, Semicond. Sci. Technol., 19, R75–R108 (2004)

SiGe growth stable region

- **Reduced SiGe thickness**

→ Stricter control of the TTV before the Si etch soft landing step with dil NH_4OH.

- **Reduced Ge concentration**

→ Less process window for the Si etch soft landing step with dil NH_4OH (= more SiGe loss).

30' @ 500°C + 30' @ 700°C

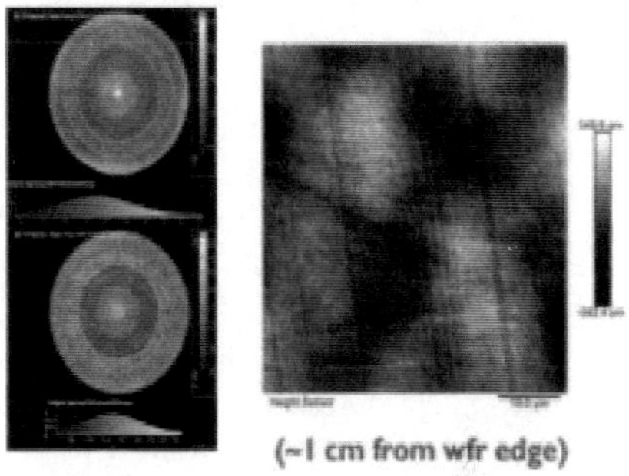

(~1 cm from wfr edge)

SP5
2.74 ppm
3.80 ppm

→ Still some misfits at the edge after EPI anneal

imec

Challenge for the back side silicon thinning (2)

TTV control during Si thinning

- The TTV during the Si thinning decides the thickness of the SiGe ESL. We historically used a high $Si_{0.75}Ge_{0.25}$ thickness of 100 nm and later 50 nm to compensate the high TTV during the Si thinning.

→ For thinner SiGe ESL, lowering the TTV specifications after grinding + CMP + dry etch + wet etch is essential.

High TTV before dil NH_4OH | Thinning not well stopping on 10 nm $Si_{0.75}Ge_{0.25}$ ESL

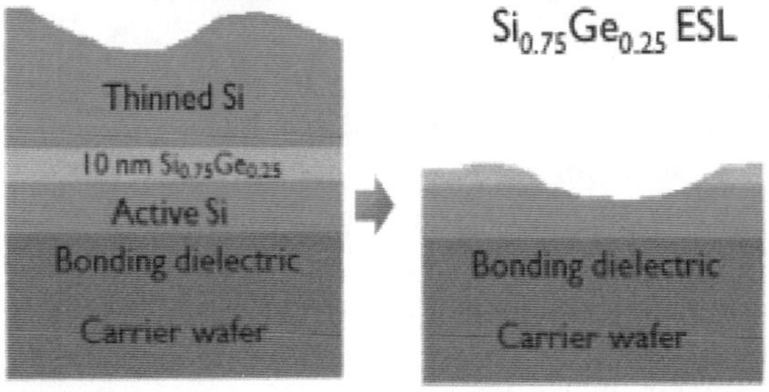

Low TTV before dil NH_4OH | Thinning well stopping inside the 10 nm $Si_{0.75}Ge_{0.25}$ ESL

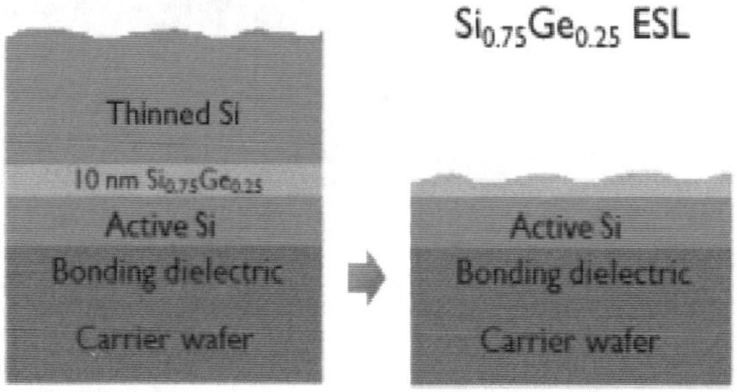

imec

- Introduction
- Challenge for the current back side silicon thinning
- **Back side silicon thinning with $Si_{075}Ge_{0.25}$:B ESL**
- Back side silicon thinning with $Si_{075}Ge_{0.25}$ ESL
- Summary

imec

Back side silicon thinning with $Si_{0.75}Ge_{0.25}$:B ESL (1)

SiGe wet etch rate reduction with boron doping

→ Tracking the mass evolution on Si_$Si_{0.75}Ge_{0.25}$ stack wafers during the wet etching with dil NH_4OH.

SiGe ESL conditions:

- 10nm
- $Si_{0.75}Ge_{0.25}$ vs $Si_{0.75}Ge_{0.25}$:B ($10^{19}/cm^3$)
- Anneal* vs no anneal

* 1000 °C_He_1.5s (activation anneal)
\+ 500 °C_steam_60min + 700C_N_2_20 min
\+ 420C_N_2/H_2_20 min (STI)

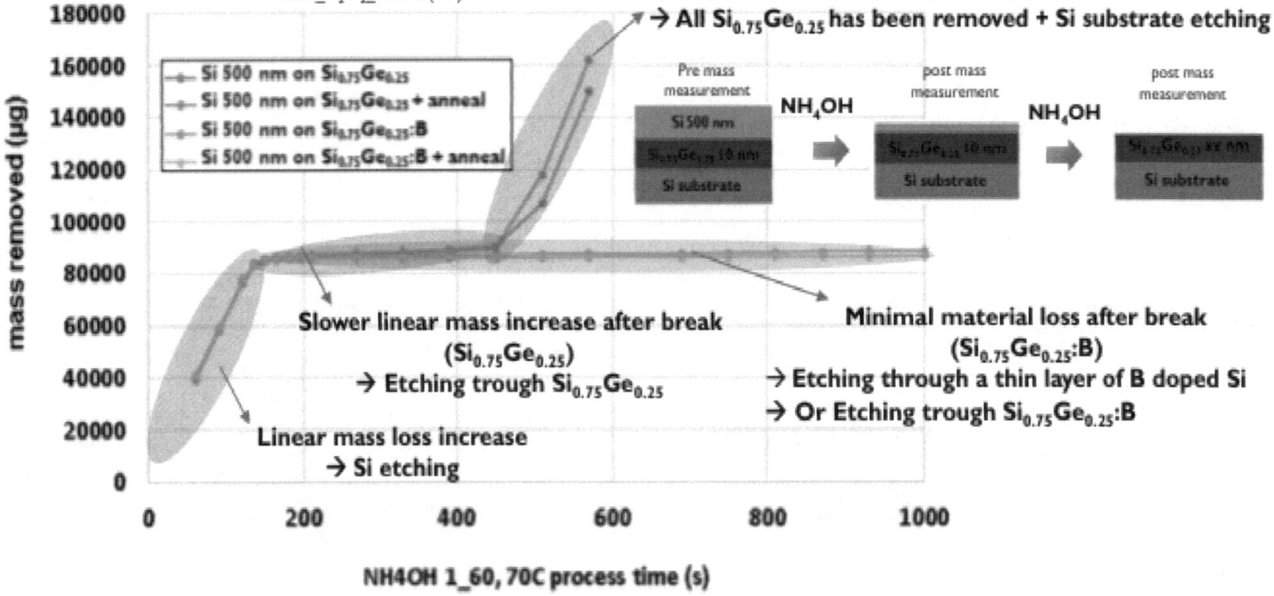

imec

Back side silicon thinning with $Si_{0.75}Ge_{0.25}$:B ESL (2)

SiGe wet etch rate reduction with boron doping

→ Tracking the mass evolution on Si_$Si_{0.75}Ge_{0.25}$ stack wafers <u>during the wet etching with dil NH4OH.</u>

- 500 nm Si on 10 nm $Si_{0.75}Ge_{0.25}$:B + anneal + **1000s Dil NH_4OH 70C (High over etch)**
- 500 nm Si on 10 nm $Si_{0.75}Ge_{0.25}$:B + anneal + **210s Dil NH4OH 70C (low over etch)**

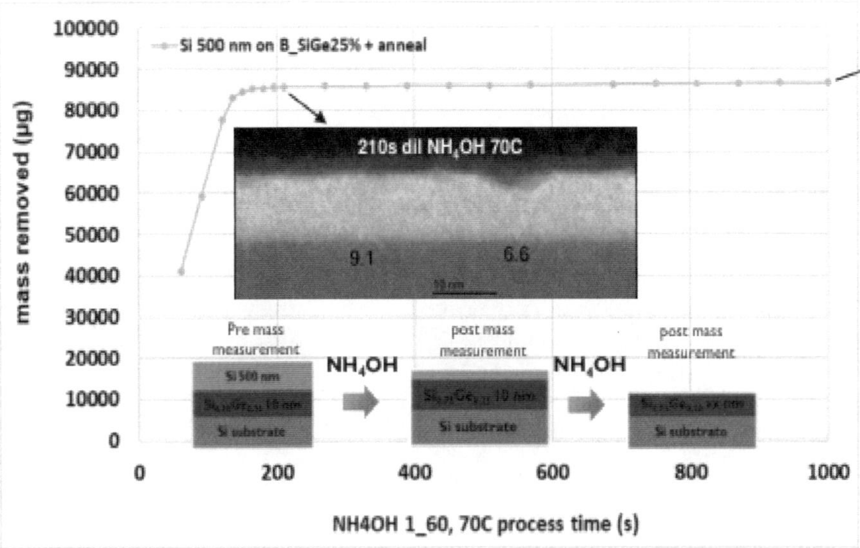

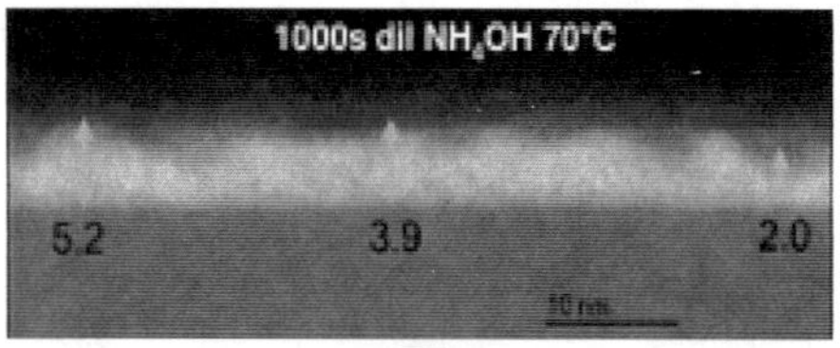

- **Dil NH4OH at 70C etches through the $Si_{0.75}Ge_{0.25}$:B with a very slow etching rate (defect etching).**

→ Selectivity Si vs $Si_{0.75}Ge_{0.25}$:B at least > 500_1

imec

Back side silicon thinning with $Si_{0.75}Ge_{0.25}$:B ESL (3)

Discussion: impact of B diffusion into the EPI Si

30 nm of B diffusion on a 500 nm EPI Si channel

30 nm B diffusion

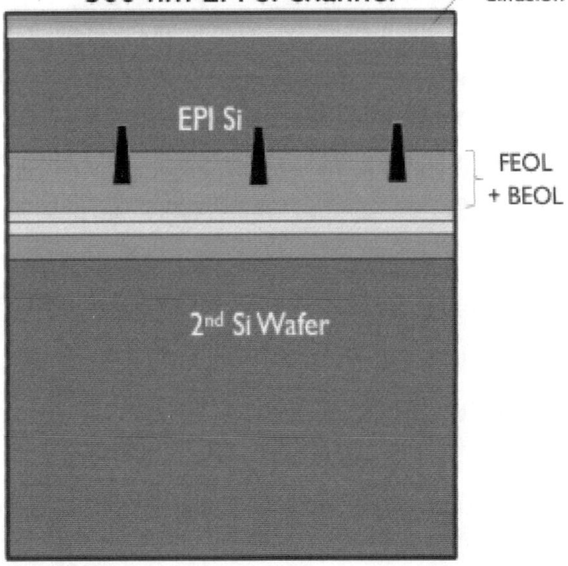

B still far from the transistors ➡ **$Si_{0.75}Ge_{0.25}$:B as an ESL seems safe enough**

imec

30 nm of B diffusion on a 50 nm EPI Si channel (= possible integration involving thinner Si channel)

30 nm B diffusion

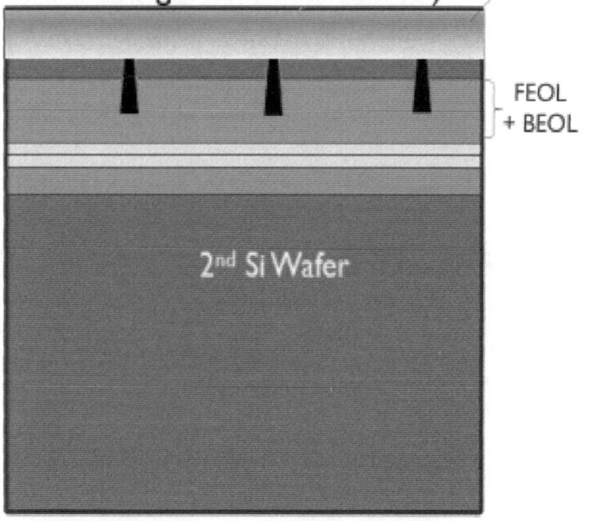

B much closer from the transistors ➡ **$Si_{0.75}Ge_{0.25}$:B as an ESL seems not safe enough**

- Introduction
- Challenge for the current back side silicon thinning
- Back side silicon thinning with $Si_{075}Ge_{0.25}$:B ESL
- **Back side silicon thinning with $Si_{075}Ge_{0.25}$ ESL**
- Summary

imec

15

Back side silicon thinning with $Si_{0.75}Ge_{0.25}$ ESL (1)

TTV control during Si thinning

- Si thinning stopping on $Si_{0.75}Ge_{0.25}$ ESL = strict TTV control during the thinning process.

- TTV optimization of the steps that constitutes the thinning process: grinding + CMP + wet etch.

- Si thinning studied case:

1/ Improved grinding + CMP → down to 50 μm.
2/ Bulk Si removal: acidic etching with HF_HNO_3 → down to 10 μm.
3/ Soft landing: improved TMAH and dil NH_4OH → down to $Si_{0.75}Ge_{0.25}$ ESL.

imec

Back side silicon thinning with $Si_{0.75}Ge_{0.25}$ ESL (2)

DISCO

TTV control after grinding + CMP

Tool: DGP8761HC from DISCO
Measurement: SUSS TTV (= interferometer, measurement on the Si thickness at the FS)

50 µm remaining Si
Si TTV: ~ 1.4 µm (100%)

Rough grinding: 681 µm of removed Si
Fine grinding: 40 µm of removed Si
CMP: 40 µm of removed Si

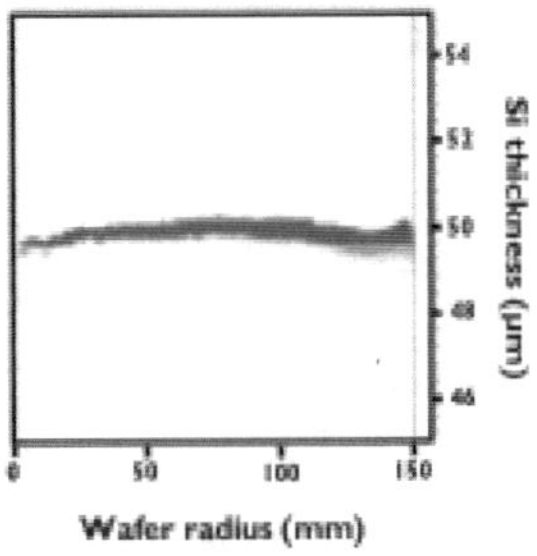

10 µm remaining Si
Si TTV: ~ 1.5 µm (100%)

Rough grinding: 721 µm of removed Si
Fine grinding: 40 µm of removed Si
CMP: 40 µm of removed Si

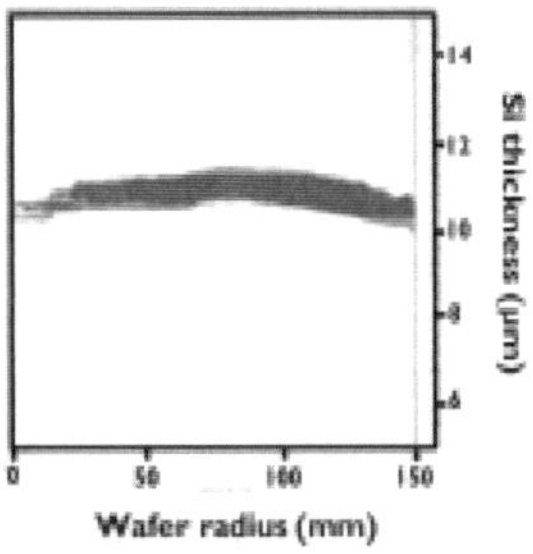

3 µm remaining Si
Si TTV: ~ 1.2 µm

Rough grinding: 728 µm of removed Si
Fine grinding: 40 µm of removed Si
CMP: 40 µm of removed Si

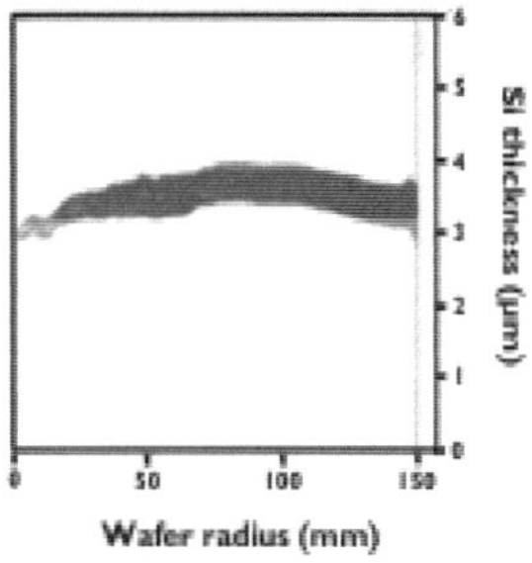

- Ability in grinding down to remaining a Si thickness < 50 µm while keeping a decent TTV.

- New thinning process sequences are being investigated depending on the remaining thickness left after grinding.
  - Grinding stops at 50 µm remaining Si: HF_HNO_3 + TMAH + dil NH_4OH.
  - Grinding stops at 10 µm remaining Si: TMAH + dil NH_4OH.
  - Grinding stops at 3 µm remaining Si: dil NH_4OH.

imec

Back side silicon thinning with $Si_{0.75}Ge_{0.25}$ ESL (3)

TTV control after wet removal (before diluted NH_4OH)

SCREEN

Grinding + CMP → 50 µm Si

+ wet thinning: HF_HNO3 + TMAV MV scan → 3 µm Si

TTV = 1.27 µm

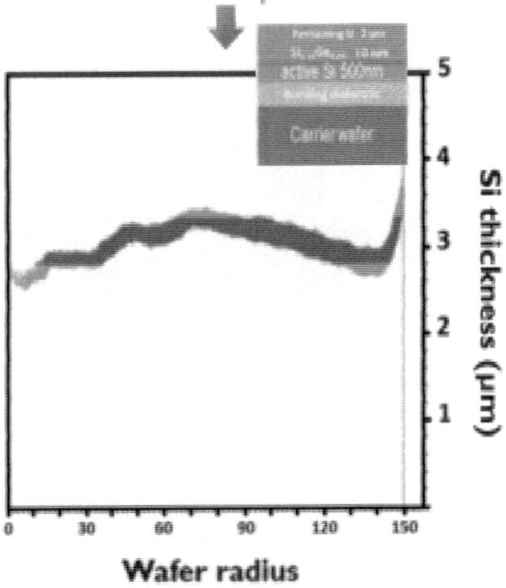

Grinding + CMP → 10 µm Si

+ wet thinning: TMAH MV scan → 3 µm Si

TTV = 1.1 µm

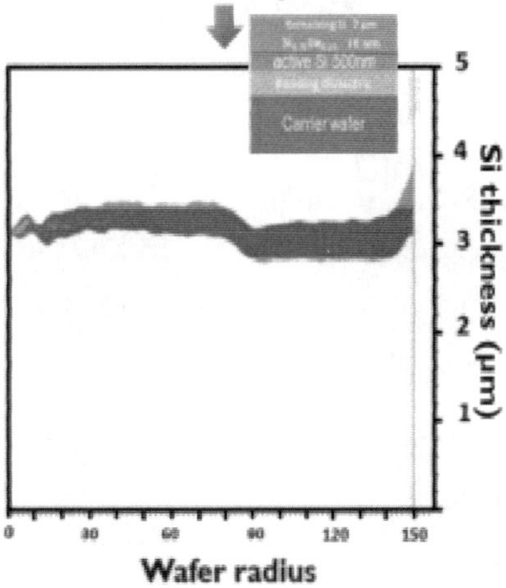

Grinding + CMP → 3 µm Si

TTV = 1.16 µm

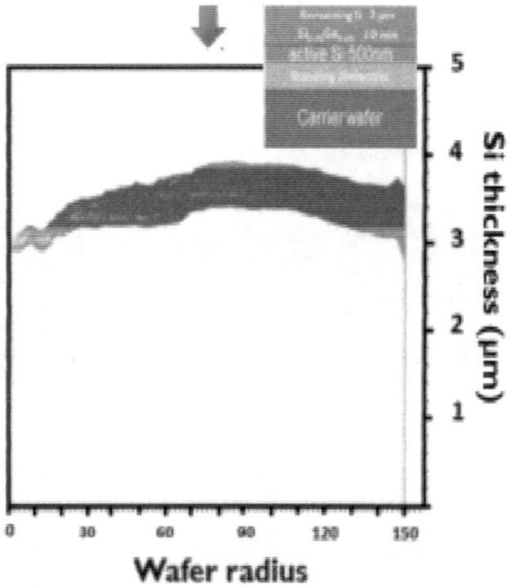

imec

Back side silicon thinning with $Si_{0.75}Ge_{0.25}$ ESL (4)

Stopping on SiGe ESL for the 3 different thinning scenarios
→ XRR

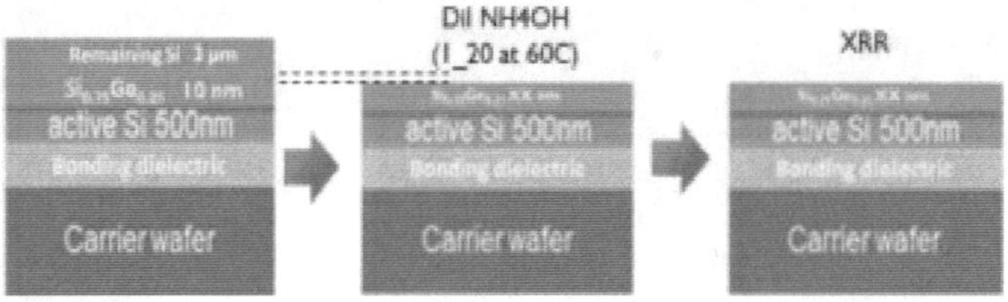

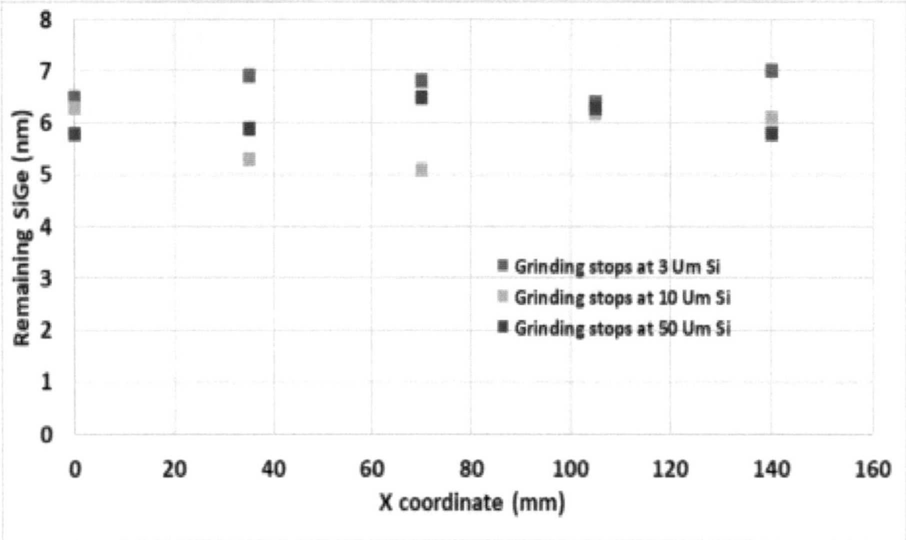

Decent remaining $Si_{0.75}Ge_{0.25}$ ESL for the 3 different thinning scenarios based on XRR measurement.

imec

Back side silicon thinning with $Si_{0.75}Ge_{0.25}$ ESL (5)

Stopping on SiGe ESL for the 3 different thinning scenarios(after diluted NH_4OH)
→ TEM

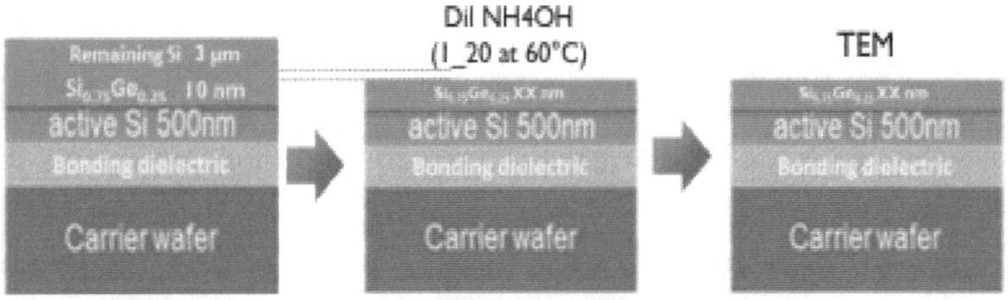

Decent remaining $Si_{0.75}Ge_{0.25}$ ESL for the 3 different thinning scenarios based on TEM as well.

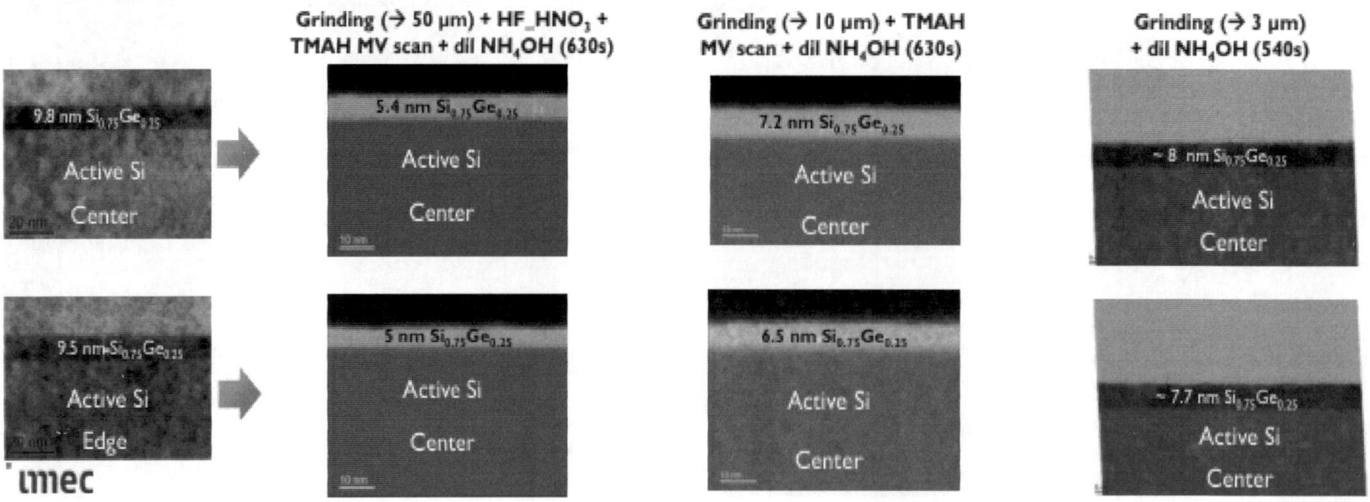

imec

- Introduction
- Challenge for the current back side silicon thinning
- Back side silicon thinning with $Si_{0.75}Ge_{0.25}$:B ESL
- Back side silicon thinning with $Si_{0.75}Ge_{0.25}$ ESL
- **Summary**

imec

21

Summary

- **In this presentation we showed you:**

➢ A description of our thinning module to enable the BSPDN process integration.

➢ A proposal for using $Si_{0.75}Ge_{0.25}$: B as an etch stop layer to enhance the selectivity of the Si etch during the last step of our thinning module with diluted NH_4OH.

➢ Some process optimization efforts to control the Si removal uniformity during the entire thinning module in case of using 10 nm $Si_{0.75}Ge_{0.25}$ as an etch stop layer.

Thanks for you for your attention !!!

farid.sebai@imec.be

imec

High Selective Wet Etching of different SiGe Layers for CFET Structures with Selectipur E-2504 Solution

SPCC Conference 2024

Francisco Javier Lopez Villanueva: francisco.lopez@basf.com
Andreas Klipp: andreas.klipp@basf.com
Farid Sebaai: farid.Sebai@imec.be
Electronic Materials
BASF SE

SPCC Conference 2024; Chandler AZ: April 24th to April 25th

BASF
We create chemistry

Overview

- Introduction
- Formulation Design Principles
- Target application
  - ▶GAA
  - ▶CFET
  - ▶3D-DRAM
- Summary

BASF
We create chemistry

Transistor Roadmap

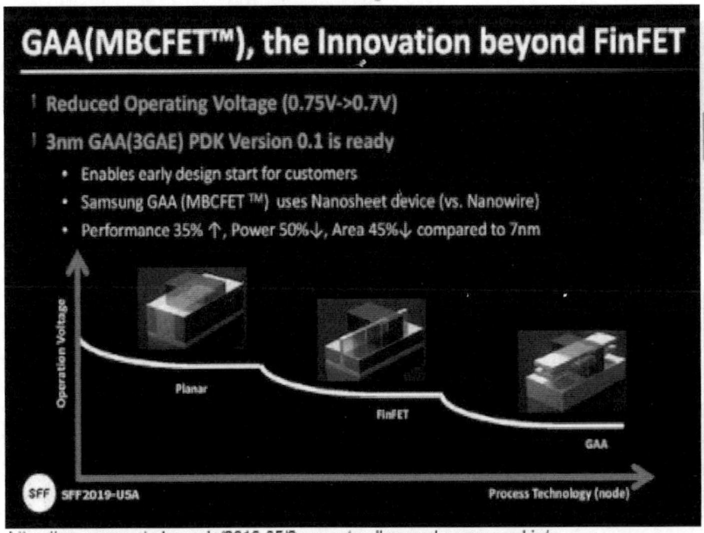

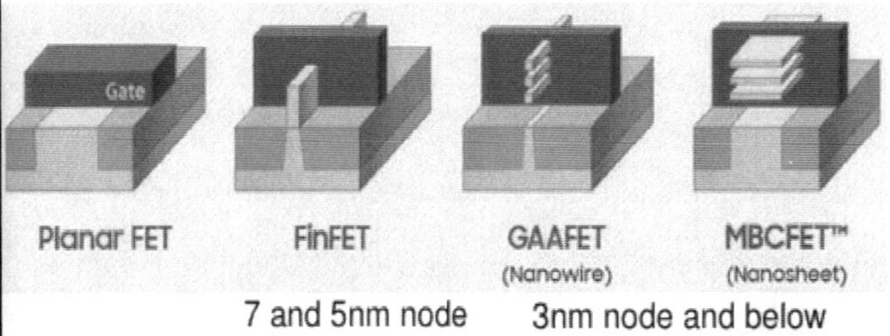

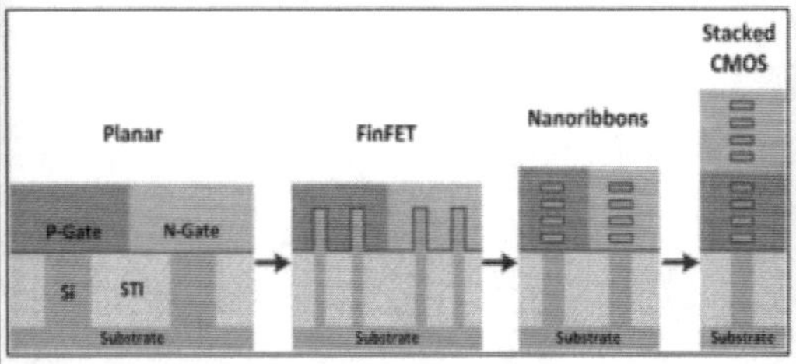

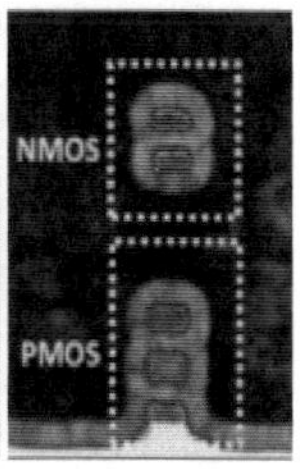

https://www.computerbase.de/2019-05/3-nm-gate-all-around-samsung-chip/

Intel

- by the introduction of the GAA FET's significant performance increase, less chip area and reduced power consumption of 50% compared to the 7nm products
- Cleaning and etching becomes a challenge as more materials and surfaces are exposed in GAA and CFET transistors

BASF
We create chemistry

Formulation Design Principles

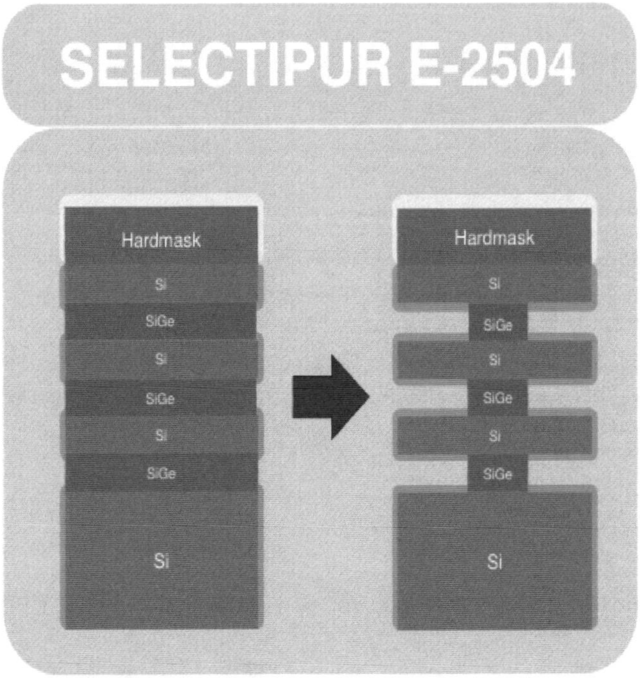

- Formulation:
  - ▶ water based
  - ▶ mixed with H_2O_2
- Hardmask Protection Additive (HPA) prevents SiO_2 from etching
- Silicon Protection Additive (SPA) prevents Silicon from etching
- Both additives protect independently

BASF
We create chemistry

Etch rate and Selectivity tuning

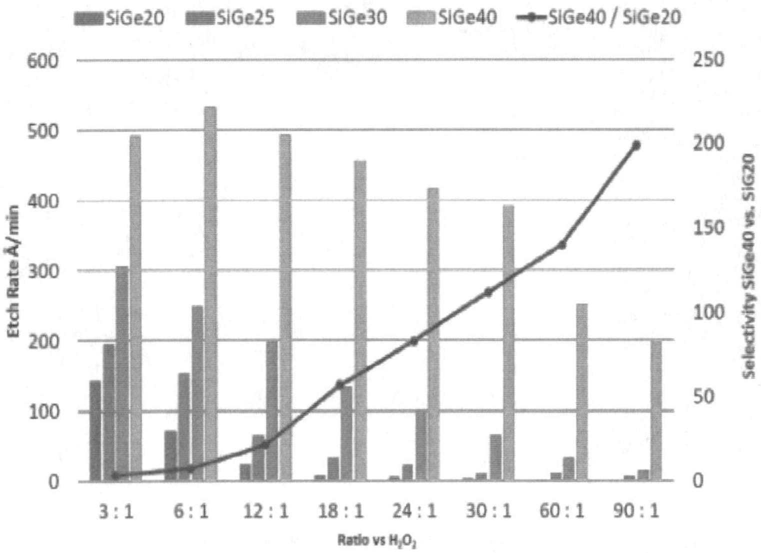

Relevant Application **SiGe vs Si** **(GAA & 3D-DRAM)** **SiGe40 vs SiG20** **(CFET)**

Tunable Formulation

Etch rate ~ $[H_2O_2]$ conc. & [Ge] conc.

formulation ($[H_2O_2]$ conc.) SiGe layer ([Ge] conc.)

- Ratio E-2504 vs. H_2O_2 variable concerning application
- Etching stops below a threshold

BASF We create chemistry

Target application: GAA

BASF We create chemistry

Transistor Roadmap

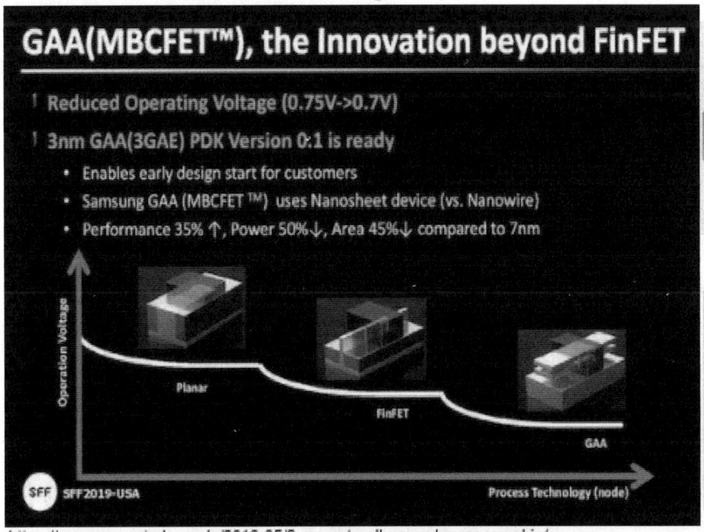

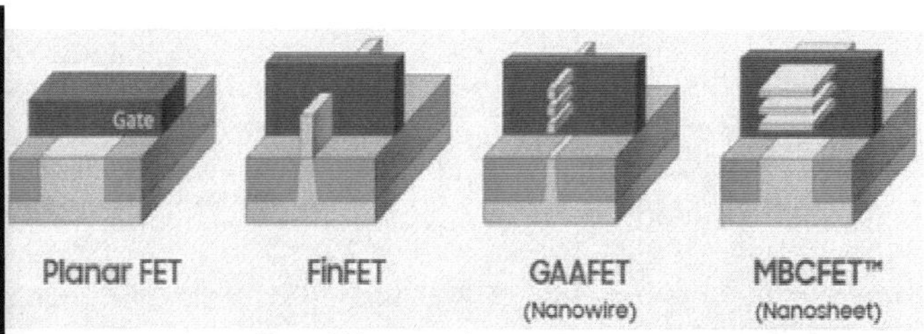

7 and 5nm node 3nm node and below

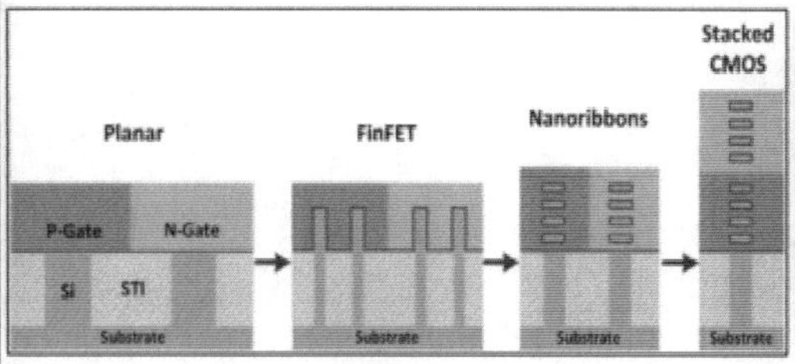

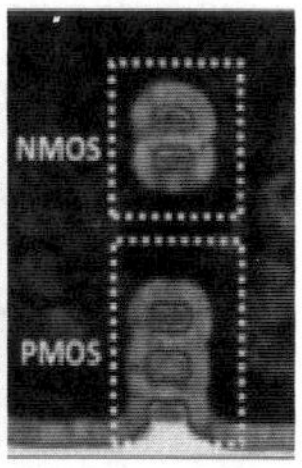

https://www.computerbase.de/2019-05/3-nm-gate-all-around-samsung-chip/

Intel

- by the introduction of the GAA FET's significant performance increase, less chip area and reduced power consumption of 50% compared to the 7nm products
- Cleaning and etching becomes a challenge as more materials and surfaces are exposed in GAA and CFET transistors

BASF
We create chemistry

3 SPCC Conference 2024; Chandler AZ; April 24th to April 25th

Formulation Design Principles

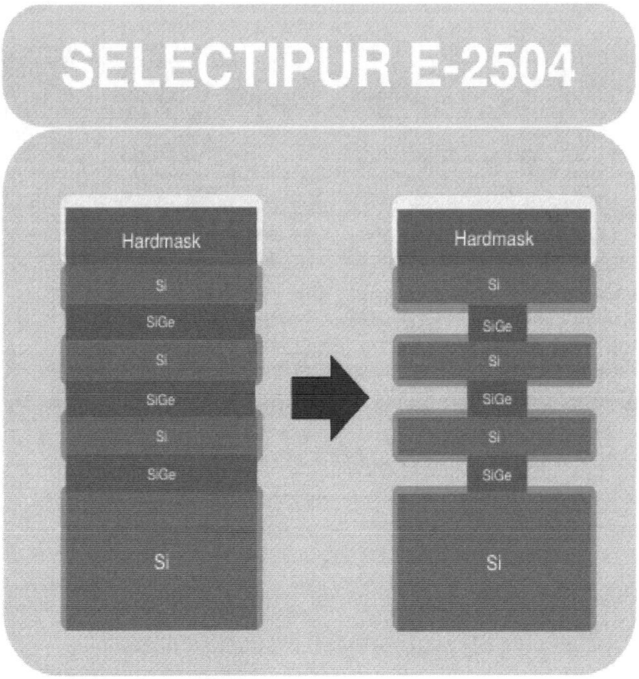

- Formulation:
  - ▶ water based
  - ▶ mixed with H_2O_2
- Hardmask Protection Additive (HPA) prevents SiO_2 from etching
- Silicon Protection Additive (SPA) prevents Silicon from etching
- Both additives protect independently

BASF
We create chemistry

4 SPCC Conference 2024; Chandler AZ; April 24th to April 25th

Results with different H_2O_2 content on Salsa 3

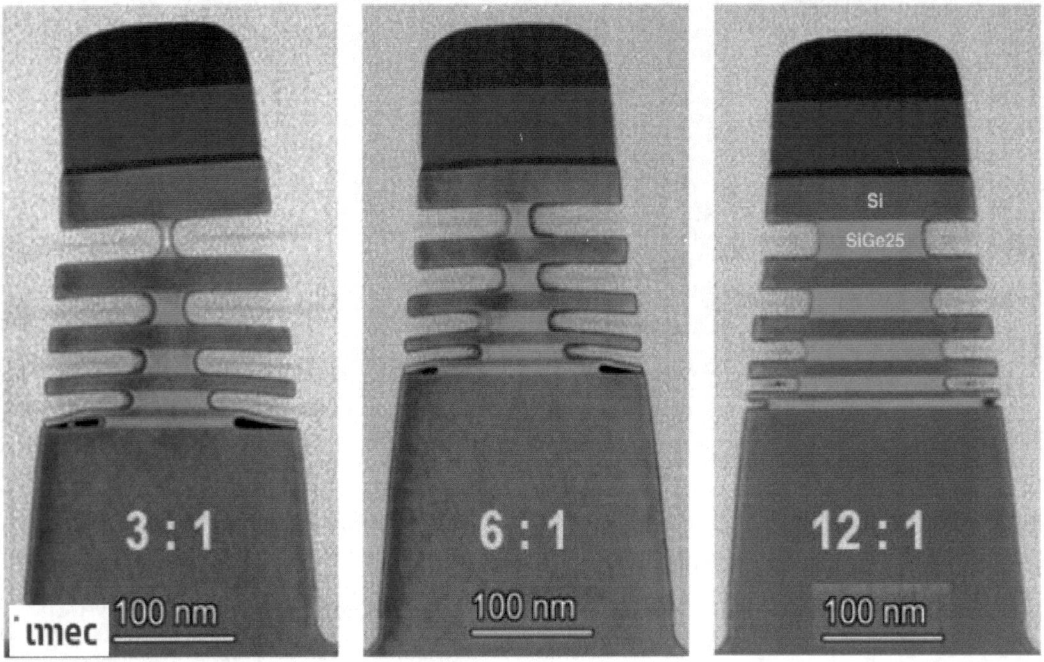

IMEC SALSA 3 test structures for GAA

Varying the ratio vs. H_2O_2 ratio does not impact the Hardmasks!

Results with different H_2O_2 content on Salsa 3

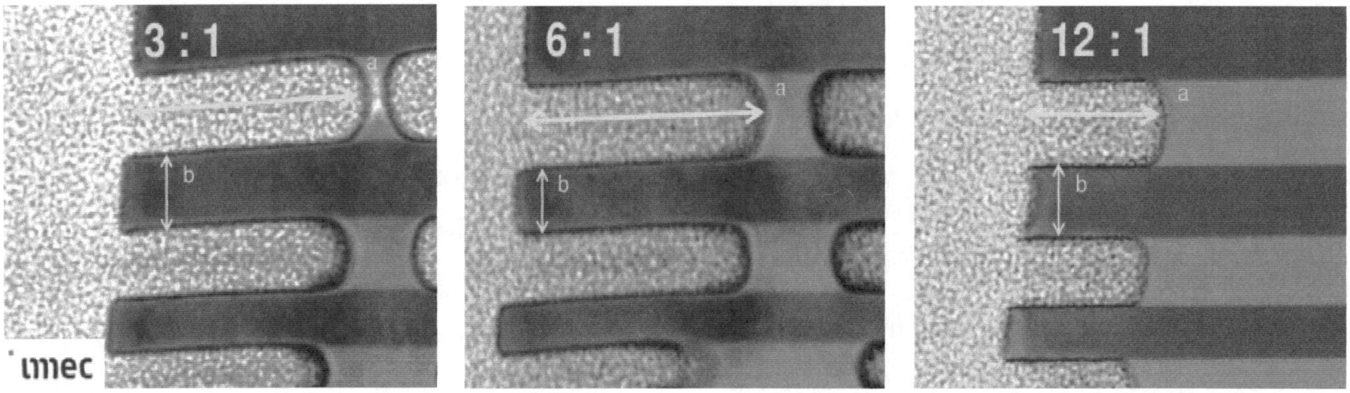

3 : 1	6 : 1	12 : 1
a = 88.3nm	a = 82.0nm	a = 46.9nm
b = 21.1nm (88.3%)	b = 16.6nm (86.9%)	b = 20.8nm (97.7%)

- Sheet thinning because after Germanium migration into the Silicon layer
  ▶ this Interlayer becomes etchable at higher H_2O_2 concentrations
- Based on H_2O_2 formulation can be used for full release as well as for the high precise recess process

Target application: CFET

Target application: CFET

Middle bottom isolation (MDI or BDI)
SiGe high Ge content vs SiGe lower content

MDI: Middle Dielectric Isolation
BDI: Bottom Dielectric Isolation
ISP: Inner spacer
NS: Nanoheet

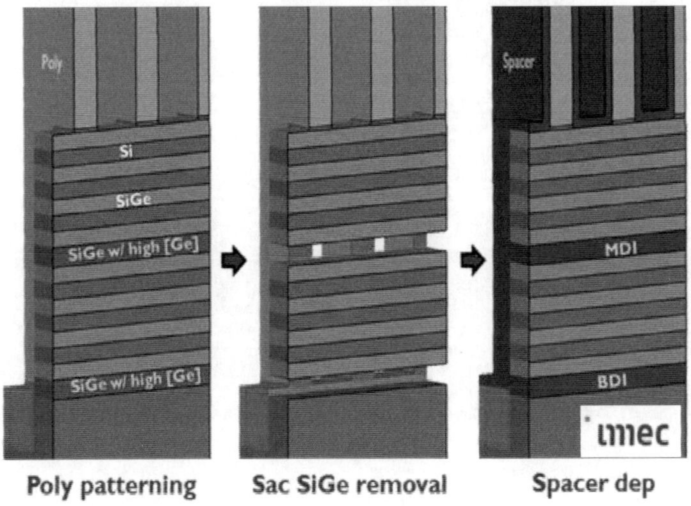

- SiGe1 – 20%
- SiGe2 – 40%

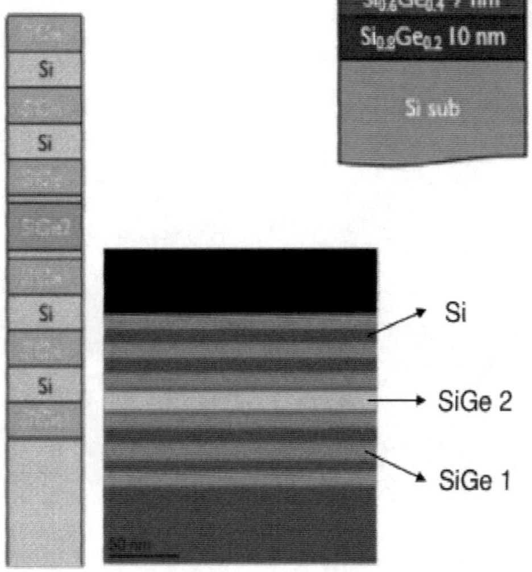

IMEC BDI Test Structure

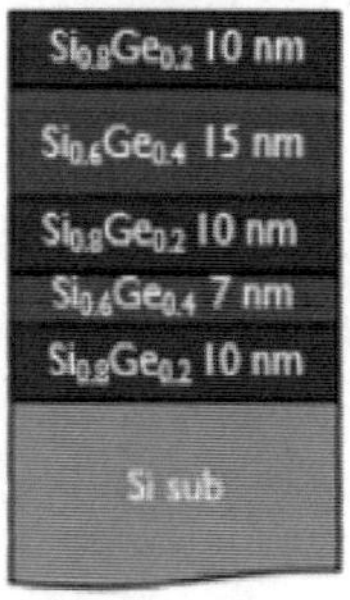

Mechanistic study for CFET (90 : 1)

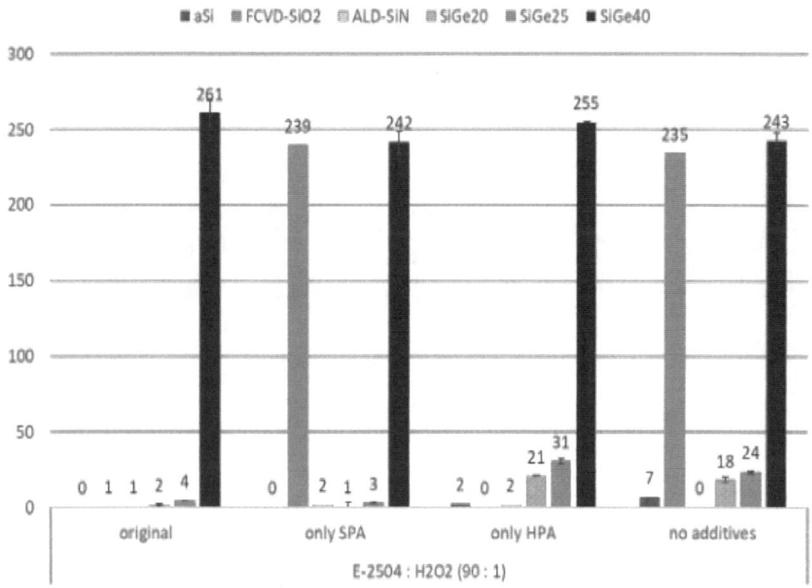

HPA = Hardmask Protection Additive
SPA = Silicon Protection Additive

- Low H_2O_2 concentration
  ▶ Ratio E-2504 : H_2O_2 = 90 : 1
- Lowering the H_2O_2 concentration lowers all etch rates
- HPA at low H_2O_2 levels also improves selectivity vs. Silicon
- SPA protection vs. Silicon gets so strong that Si etching on blanket is not observed anymore

BASF
We create chemistry

13 SPCC Conference 2024; Chandler AZ; April 24th to April 25th

Mechanistic study for CFET

Mixture: E-2504 : H_2O_2 = 90 : 1

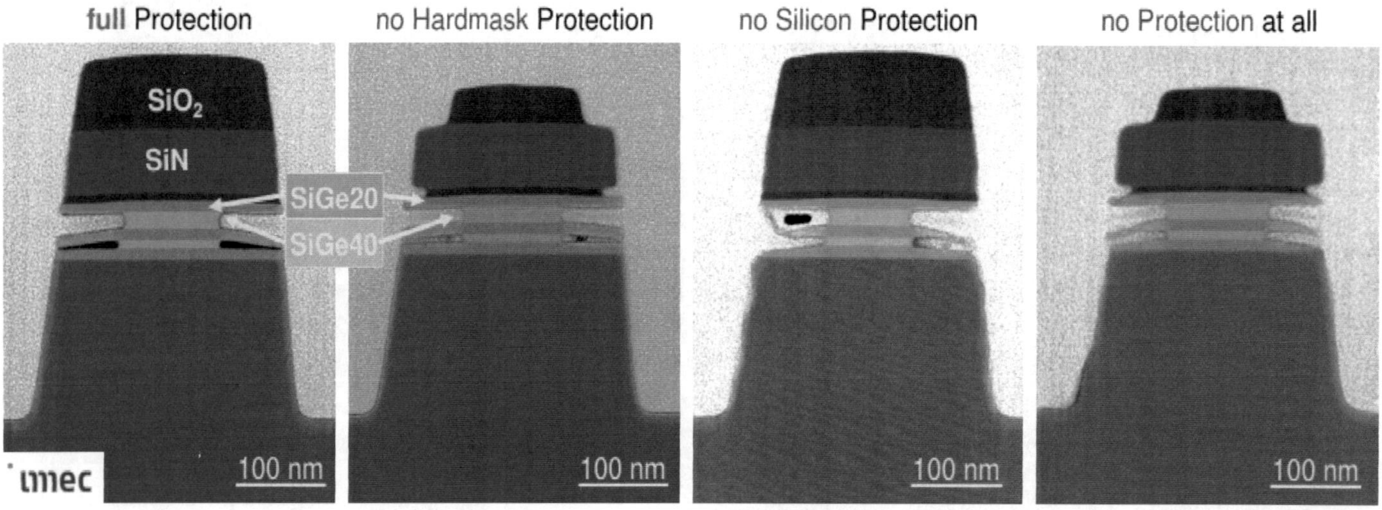

BASF
We create chemistry

14 SPCC Conference 2024; Chandler AZ; April 24th to April 25th

Performance on IMEC BDI Test structure

Mixture: E-2504 : H_2O_2 = 90 : 1

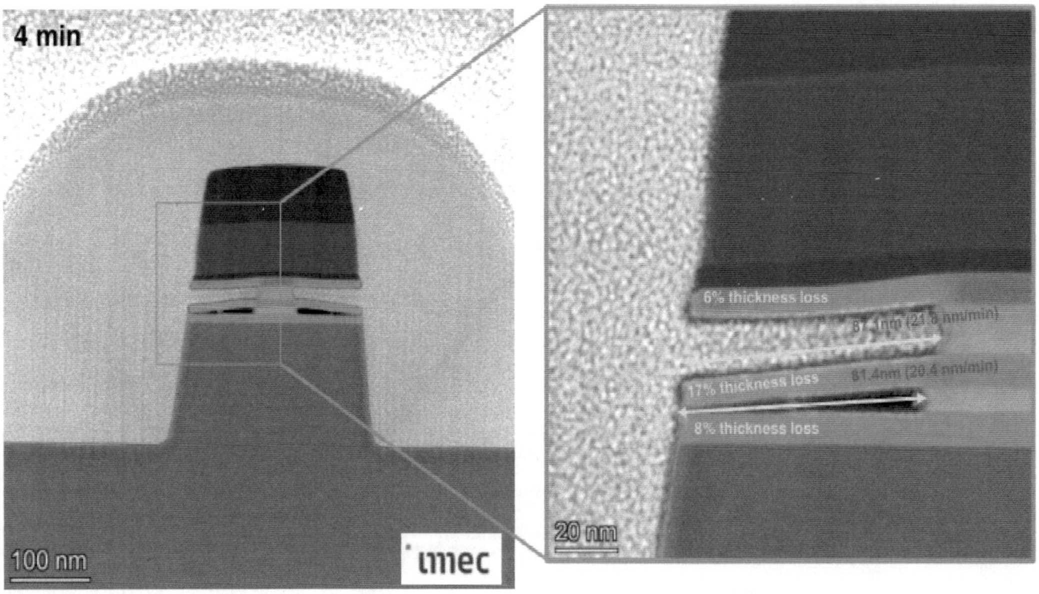

IMEC BDI Test Structure

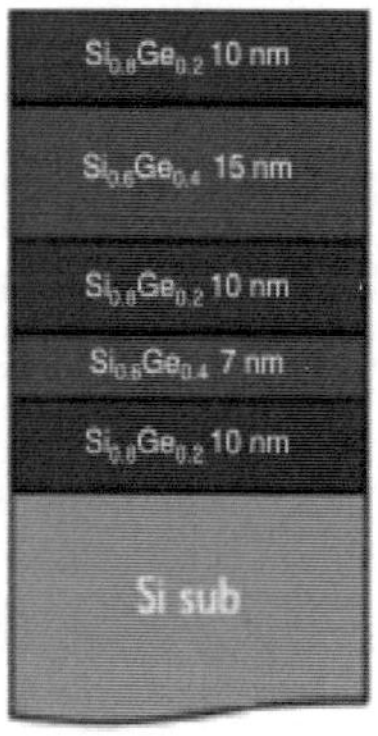

BASF
We create chemistry

SiGe etch for MDI/BDI: tool test results

IMEC CFET Test Structure

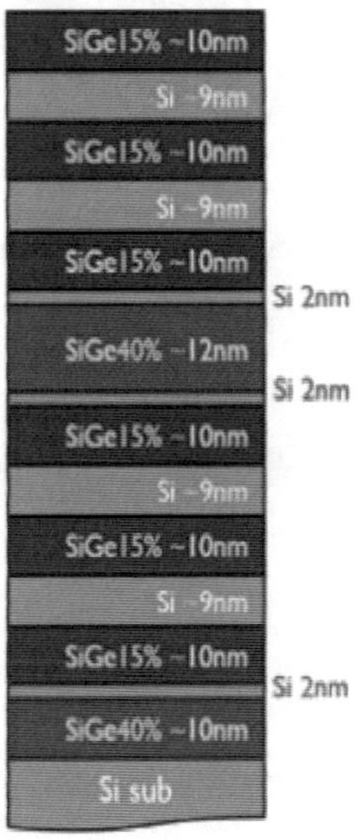

BASF tool demo @ IMEC, E-2504, center

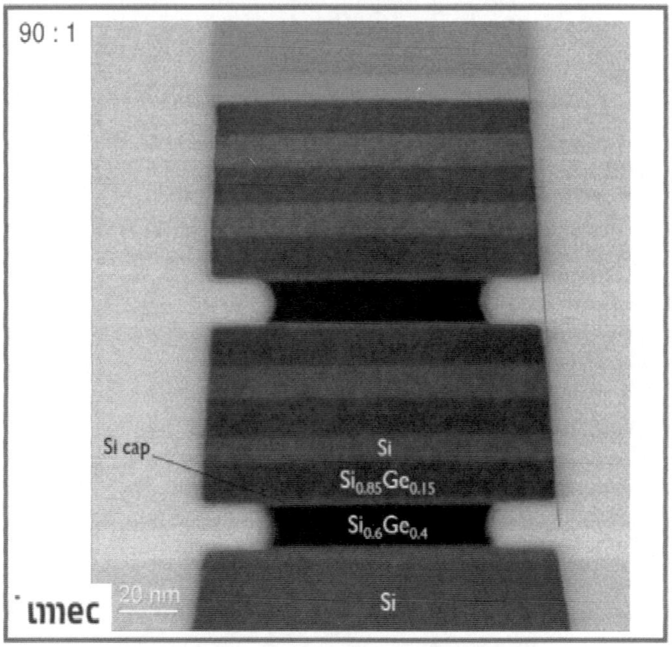

- Perfect etch of the SiGe40% layer under tool conditions
- Silicon and SiGe15% not etched at all

Selectivity:
no SiGe15% loss visible!

BASF
We create chemistry

Target application: 3D-DRAM

BASF
We create chemistry

IMEC Integration Plans on 3D-DRAM

Thick SiGe

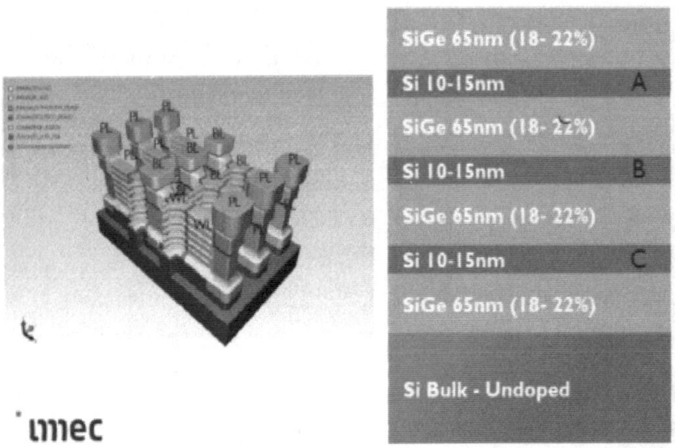

Thick Si

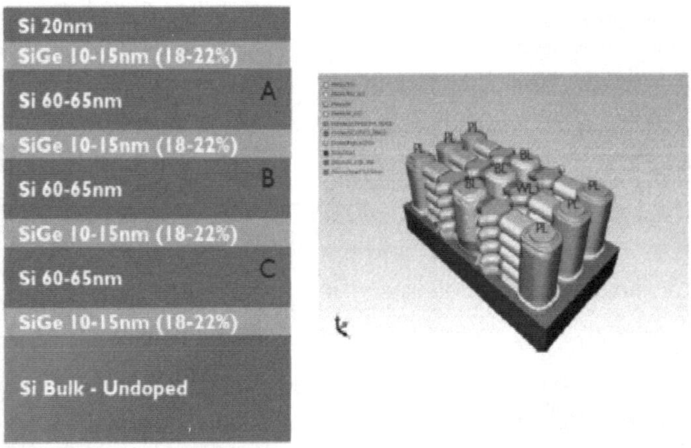

Solution for this feasible

BASF
We create chemistry

E-2504 on Salsa 3 based test structure from IMEC

Mixture: E-2504 : H_2O_2 = 6 : 1

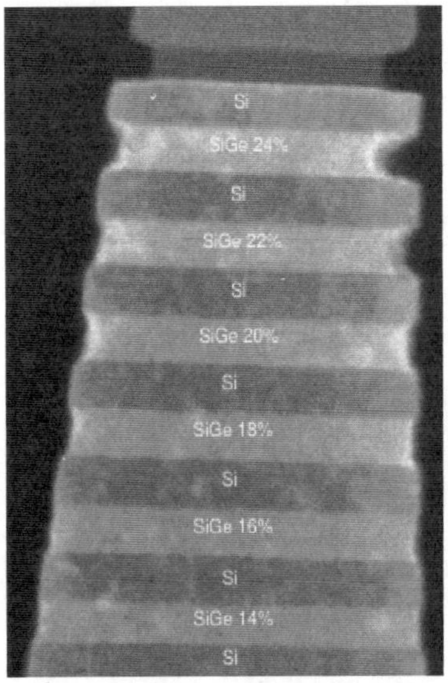

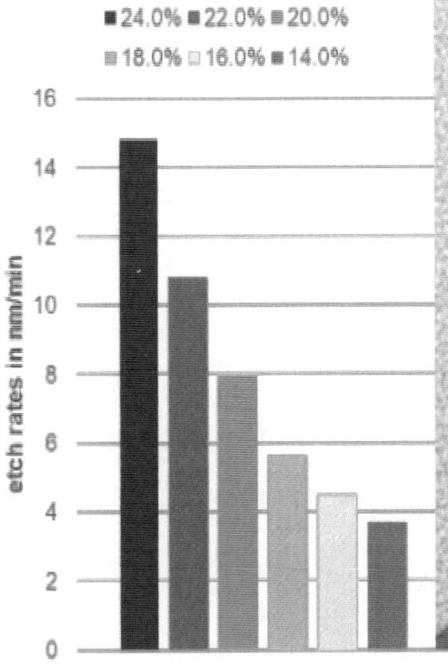

- Possible using a batch process (due to the long etching times)

BASF
We create chemistry

Summary

SELECTIPUR E-2504 is **THE** multitool for the selective etching of SiGe layers

- Simple tuning by H_2O_2 ratio
- Easy handling
- Suitable for
  - ▶CFET (MDI/BDI)
  - ▶GAA
  - ▶3D-DRAM

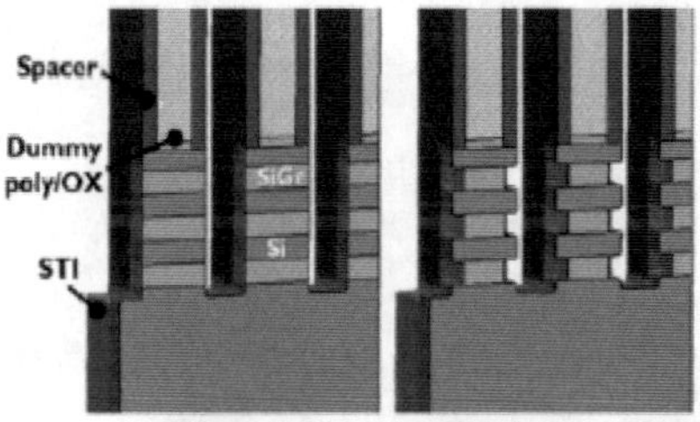

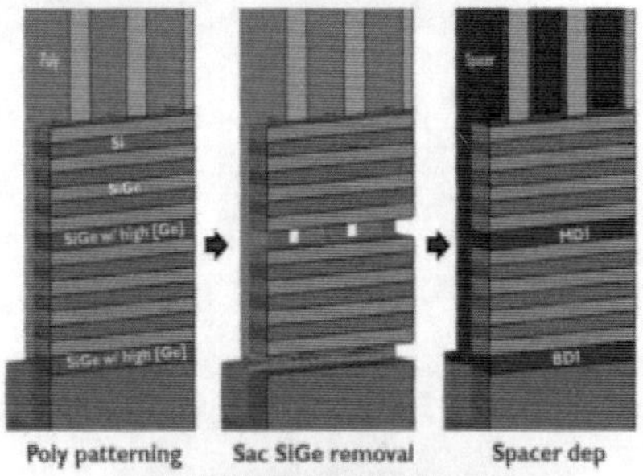

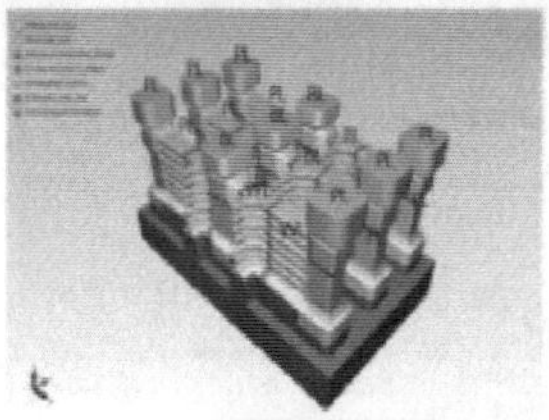

BASF
We create chemistry

Acknowledgements

- Global Semiconductor Application team BASF Electronic Materials
- Application Development Lab team BASF Electronic Materials Taiwan
- Analytical department from BASF Central Research
- Collaboration partners at IMEC for CFET Development:
  - ▶ Farid Sebaai, Efrain Altamirano Sanchez, Antoine Pacco, Jana Loyo Prado

BASF

We create chemistry

Wet Etching Mechanism of Ruthenium : Effect of Oxidants and Inhibitors

Hwan Park[1], Heeju Ha[1], Taeseok Jung[2], Yongsun Ko[1], Yunhee Cho[3], Wonseob Cho[3], Andreas Klipp[4] and Heeyeop Chae[1,2,*]

[1]School of Chemical Engineering, SKKU, Republic of Korea
[2]Department of Semiconductor Convergence Engineering, SKKU, Republic of Korea
[3]BASF Company Ltd., Research Center No.1, SKKU, Republic of Korea
[4]BASF SE, G-EDM/SA – J534 Office 541, Germany

E-mail: hpark1357@gmail.com
** E-mail: hchae@skku.edu*

Ruthenium: Alternative of Copper

- **Scaling down of semiconductor devices**

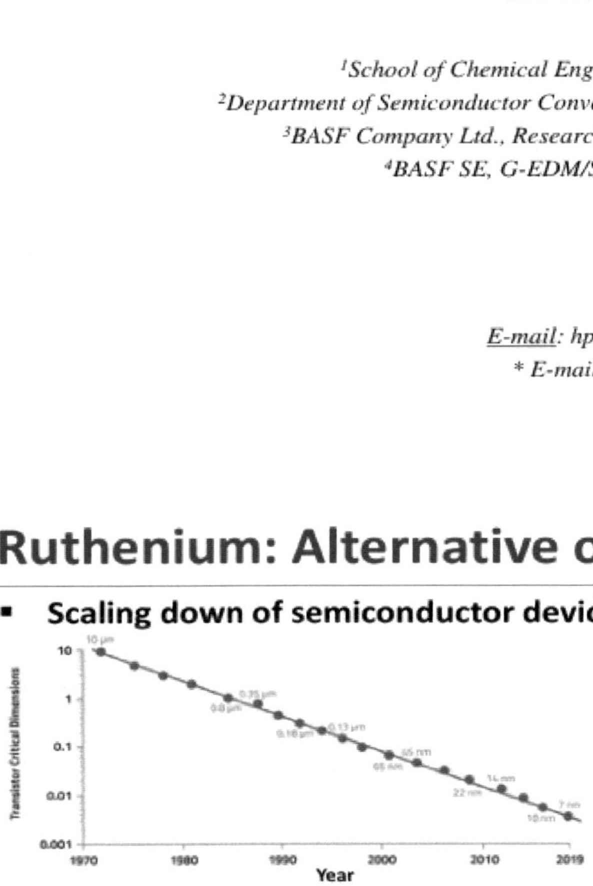

Semiconductor engineering (2019)

- **Limitation of copper interconnect**

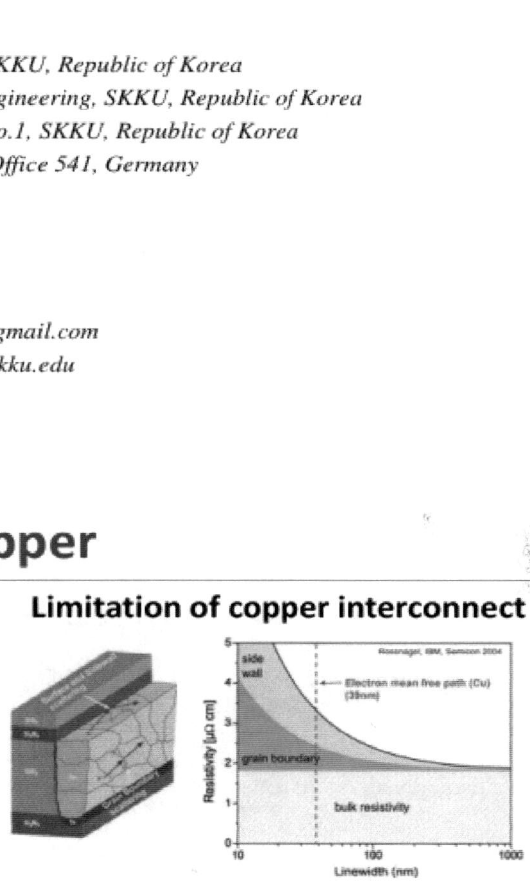

Semiconductor-digest (2020)

- **Alternative metal**

Material	Bulk resistivity ($\mu\Omega$-cm)	Electron mean free path (nm)
Cu	1.7	39.9
Co	6.2	11.8
Ru	7.8	6.6

J. Appl. Phys. 119, 085101 (2016)

- **Resistivity of copper & ruthenium**

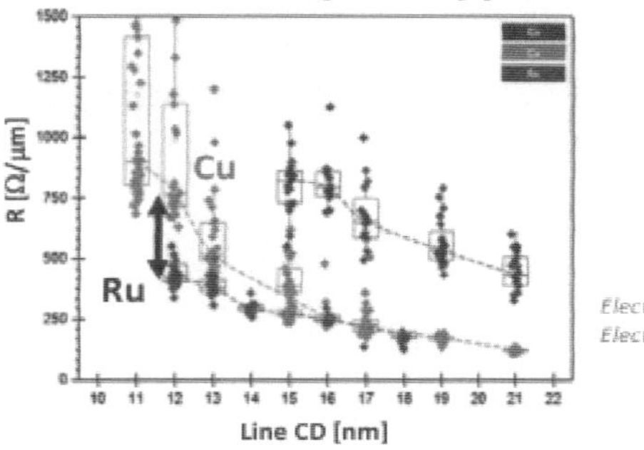

IITC (2018)
Electrochim. Acta. 274, 306 (2018)
Electrochim. Acta. 306, 285 (2019)

- Ruthenium has a lower resistance than copper as the line width decreases.
- Ruthenium could be an alternative metal to copper.

2

Application of Ruthenium

- **Interconnect structure**

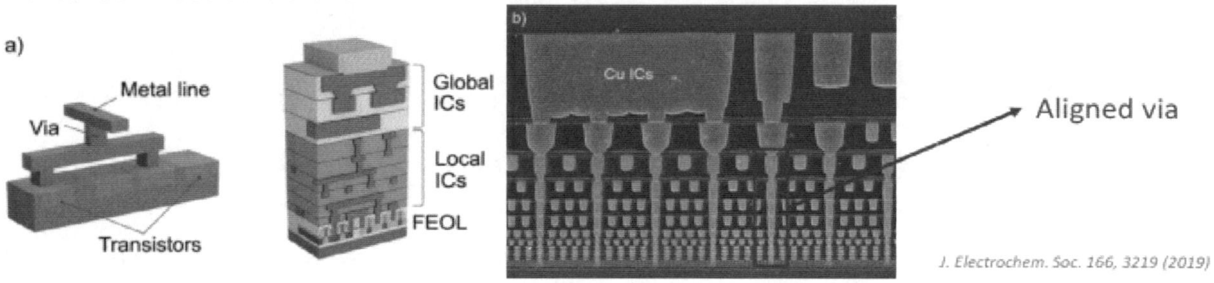

J. Electrochem. Soc. 166, 3219 (2019)

- **Aligned via**

a) Self Aligned Via(SAV)

b) Fully Self Aligned Via(FSAV)

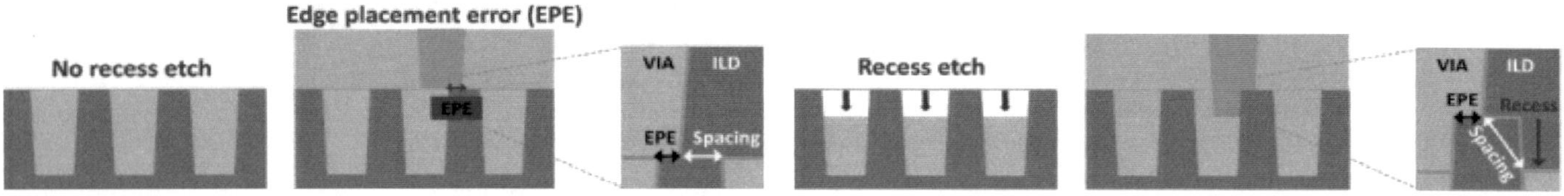

- FSAV structure prevents leakage current by securing space with adjacent lines.
- Recess etch is required to create the FSAV structure.

3

Metal Recess Etch

- **Recess etch for FSAV application**

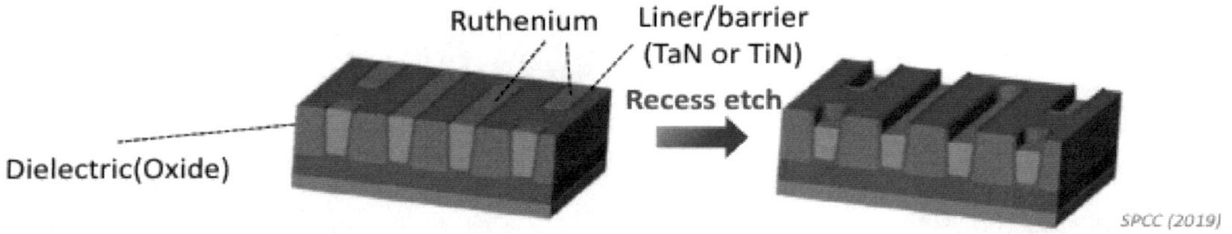

SPCC (2019)

- **Wet etching vs Dry etching**

	Wet etching	Dry etching
Etch profile	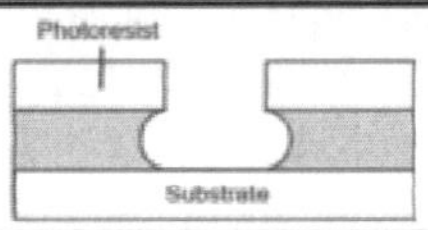	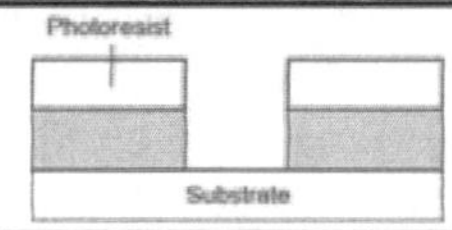
Method	Chemical reaction	Physical/ Chemical reaction
Etch rate	High	Low
Selectivity	High	Low

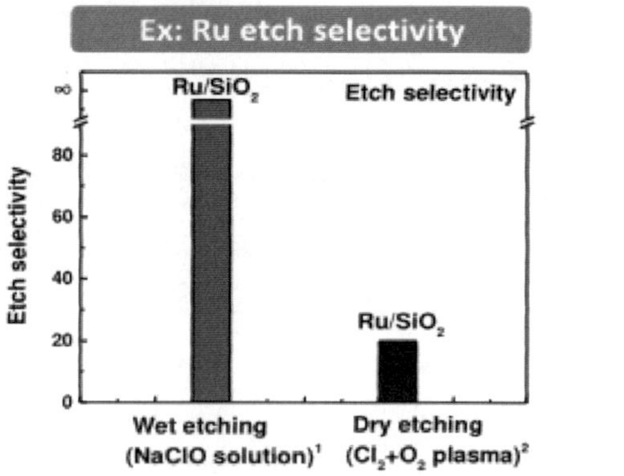

1: This work, 2: J. Vac. Sci. Technol. B. 36(3), E103 (2018)

- Recess etch requires a method to etch only Ru among various materials. (high selectivity)
- Wet etching has higher selectivity than dry etching, research on wet etching is needed.

4

Approach: Ruthenium Oxidation

- **Schematic of process**

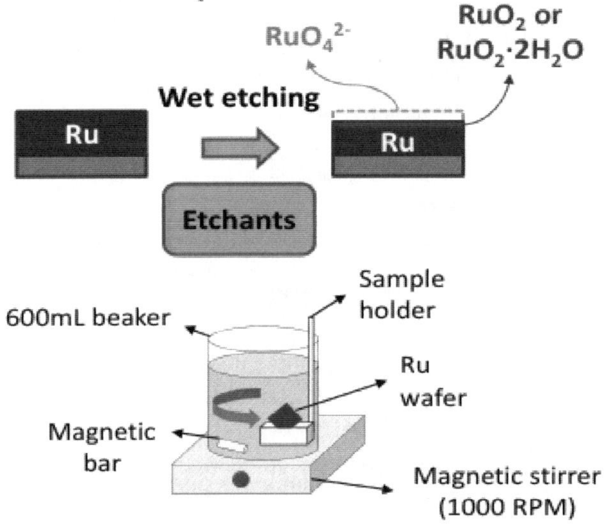

- **Ruthenium Pourbaix diagram**

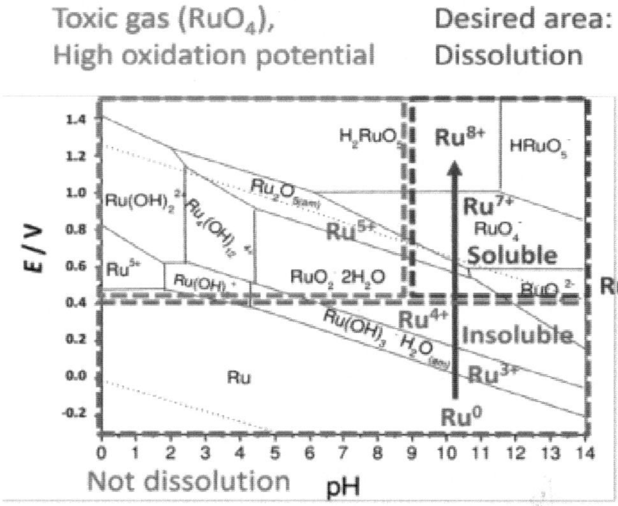

J. Electrochem. Sci. Eng. 6, 145 (2016)

- Ruthenium is soluble when it becomes an oxide with a high oxidation number.
- RuO_4, a ruthenium oxide produced at low pH, is a toxic gas, so reacting at high pH appears to be the desired area.

5

Candidates of Ru Etchants

Criterion 1: Similar Chemical Structure to Known Etchants
Criterion 2: Oxidizer used in Ru CMP papers
Criterion 3: High standard reduction potential (E°)

Criterion	Chemical Formula (Criterion 1)		Chemical Structure	Ru Etching (Criterion 2)	Standard reduction potential (E°) (Criterion 3)	pH
1) Similar structure	Cl-O ion	**NaClO**		Immersion	**1.630**	12
		$NaClO_2$	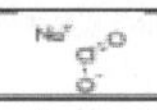	-	0.93 - 1.5	10.9
		$NaClO_3$	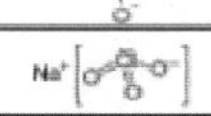	-	1.15 - 1.47	5.4
		$NaClO_4$	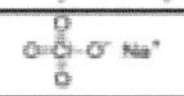	-	1.204 - 1.42	5.7
	Br-O ion	$NaBrO_3$		CMP	1.44	5.5
	I-O ion	**H_5IO_6 (HIO_4)**	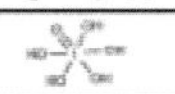	Immersion	**1.603**	1.4
		$NaIO_4$	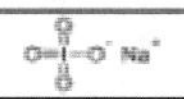	CMP	**1.60**	4.7
		$NaIO_3$		-	1.195	5.9
		KIO_4	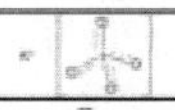	CMP	1.60	4.5
		KIO_3	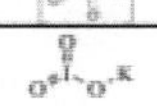	-	1.195	5.7
2) CMP oxidizer	Ce^{4+} ion	**$(NH_4)_2[Ce(NO_3)_6]$**	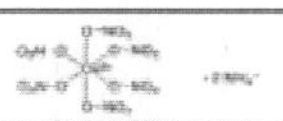	CMP	**1.47 - 1.72**	0.8
3) Strong oxidant	MnO_4^- ion	$KMnO_4$		CMP	1.51 - 1.70	9.9
	$S_2O_8^{2-}$ ion	$K_2S_2O_8$	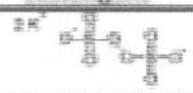	CMP	2.01	3.7
		$Na_2S_2O_8$	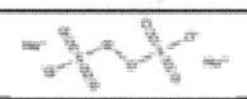	-	2.01	4.4
	H_2O_2			CMP	1.763	3.9

- 15 etchants were selected by considering 3 criteria.

6

Screening of Etchants

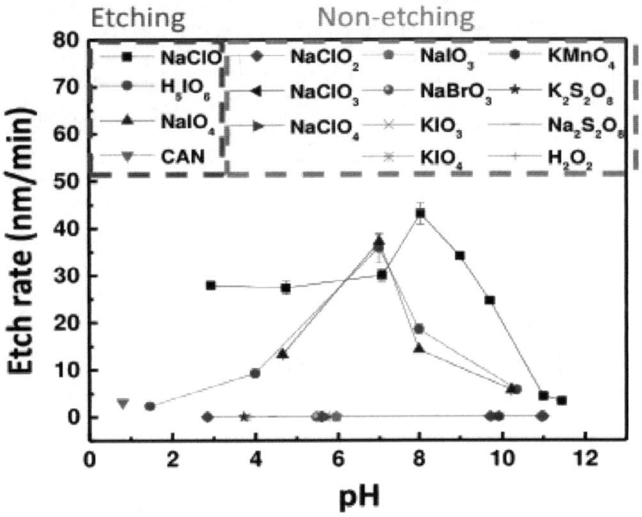

- ✓ Same molality
- ✓ pH adjuster: HNO_3, NaOH
- ✓ Room temperature

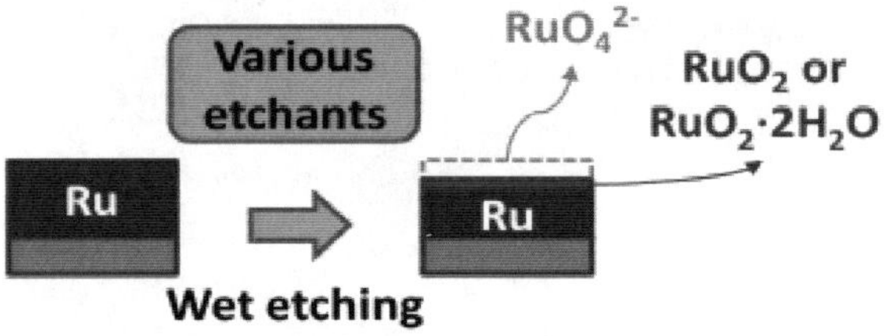

➢ **Known reaction mechanism**

✓ $Ru + 3OCl^- + 2OH^- \rightarrow RuO_4^{2-} + H_2O + 3Cl^-$

✓ $2Ru + 2IO_4^- + 4OH^- \rightarrow 2RuO_4^{2-} + 2H_2O + 3O_2 + 2I^-$

❖ CAN: Ceric Ammonium Nitrate

❖ KIO_4: not dissolved in water

- Etching reaction occurs with only 4 etchants(NaClO, H_5IO_6, $NaIO_4$, Ceric Ammonium Nitrate)
- The etching reaction seems to be possible only when a reaction path exists in which Ru is directly oxidized to RuO_4^{2-}, since RuO_2 is a very stable.

(The Chemistry of RUTHENIUM, RHODIUM, PALLADIUM, OSMIUM, IRIDIUM and PLATINUM, PERGAMON PRESS (1973))

7

NaClO: Passivation Layer Formation

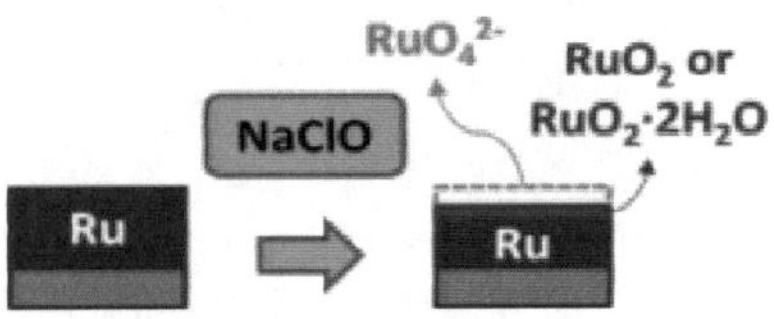

- ✓ Same molality
- ✓ pH adjuster: HNO_3
- ✓ Room temperature
- ✓ Same etch depth

▪ **Depth profile (SIMS)**

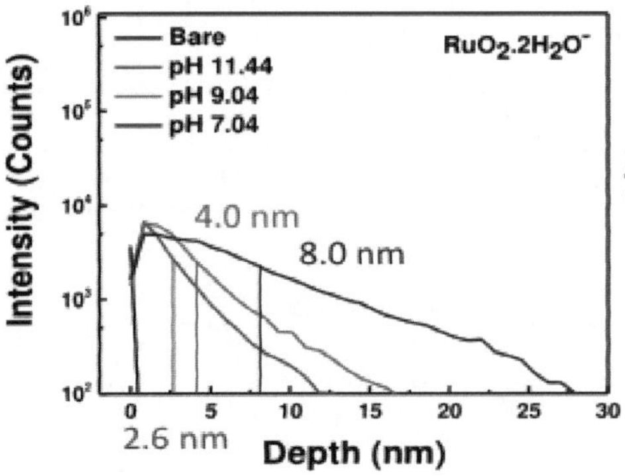

▪ **Pourbaix diagram**

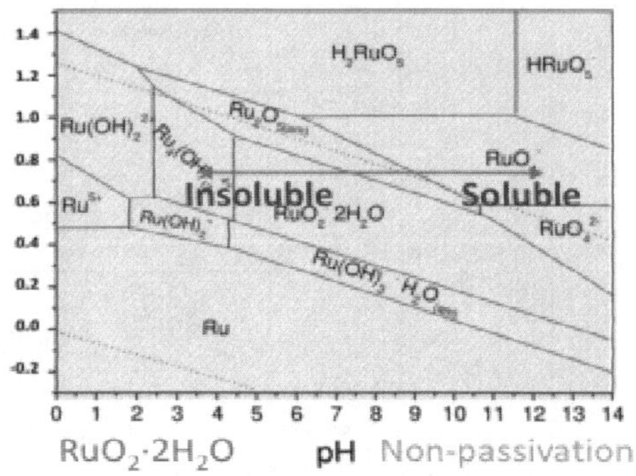

Atlas of Electrochemical Equilibria in Aqueous Solutions, National Association of Corrosion (1974)

➢ **Known etching mechanism**

✓ Rxn 1: $Ru + 3OCl^- + 2OH^- \rightarrow RuO_4^{2-}(aq) + H_2O + 3Cl^-$

(Red: Soluble) (Blue: Insoluble)

➢ **Known reaction mechanism at low pH**

✓ Rxn 2: $3RuO_4^{2-} + 4H^+ \rightarrow RuO_2 \cdot 2H_2O(s) + 2RuO_4^-$

✓ Rxn 3: $4RuO_4^- + 4H^+ \rightarrow RuO_2 \cdot 2H_2O(s) + 3RuO_4$

J. Electrochem. Soc. 156, 188 (2009)
ECS Trans. 25, 329 (2009)

- The thickness of the $RuO_2 \cdot 2H_2O$ layer increased, at pH 7.
- $RuO_2 \cdot 2H_2O$ passivation layer is formed at pH ≤ 7, and it reduces the etch rate.

8

NaClO: Etch Reaction by ClO^- Ion

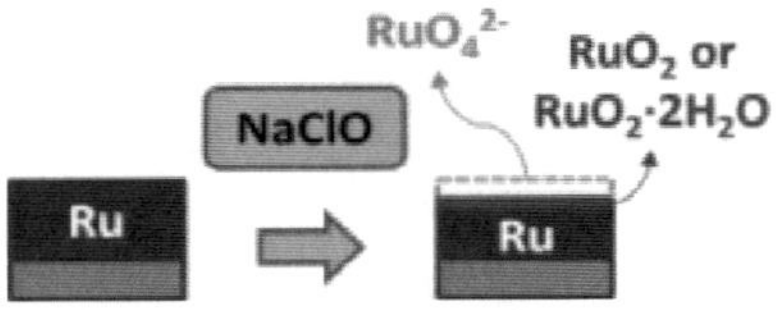

- ✓ Same molality
- ✓ pH adjuster: HNO_3
- ✓ Room temperature
- ✓ Same etch depth

▪ **Etch rate**

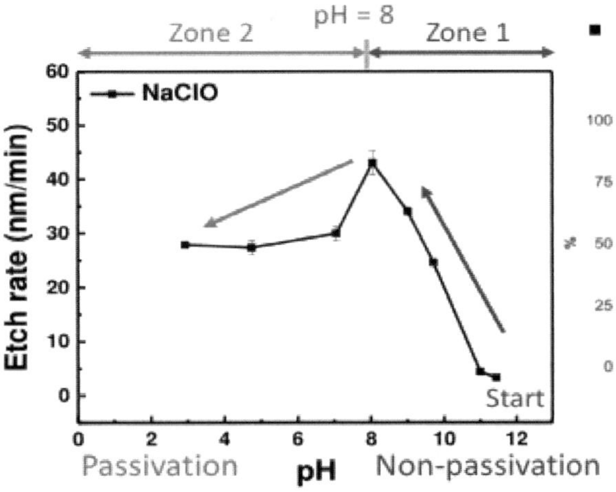

▪ **Stability of OCl^- ion**

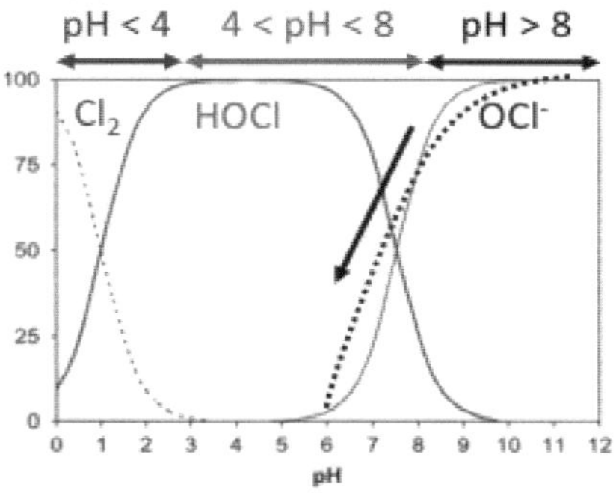

Water Res. 42, 13 (2008)

➢ **Known etching mechanism**

✓Rxn 1: $Ru + 3OCl^- + 2OH^- \rightarrow RuO_4^{2-}(aq) + H_2O + 3Cl^-$

(Red: Soluble) (Blue: Insoluble)

➢ **Known reaction mechanism at low pH**

✓Rxn 2: $3RuO_4^{2-} + 4H^+ \rightarrow RuO_2{\cdot}2H_2O(s) + 2RuO_4^-$

✓Rxn 3: $4RuO_4^- + 4H^+ \rightarrow RuO_2{\cdot}2H_2O(s) + 3RuO_4$

- Zone 1: pH ↓, etch rate ↑: H^+ ion increased and RuO_4^{2-} ion stabilized by reaction 2 and 3.
- Zone 2: pH ↓, etch rate ↓: $RuO_2{\cdot}2H_2O$ passivation layer is formed, OCl^- ion unstable below pH 8.

9

$H_5IO_6(HIO_4{\cdot}2H_2O)$, $NaIO_4$: Etch Reaction by IO_4^- Ion

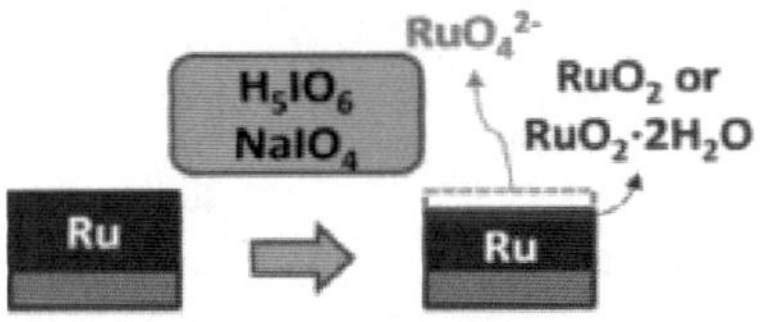

- ✓ Same molality
- ✓ pH adjuster: NaOH
- ✓ Room temperature
- ✓ Same etch depth

▪ **Etch rate**

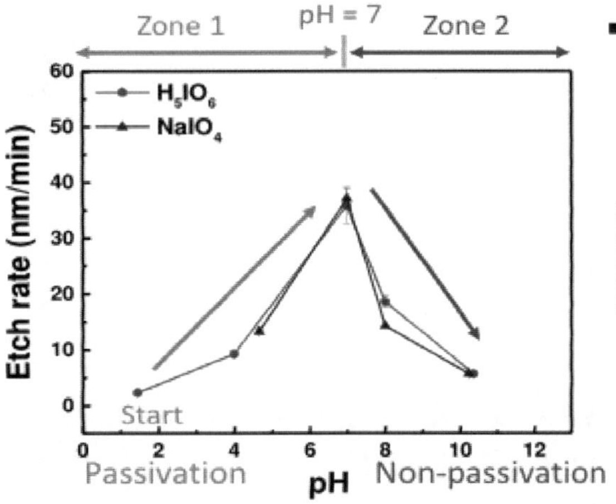

▪ **Stability of IO_4^- ion**

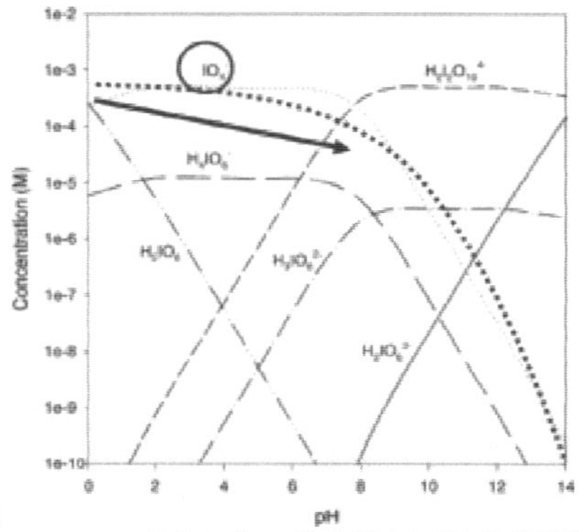

J. Photochem. Photobiol. A. 165, 35 (2004)

➢ **In acidic and neutral regions (pH ≤ 7)**

✓$2Ru + IO_4^- + 4H_2O \rightarrow 2RuO_2{\cdot}2H_2O(s) + I^-$

✓$Ru + IO_4^- \rightarrow RuO_4(g) + I^-$

✓$RuO_4 + H_2O + 4H^+ + 4e^- \rightarrow RuO_2{\cdot}2H_2O(s) + 2H_2O$

➢ **In alkaline regions (pH > 7)**

(Red: Soluble)
(Blue: Insoluble)

✓$2Ru + IO_4^- \rightarrow 2RuO_2 + I^-$

✓$RuO_2 + 4OH^- \rightarrow RuO_4^{2-}(aq) + 2H_2O + 2e^-$

✓$RuO_2 + 4OH^- \rightarrow RuO_4^-(aq) + 2H_2O + 3e^-$

- Zone 1: pH ↑, etch rate ↑: H^+ ion reduced, and less $RuO_2{\cdot}2H_2O$ layer is formed.
- Zone 2: pH ↑, etch rate ↓: IO_4^- ion unstable above pH 7.

10

NaClO, $H_5IO_6(HIO_4{\cdot}2H_2O)$, $NaIO_4$: Etch Rate

✓ Same molality
✓ pH adjuster: HNO_3, NaOH
✓ Room temperature

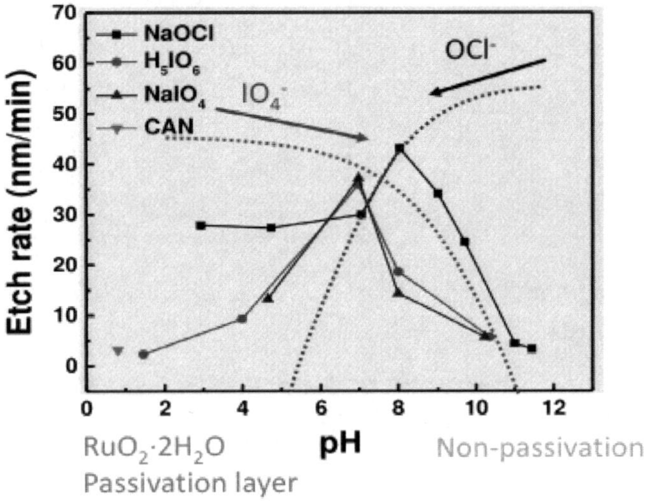

- **Stability of OCl^-, IO_4^- ion**

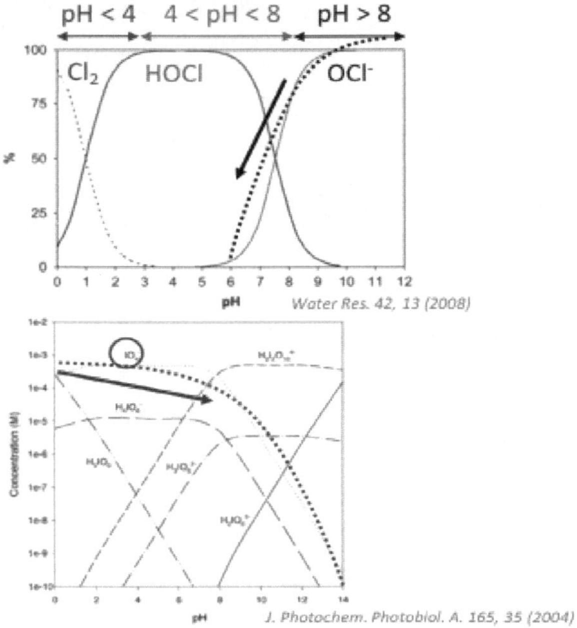

- At pH ≤ 7, a $RuO_2{\cdot}2H_2O$ passivation layer is formed, which reduces the etch rate.
- OCl^- ions are unstable at pH ≥ 8 and IO_4^- ions are unstable at pH ≤ 7, the etch rate decreases.

11

Surface Compositions & Roughness

✓ Same molality
✓ pH adjuster: HNO_3, NaOH
✓ Room temperature
✓ Same etch depth

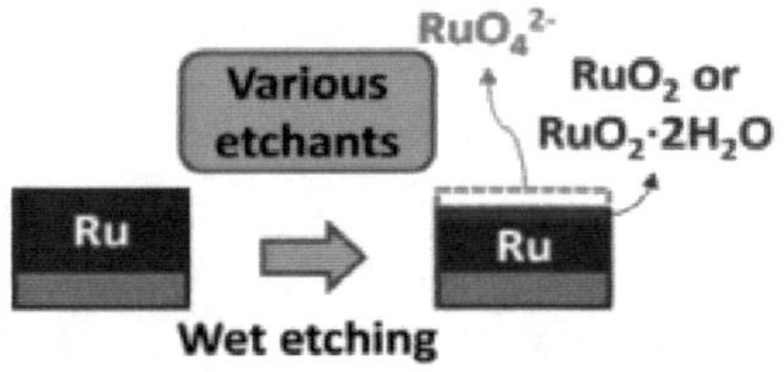

- **Ru oxide percentage**

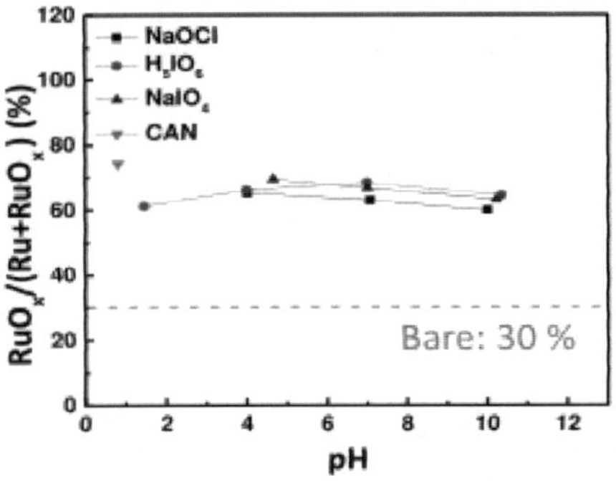

- **Surface roughness**

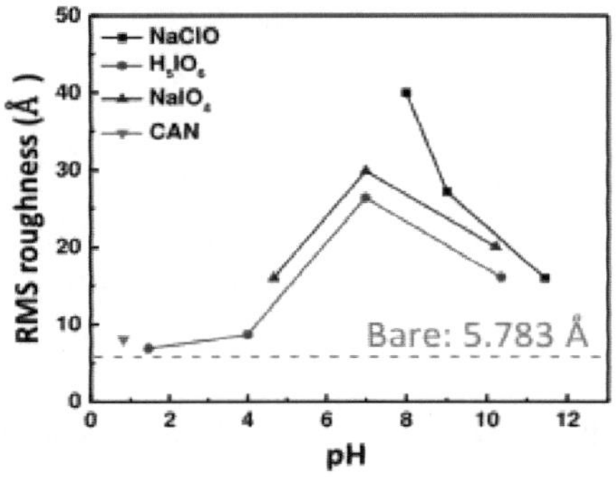

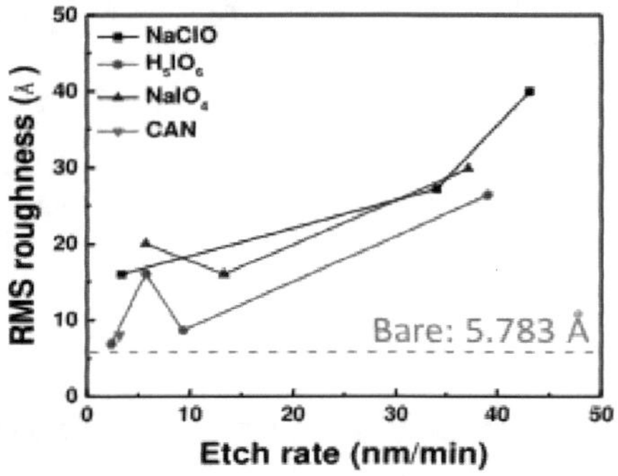

- The $RuO_x/(Ru+RuO_x)$ ratio is almost the same regardless of etchant type and pH, if the etch depth is the same.
- Surface roughness is assumed to be affected by etch rate even if the etch depth is the same.

12

Approach: Inhibitor Effects in Wet Etching

- **Schematic of process**

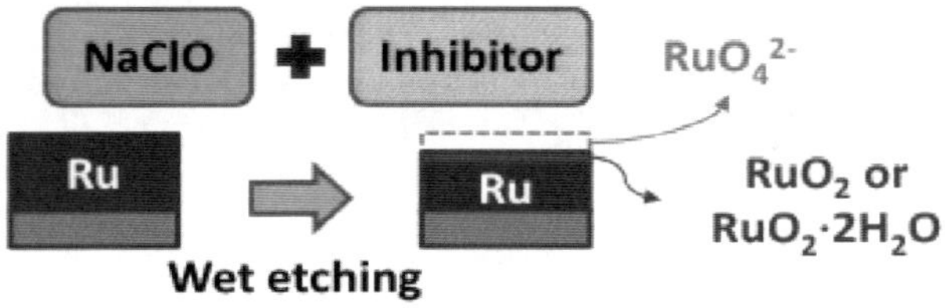

- **Mechanism of etching inhibitors**

Group	**Zwitterion group**	**Azole group**	**Acid group**
Name	Glycine	Triazole (TAZ)	L-ascorbic acid
Structure	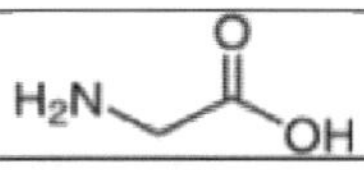	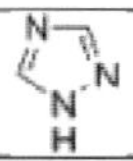	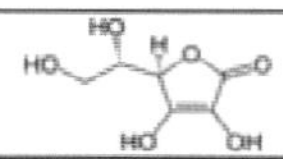
Mechanism	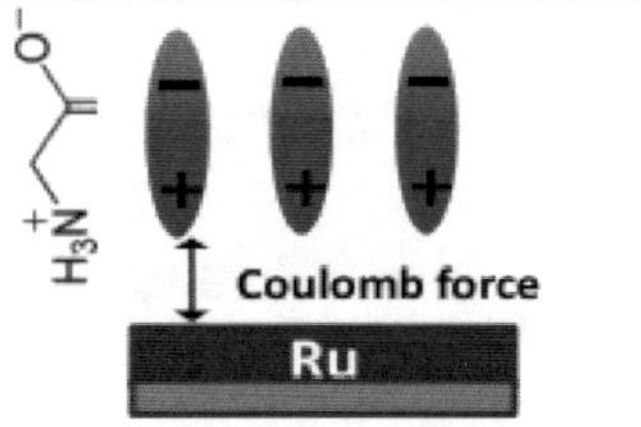 **Coulomb force** *Appl. Surf. Sci. 506, 144976 (2020)*	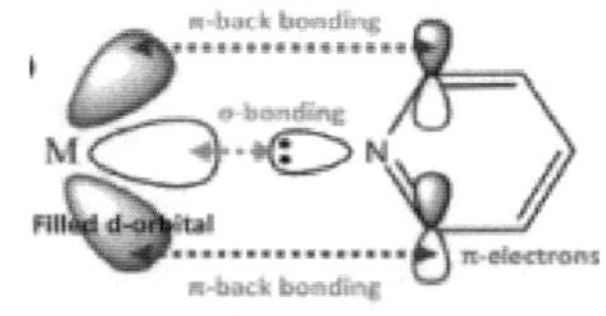 **Coordinate bond (M-N bonding)** *Sci. Rep. 11, 21214 (2021)*	Dehydrogenation 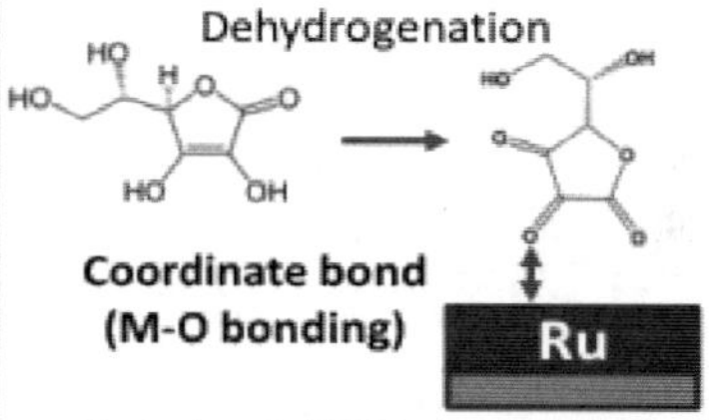 **Coordinate bond (M-O bonding)** *Electrochem. Solid-State Lett. 14, 306 (2011)*

- Compare the Ru etching reaction depending on the inhibitor mechanism.

13

Etch Rate with Inhibitor

✓Inhibitor concentration : Same molality
✓pH adjuster: HNO_3, NaOH
✓Room temperature

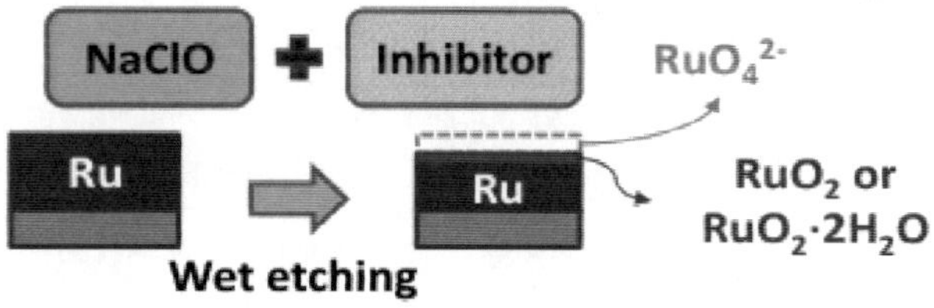

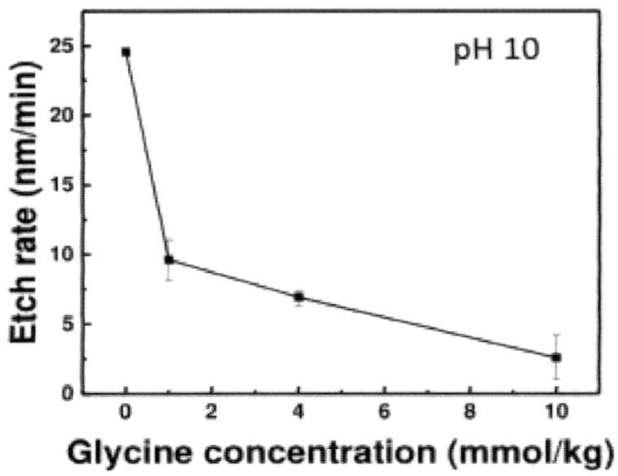

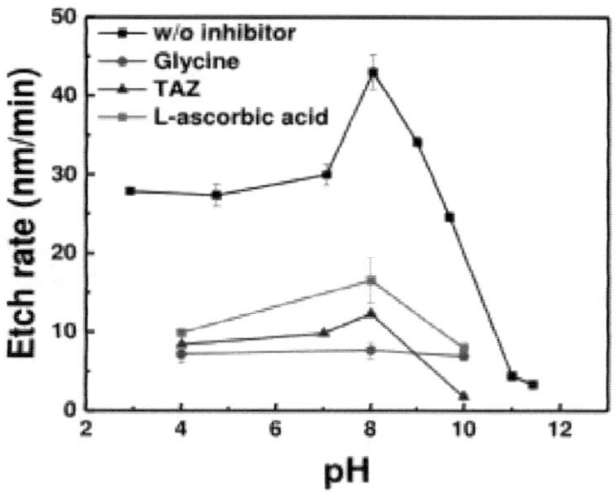

- The etch rate decreases as the concentration of glycine increases.
- The inhibitor effect was in the order of Glycine > TAZ > L-ascorbic acid in most areas, but at pH 10, when TAZ was added, the etching rate was greatly reduced.

14

Surface Composition & Roughness with Inhibitor

- ✓Inhibitor concentration : Same molality
- ✓pH adjuster: HNO_3, NaOH
- ✓Room temperature
- ✓Same etch depth

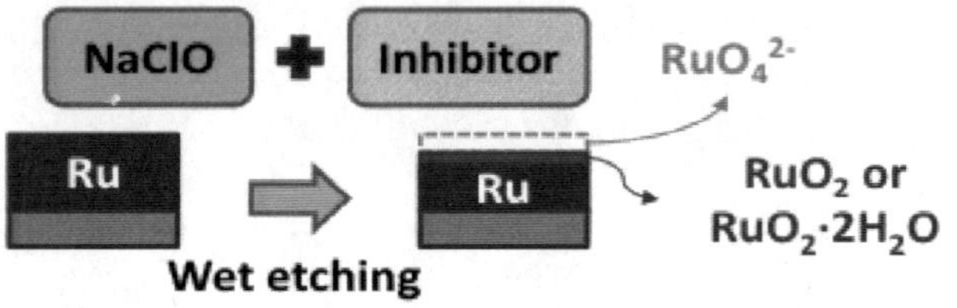

- **Ru oxide percentage**

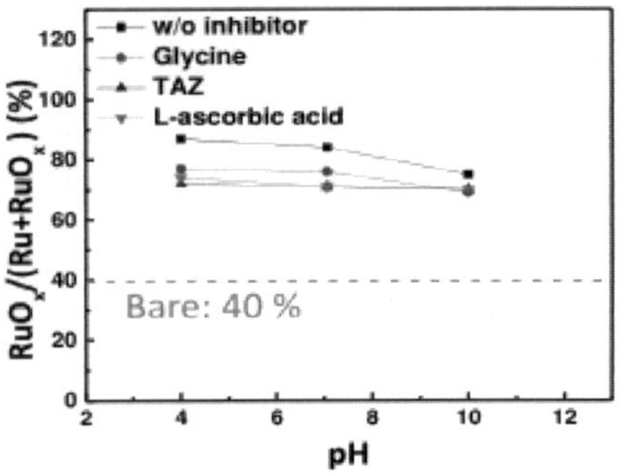

- **Surface roughness**

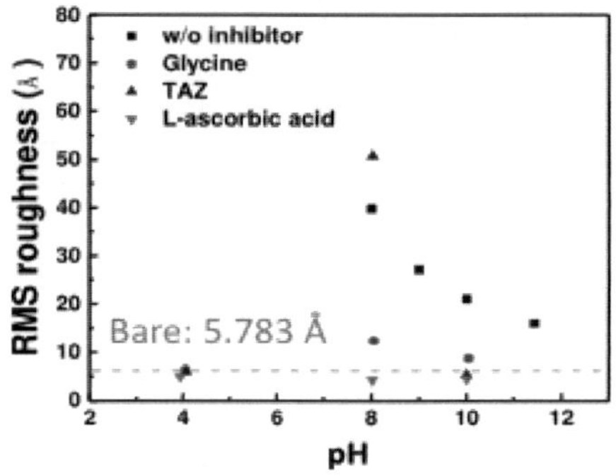

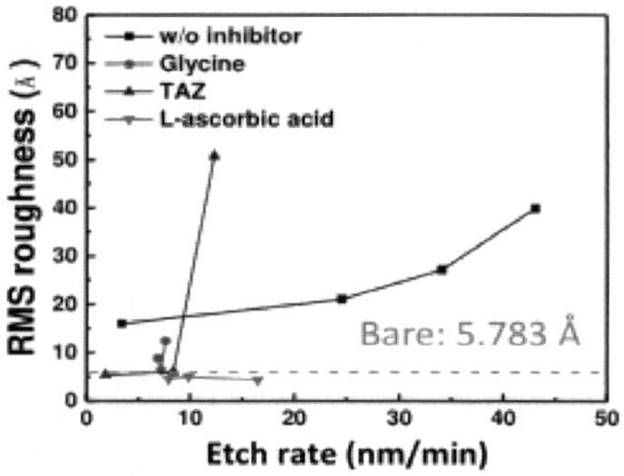

- The $RuO_x/(Ru+RuO_x)$ ratio is almost the same regardless of inhibitor type and pH, if the etch depth is the same.
- Surface roughness decreased in most areas with addition of inhibitor, regardless of etch rate.

15

Conclusion

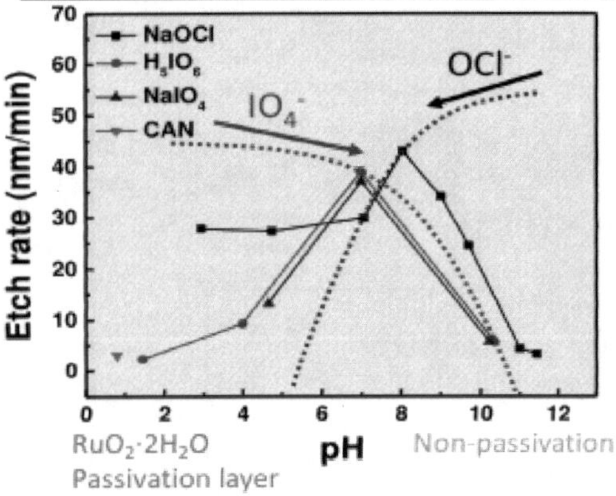

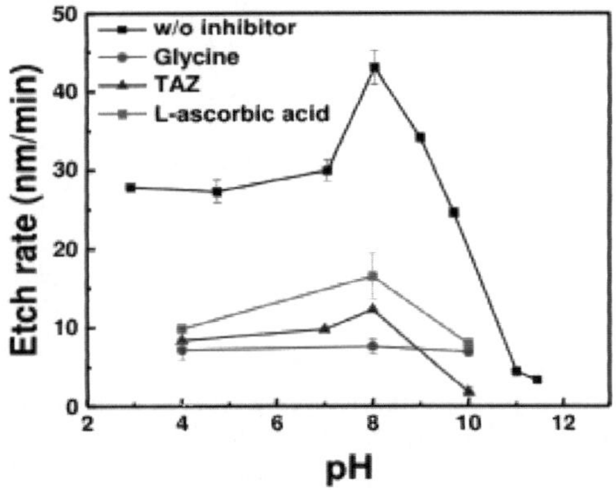

- In this study, a wet etching process for ruthenium was studied.
- Etching reactions occurred in 4 of the 15 etchants selected based on 3 criteria.
- The etch rate is determined by the stability of the etchant ions and the presence or absence of a surface passivation layer.
- The etch rate and surface roughness were significantly reduced in all pH ranges when inhibitors were added.

Acknowledgment

This work was supported by BASF. (s-2023-0865-004)

16

MODELING AND SIMULATION OF BORON REMOVAL FROM WATER USING N-METHYLGLUCAMINE

Ipsita Basu[1] , Nitheeshkumar Kumar[1] , Naveen Kulkarni[1] , Ashutosh Bhabhe[2]

1 quantumzyme

2 SILICON SOLUTIONS

Outline

- Introduction Part 1: Importance of Modeling & Simulation
- Introduction Part 2: Problem Statement Overview
- Modeling and Simulation Approach
- Results & Discussion
- Conclusions

2

Unified Approach to Contamination Control

MATERIAL
(Resins, Polymers, Chemicals etc.)

MANUFACTURING
(Of semiconductors, Chemicals, Components etc.)

METROLOGY
(Characterize, Characterize, Characterize!)

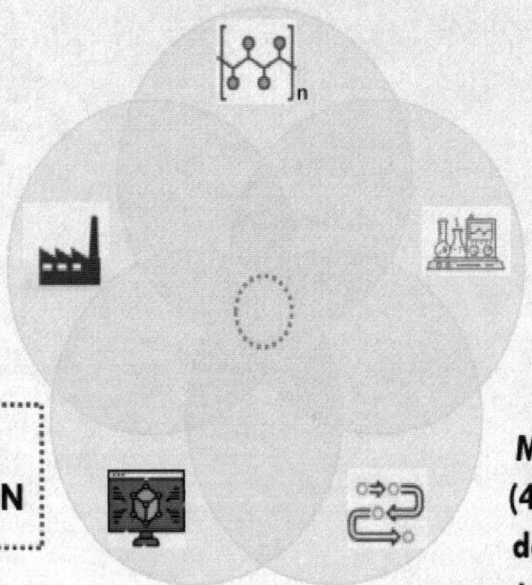

MODELING & SIMULATION

METHODOLOGY
(4W & 1H Q's - of defect formation, contaminant formation, its interaction etc.)

MITIGATION
Targeted, Optimized, & Proactive

3

A Fundamental Re-look at Contamination Removal and Generation Mechanisms is ESSENTIAL!

FUNDAMENTAL PROBLEMS REMAIN UNRESOLVED

- **SPECIFICITY**: Removal of target contaminant/s from target chemistry at target concentration/s
- **(REVERSE) SPECIFICITY**: Removal of target contaminant/s from the material of interest (Example: Filter Material) with new and/or known chemistry/ies
- **CAPACITY**: How much is too little? How much is too much?
- **DEPENDENCY**: Impact of chemical environment, concentration, application condition, mod. type

CURRENT TEST METHODOLOGY IS NOT OPTIMIZED

- **TRIAL AND ERROR**: Lack of wide enough net → better targeted, optimized approaches may be left out
- **SERIALIZED APPROACH**: Development process is long, linear, and is difficult to multi-task, parallelize due to multiple unknowns
- **RESOURCE** (Time, Personnel, Metrology) **INTENSIVE**
- **APPLICABILITY** (to multiple chemicals, multiple materials) & **SCALABILITY** (e.g.: capacity of and for removal) requires separate (and often additive) focus

4

Removal of Boron from Water (UPW) is a known challenge

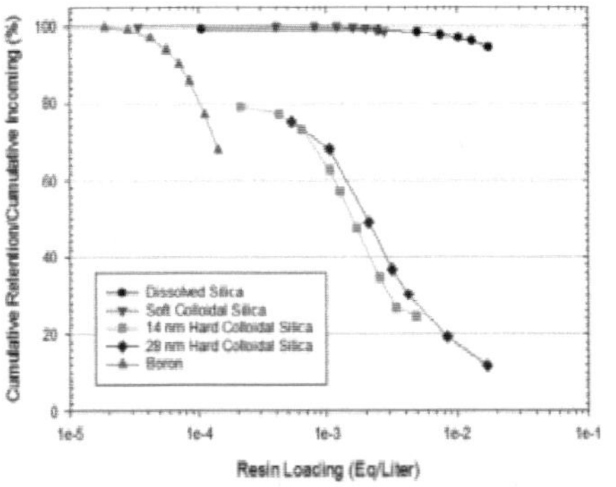

S. Libman et al. UPW Conference, Portland, US 2017

Traditional IEX Resins show very low selectivity to boron → Breakthrough of boron happens rapidly

Known Solutions:

a) AmberSep™ Resin IRA743 from Dow is an industrial grade Chelating Resin from boron removal from Water using n-methylglucamine modified styrene-divinylbenzene beads (500-700 um); Similar resin-based solutions (DIAION™ CRB03, 05) are available from Mitsubishi Chemical Company

b) Membrane-based methods for removal of boron using Amino Polyol Ligands

BUT, despite known solutions, there is **only (limited) empirical literature** on understanding exactly **HOW boron is removed from Water** by n-methylglucamine (NMG) → A Deeper Understanding is therefore required

5

What is known from Literature?

	Known Key Information
Removal of boron from Water	• NMG is a known-good chelator for boron
pH related information	• NMG removes boron via Chelation between pH of 3 – 10 • boron (often) exists as Boric Acid (at pH < 7) & Tetrahydroxy Borate (at pH > 7)
Removal Mechanism	• Boron is removed via formation of a complex with the chelating group. • Both 1(ligand) :1(boron) and 2:1 complexes have been reported in literature
Control Ligands	• Polyvinyl Alcohol (PVA) has insufficient efficacy in removal of boron → will be used as a negative control

6

Problem Statement: Goal & Objectives

Goal

- To investigate the chelation mechanism for boron removal by NMG using in-silico modeling & simulation studies

Objectives

- To obtain the structural details of NMG-boron complex at the QM level.
- To obtain the energy of complex formation at the QM level.
- To study the assembly of NMG conjugated polystyrene material using MD simulation at MM level.
- To model the early process of boron (either as Boric Acid or Tetrahydroxy Borate) binding to NMG using MD simulation.

QM: Quantum Mechanical
MD: Molecular Dynamics
MM: Molecular Mechanics

7

Modeling and Simulation Approach

- The Chelation process involves a chemical reaction and bond formation → QM calculations are needed to study the chelation
- Microscopically, NMG is the functional unit for chelation
- In the filtration matrix, NMG groups are attached to large polymers
- Polymers can be omitted in QM studies to save computational cost

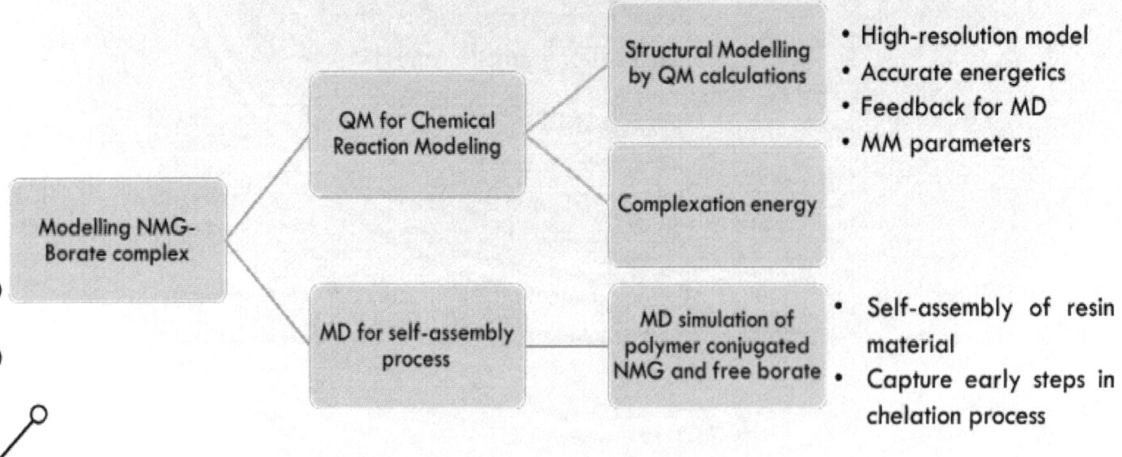

8

Boron Complexes (ligand : boron)

NMG-B⁻ (1,2) Complex

NMG-B⁻ (3,4) Complex

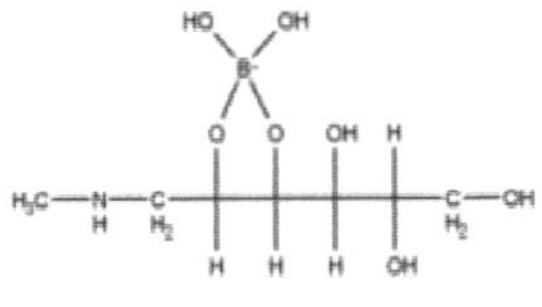

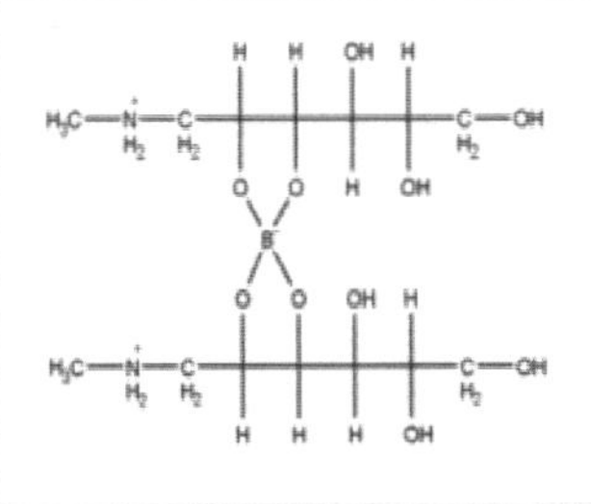

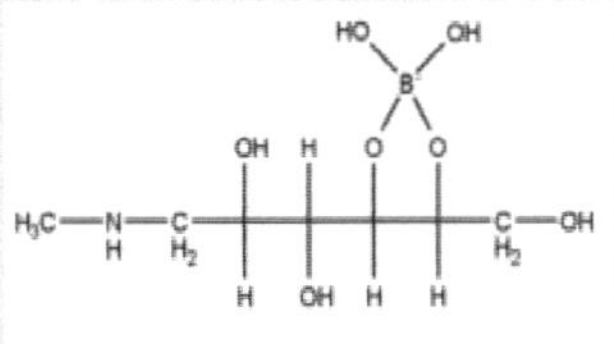

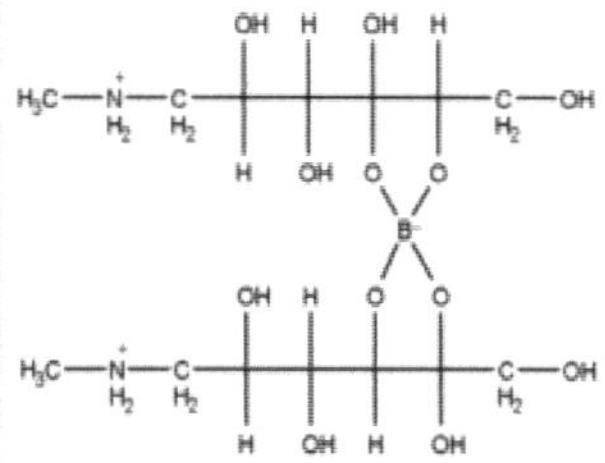

1:1 Complex

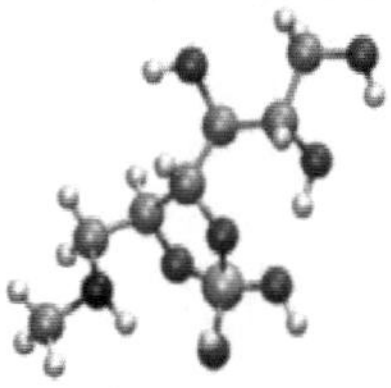

2:1 Complex

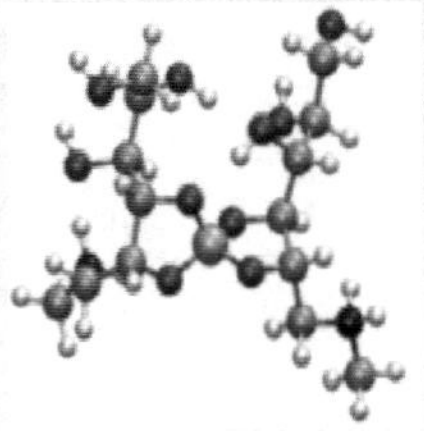

1:1 Complex

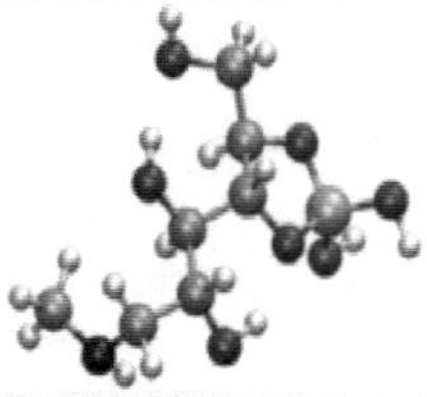

2:1 Complex

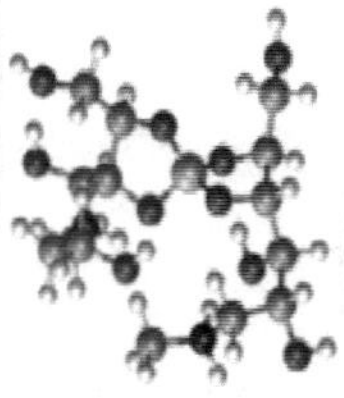

Thermodynamics: Complexation Energy Comparison

1:1 Complex

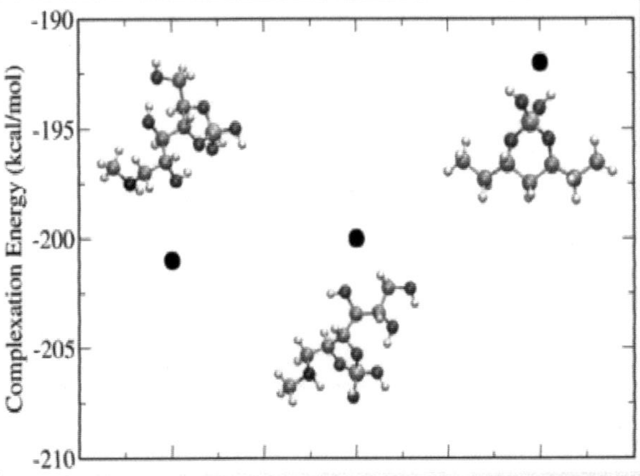

NMG-B⁻ (3,4) **NMG-B⁻ (1,2)** **PVA-B⁻**

2:1 Complex

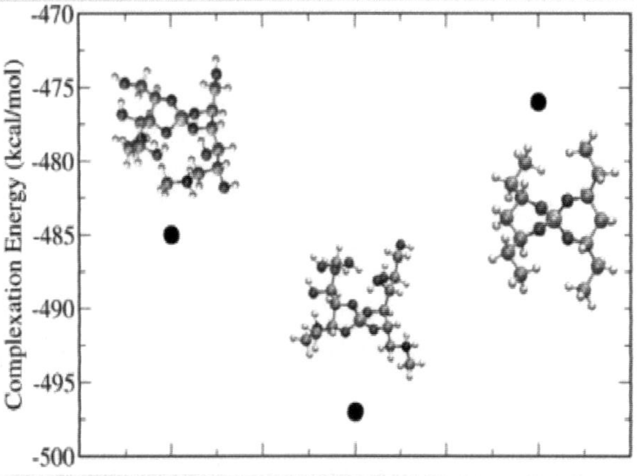

NMG-B⁻ (3,4) **NMG-B⁻ (1,2)** **PVA-B⁻**

- **2:1 (2x ligands : 1x boron) is a higher stability complex → validates findings of Okabira et al.[1]**
- **NMG is more effective in removing boron compared to PVA → validates findings of Harada et al.[2]**

[1] Okabira et al., Essential Structural Factors of Bis-1,2-diol in Polyol Molecules for Capturing Carbon, 2011

[2] Harada et al., boron adsorption mechanism on polyvinyl alcohol, 2011

10

Kinetics: Steps for Molecular Dynamics Simulation

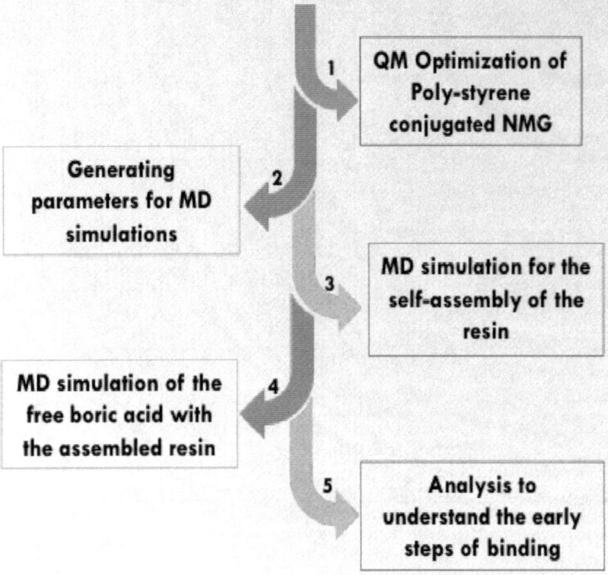

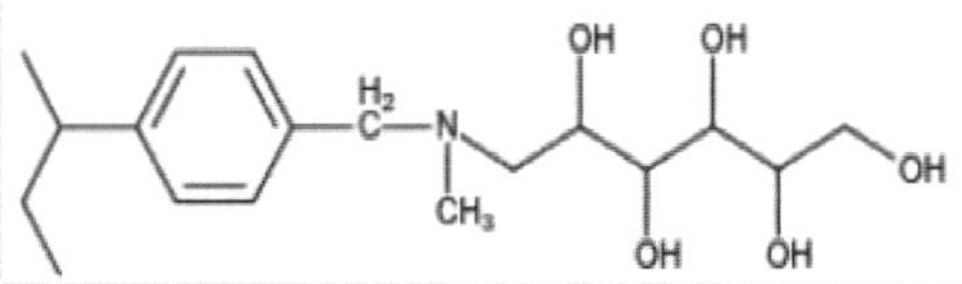

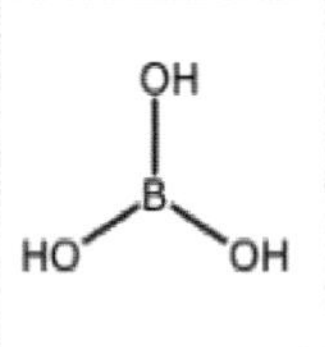

Boric acid

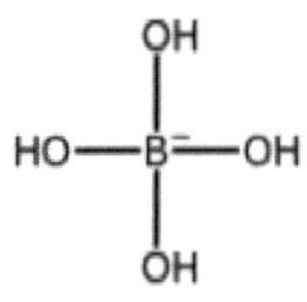

Tetra-hydroxy Borate

Reference: Kurt, Barış, and Hamdi Temel. "Parameterization of boronates using VFFDT and paramfit for molecular dynamics simulation." *Molecules* 25.9 (2020): 2196.

11

Self-Assembly Simulations of NMG-Polystyrene Resin Results in Stable Bilayer Aggregates

Initial **Final**

128 molecules

256 molecules

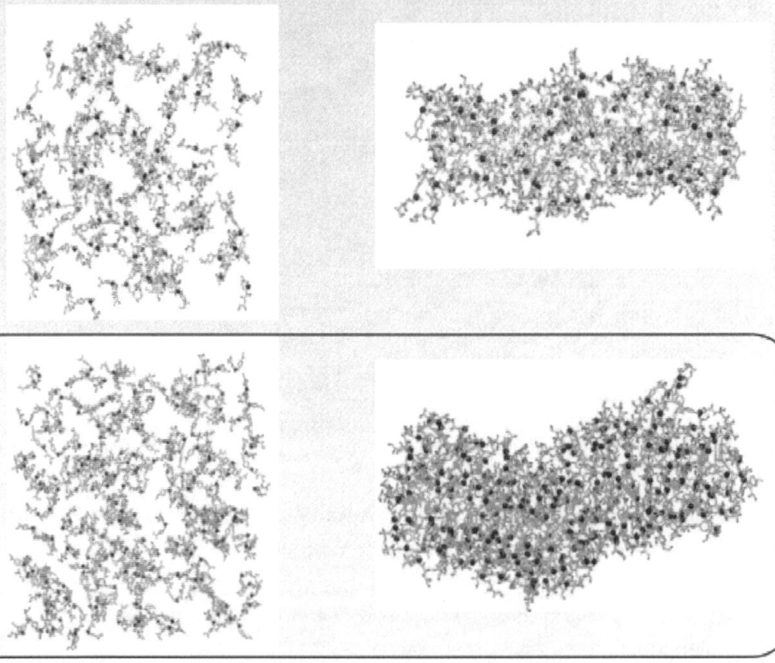

12

Boric Acid Chelation
Simulates low pH conditions

Initial

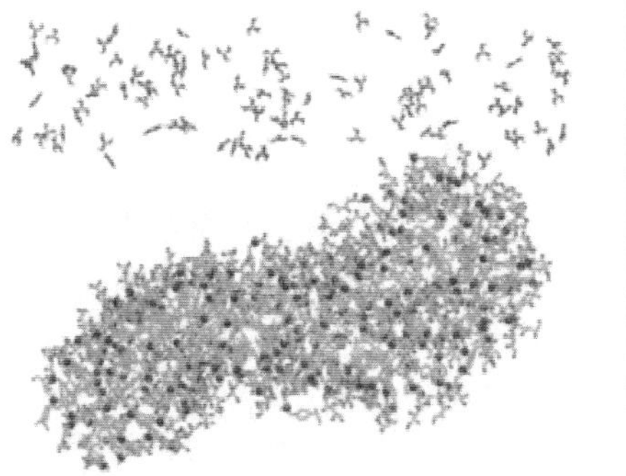

Final

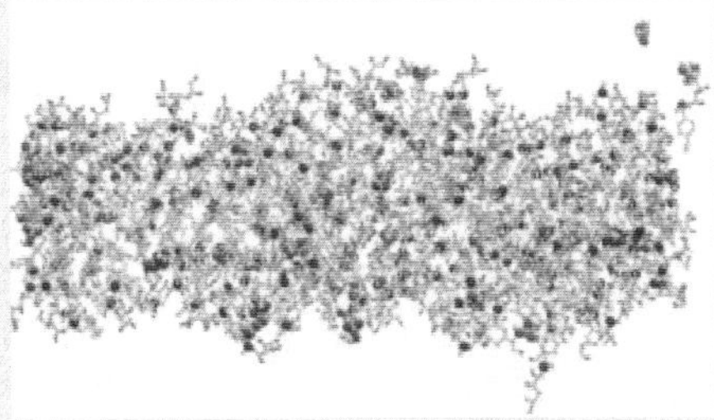

Simulation (Video)

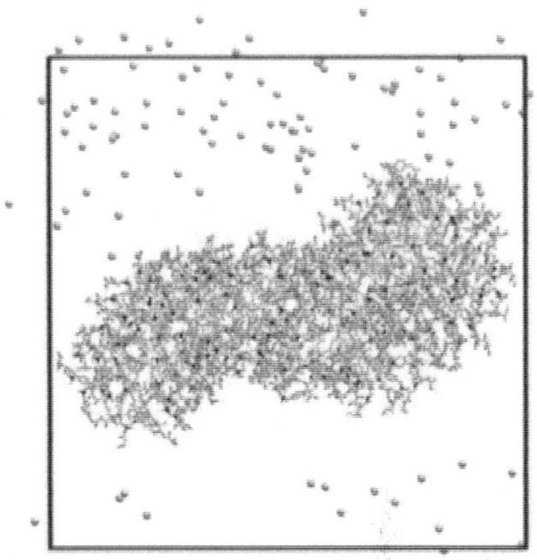

- We observe complete removal of boron (As Boric Acid) by NMG
- As demonstrated empirically, NMG is effective at completely removing boron at low pH

Ordering near NMG (w/ Boric Acid)

Boric acid B – NMG O

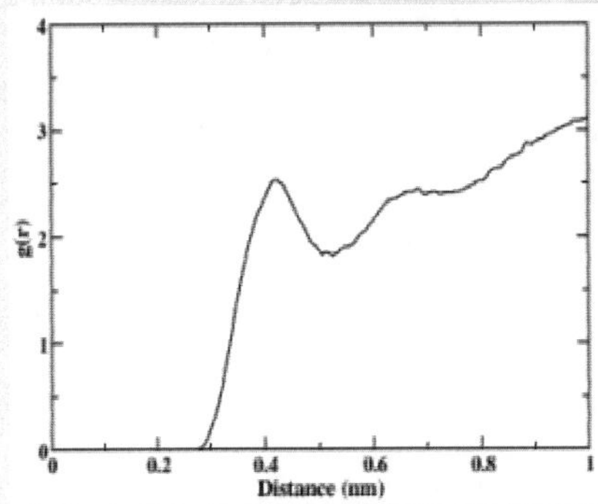

Boric acid O – NMG O

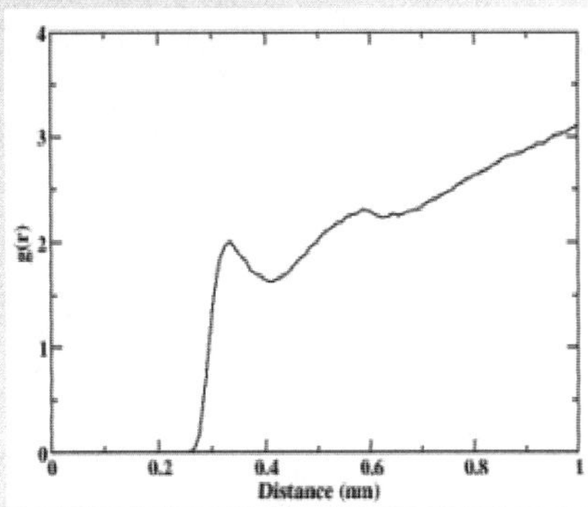

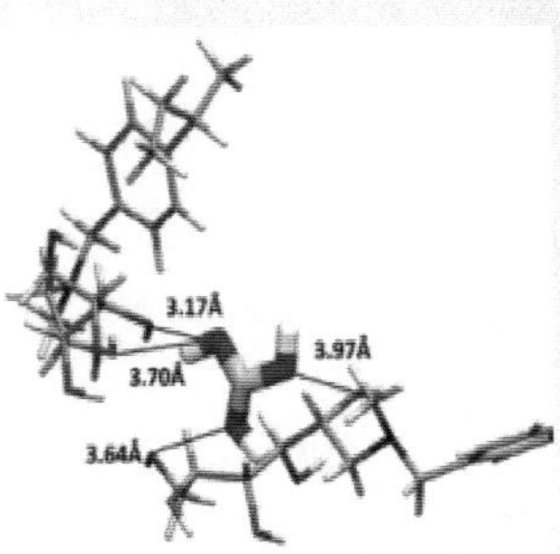

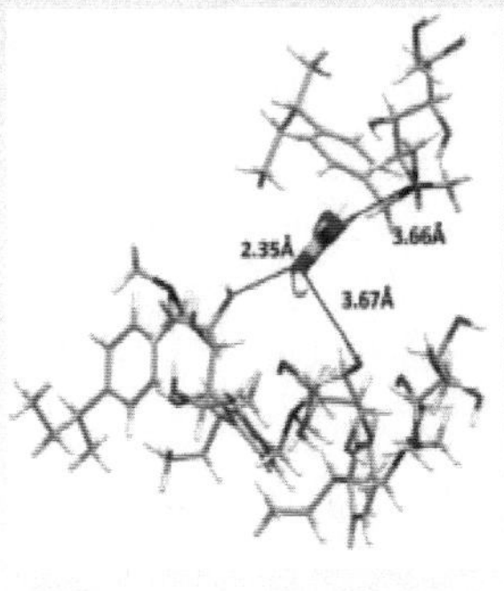

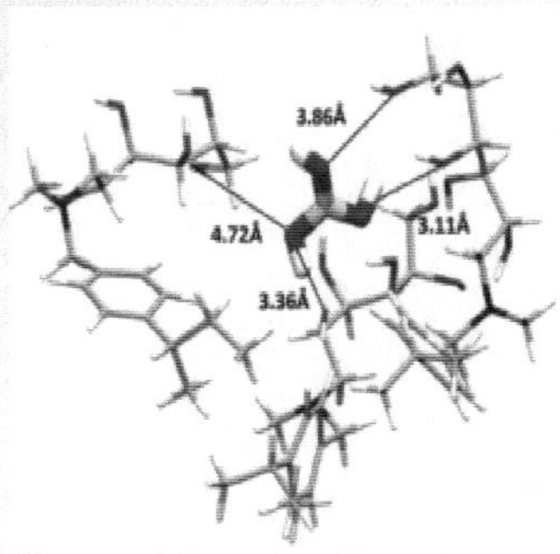

Borate Chelation
Simulates high pH conditions

Initial

Final

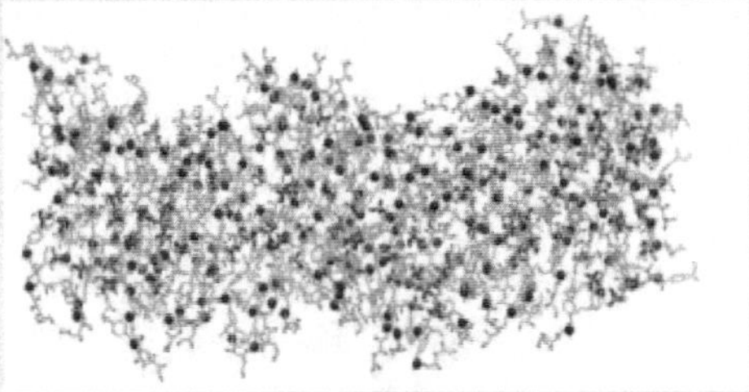

Simulation (Video)

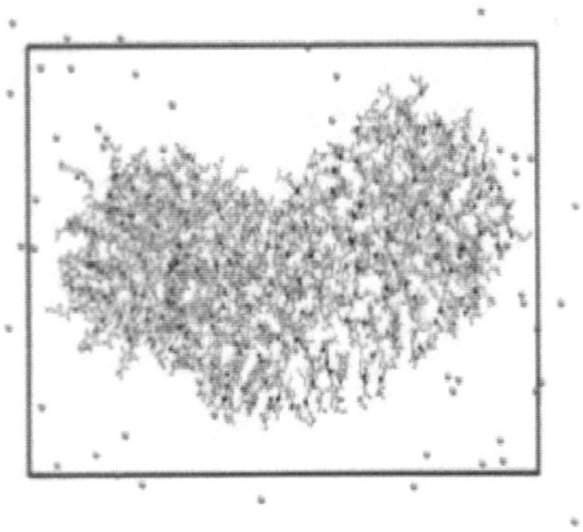

- We observe complete removal of boron (As Tetrahydroxy Borate) by NMG
- As observed empirically, NMG is effective at completely removing boron at high pH

Ordering near NMG (w/ Borate)

Borate B – NMG O

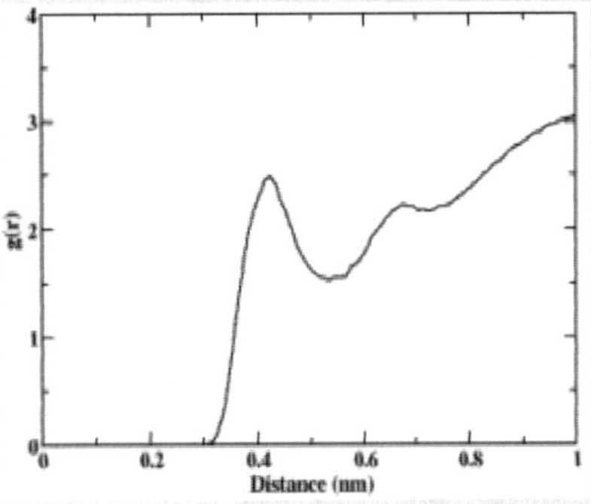

Borate O – NMG O

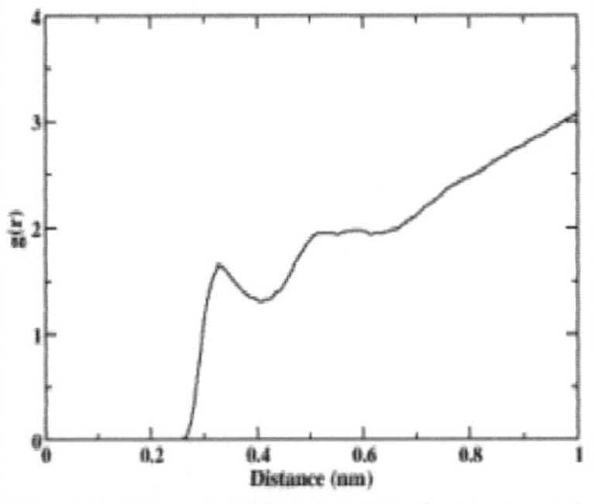

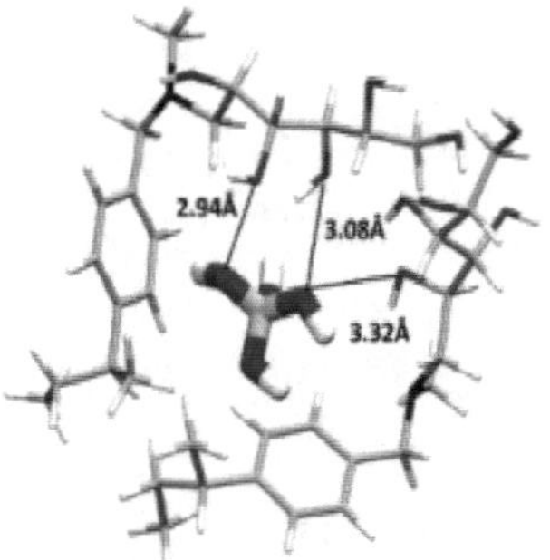

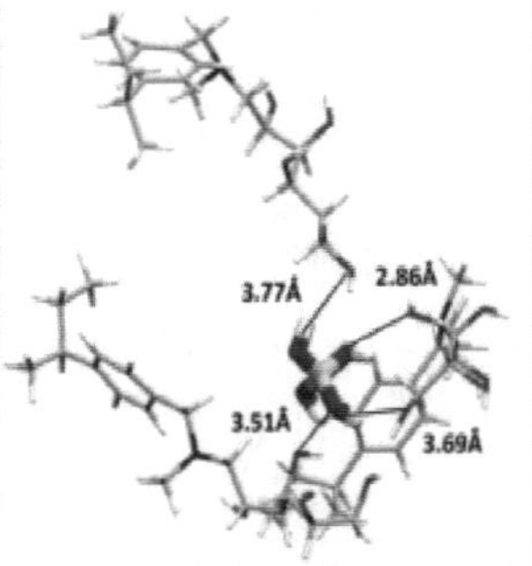

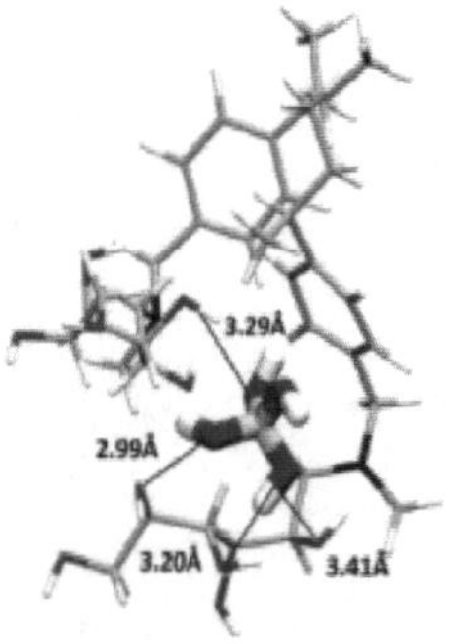

Conclusions

QM Calculations

- From complexation energy, we found NMG is better than PVA at removing boron → this is validated by empirical data.
- 2:1 Complex (2x ligand : 1x boron) is more stable than a 1:1 complex → this is validated by data available in literature.

MD Simulations

- Self-assembly of resin functionalized with NMG forms a bilayer like aggregate
- Within bounds of the pH at the lower end (pH $\geq$ 3) boron as boric acid and at the higher end (pH $\leq$ 10) boron as tetrahydroxy borate is completely removed by NMG → this is validated by empirical data from literature
- Simulations showed that spontaneous diffusion and hydrogen bonding are key steps prior to formation of a stable complex in the chelation process, across pH → this is a hitherto unknown insight into the removal of boron by NMG.

17

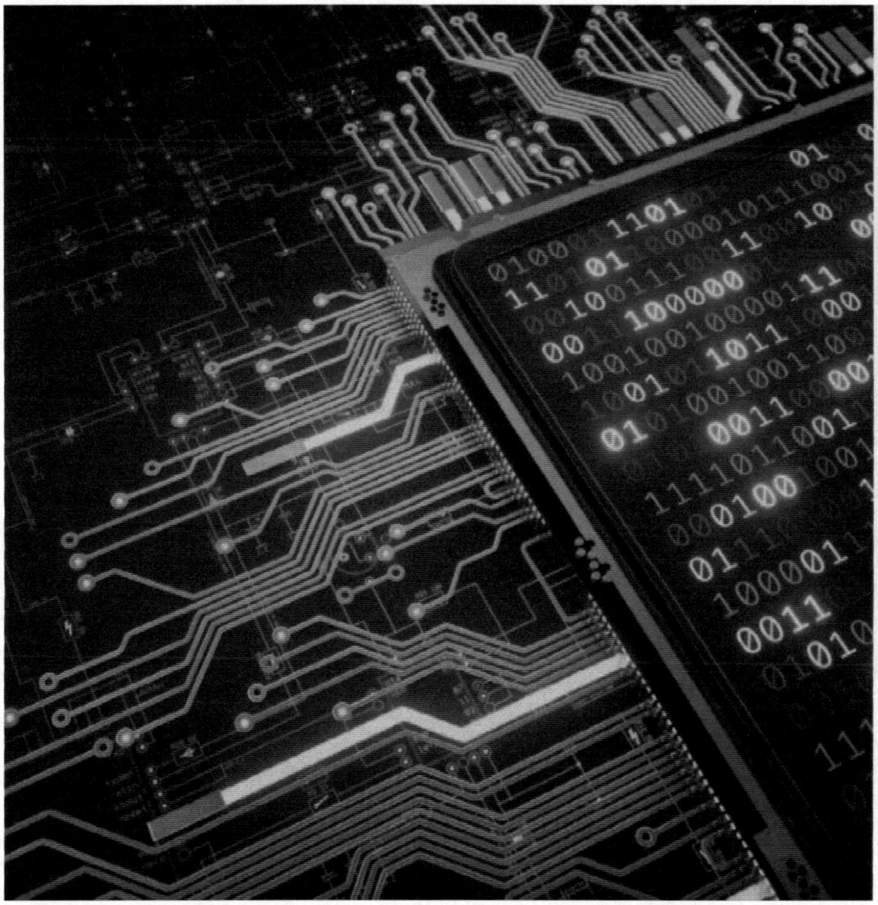

SILICON SOLUTIONS

quantumzyme

THANK YOU!

Highly Uniform PolySi Channel Etch

Jack Ke, Owen Liu, Jean Yang, Kyoung Sik Chin, Justin Kim
Advanced Cleans & Slurry Technologies (ACST)
DuPont Electronics & Industrial
Email: chunren.ke@dupont.com
Phone: (886)972016970
Address: No.2 Li-Hsin Rd.IV, Hsinchu Science Park
Hsinchu 30078, Taiwan

Oral Presentation
The Surface Preparation and Cleaning Conference (SPCC) 2024
10:00 AM, Wednesday, April 24, 2024

Outline

- ❑ **Introduction**
- ❑ **Design Criteria for Highly Uniform PolySi Channel Etch**
- ❑ **Results and Discussion**
  - Components Effects on Etch Uniformity
  - Methodology Development – Etch Product Dispersant Screening
  - Surface Roughness
  - Etch Pits
- ❑ **Conclusions**
- ❑ **Acknowledgments**

Introduction – 3D NAND Flash Memory and PolySi Channel

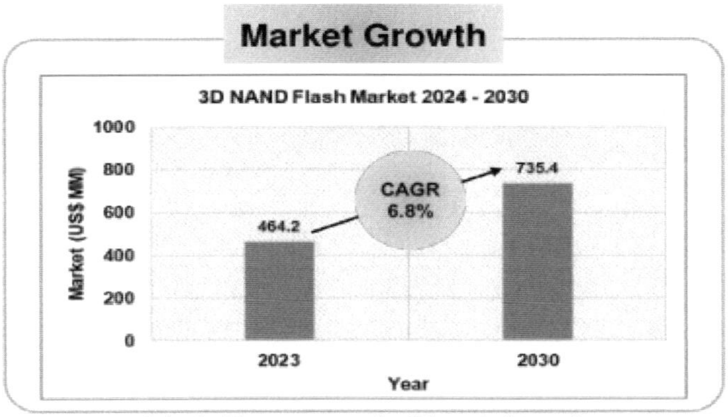

Adapted from sources by 24 market reports

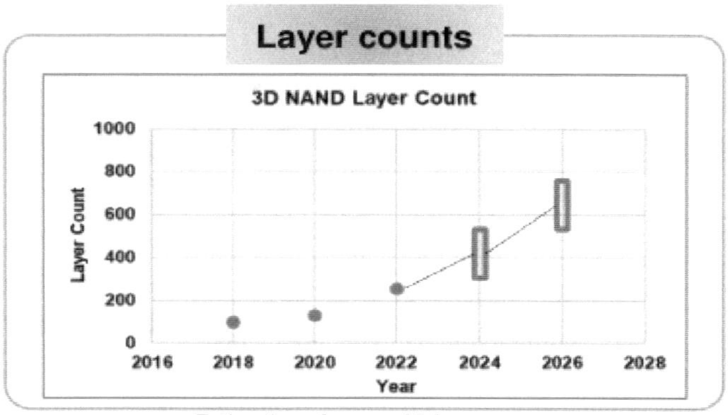

Estimations from multiple sources

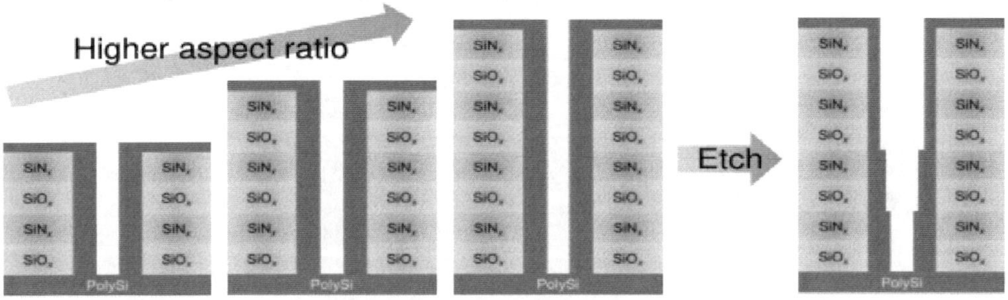

- Issue: conventional processes (dry etch or SC1 wet etch) cannot produce uniform etch in high aspect ratio (HAR) advanced structures.

One key component in 3D NAND: polySi channel

Design Criteria for Highly Uniform PolySi Channel Etch

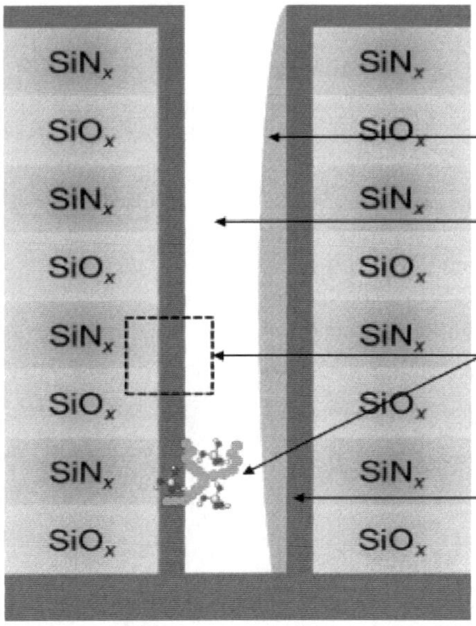

Working mechanism and key properties

1. **Uniform etch from top to bottom (T2B)**
   - Good wettability and diffusion capability to reduce T2B etch difference.
   - Suppress top etch
   - Enhance bottom etch (prevent silicic acid etch product growth)
2. **Free of etch pits**
   - Si chelator selection
   - Component effects
3. **Smooth surface**
   - Si chelator loading
   - H_2O_2 effect

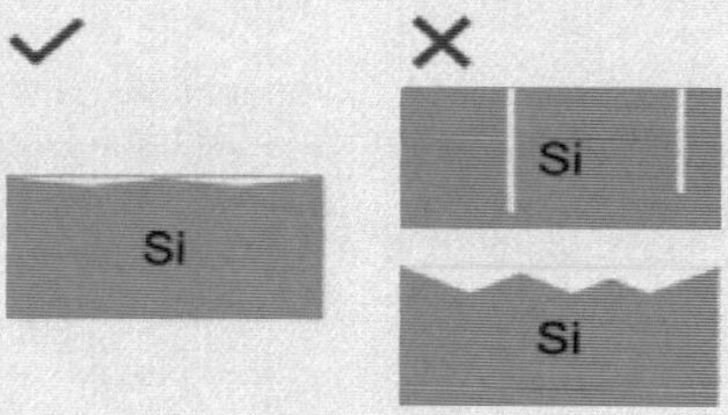

Components Effect on Etch Uniformity – An Overview

Comp.	Functionality	Etchant-1	Etchant-2	Etchant-3	Etchant-4	SC1
A	Si chelator	V	V	V	V	$NH_4OH:H_2O_2$ = 1:1
B	pH buffer	V	V	V	V	
C	Wettability improver	V	V	V	V	
H_2O_2	**Surface smoother**	-	**V**	**V**	**V**	
D	**Diffusion improver**	-	-	V	V	
E	**Uniformity adjuster**	-	-	-	V	

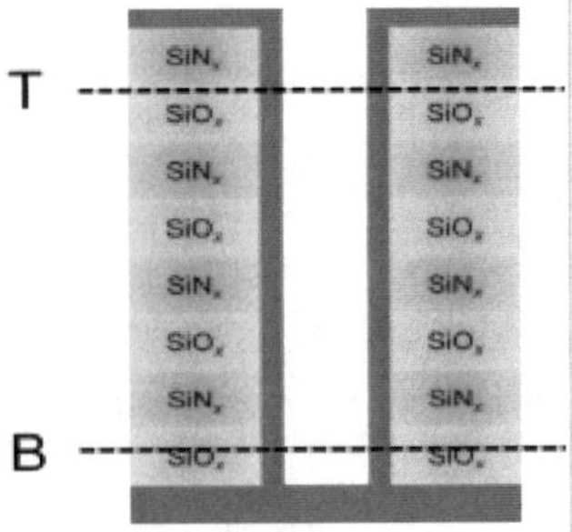

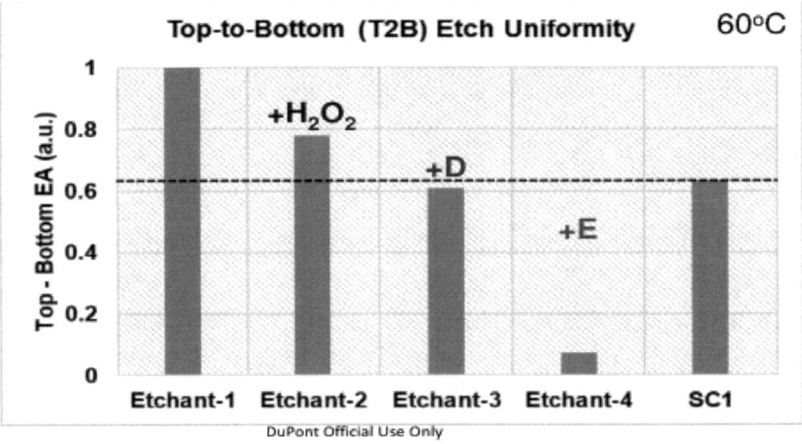

Selection of Wettability Improver (Comp.C)

Comp.	Functionality	Etchant-1
A	Si chelator	V
B	pH buffer	V
C	**Wettability improver**	**V**

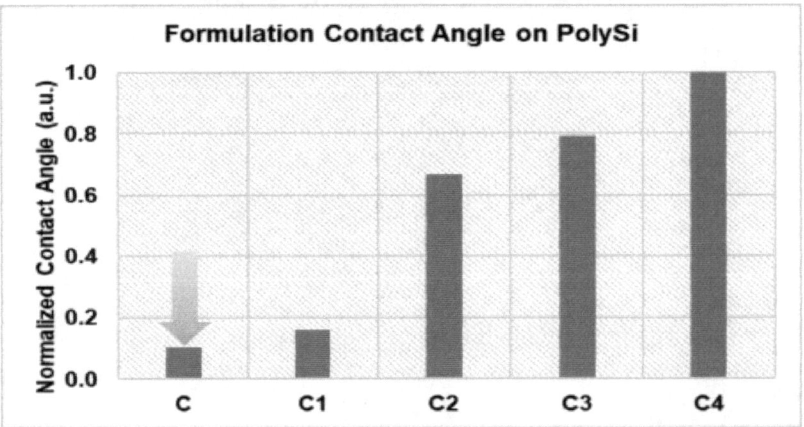

- 'C' and 'C1' offered lower formulation contact angles, suggesting better formulation spreading capability onto the polySi.

Selection of Diffusion Improver (Comp.D)

Comp.	Functionality	Etchant-3
A	Si chelator	V
B	pH buffer	V
C	Wettability improver	V
H_2O_2	Surface smoother	V
D	**Diffusion improver**	**V**

The Stokes-Einstein relation: $D_h = \frac{k_B T}{3\pi \eta D_t}$

D_h: hydrodynamic diameter
K_B: Boltzmann constant
T: temperature
η: dynamic viscosity
D_t: diffusion coefficient

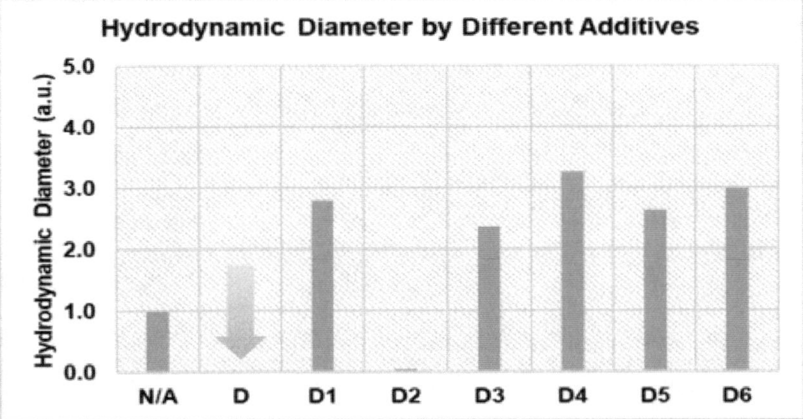

- 'D' and 'D2' reduced hydrodynamic diameters, representing better diffusion (penetration) capability into high-aspect-ratio (HAR) vias/trenches.

Comp. E Loading Effect on Etch Uniformity

Comp.	Functionality	Etchant-4
A	Si chelator	V
B	pH buffer	V
C	Wettability improver	V
H_2O_2	Surface smoother	V
D	Diffusion improver	V
E	**Uniformity adjuster**	**V**

Alkaline Si etch mechanism:

$$Si + 2OH^- + 2H_2O \rightarrow SiO_2(OH)_2^{2-} + 2H_2$$

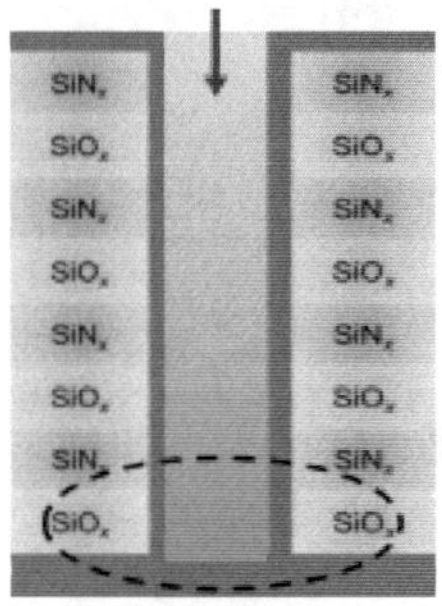

Higher concentration

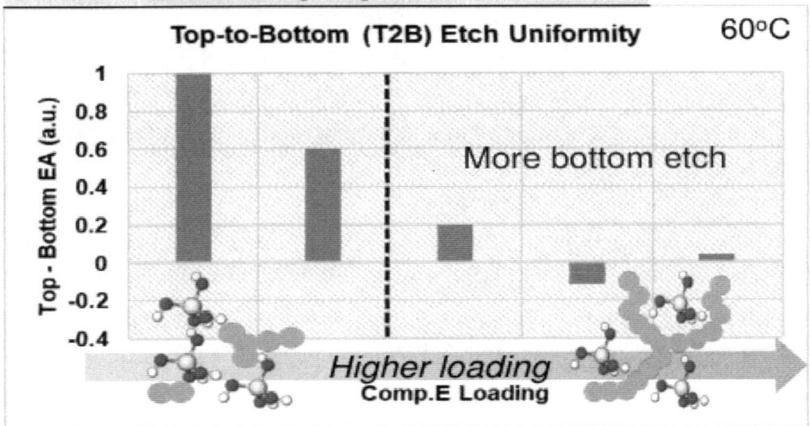

- Higher Comp. E concentration promoted better etch uniformity by minimizing silicic acid etch product growth.

Comp.E Steric Effect on Etch Uniformity

Si etchants
A+B+C+D+H_2O_2
E (different sizes)

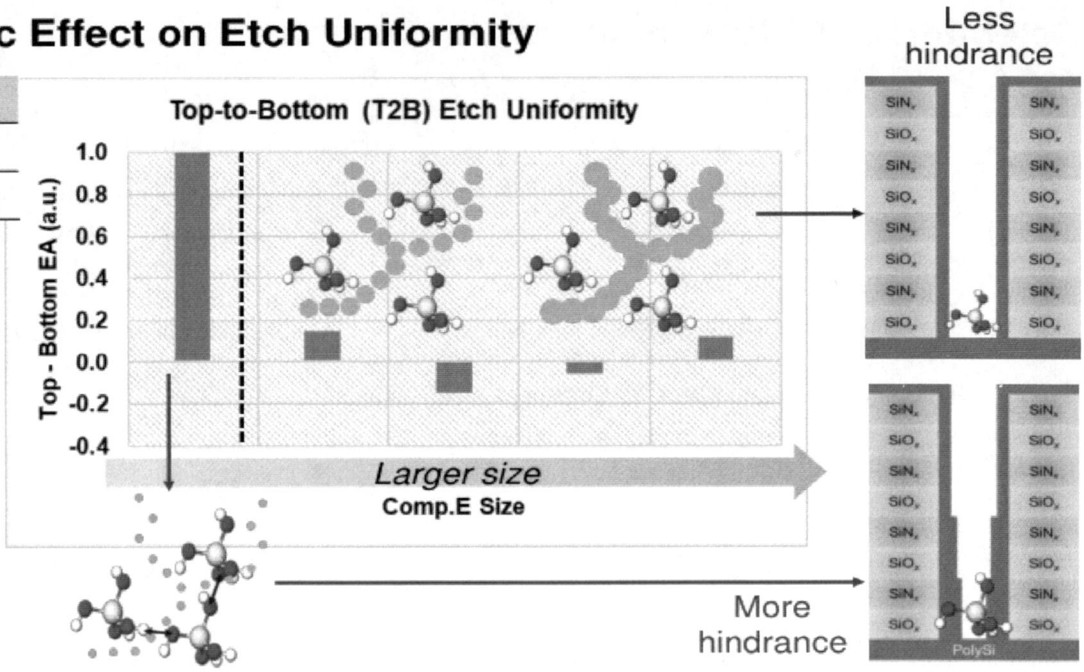

- Large size of comp. E provided good T2B etch uniformity via steric effect.
- If Comp. E is too small, silicic acid etch product growth was not effectively prevented.

Methodology Development:
Etch Product Dispersant Screening (1/2)

- ❑ Hypothesis of silicic acid etch product growth needs to be verified.
- ❑ An effective approach for screening silicic etch product dispersant requires being developed.

Step 1: activate silicic acid gelation process

Step 2: silicic acid size measured by dynamic light scattering (DLS) – hydrodynamic diameter

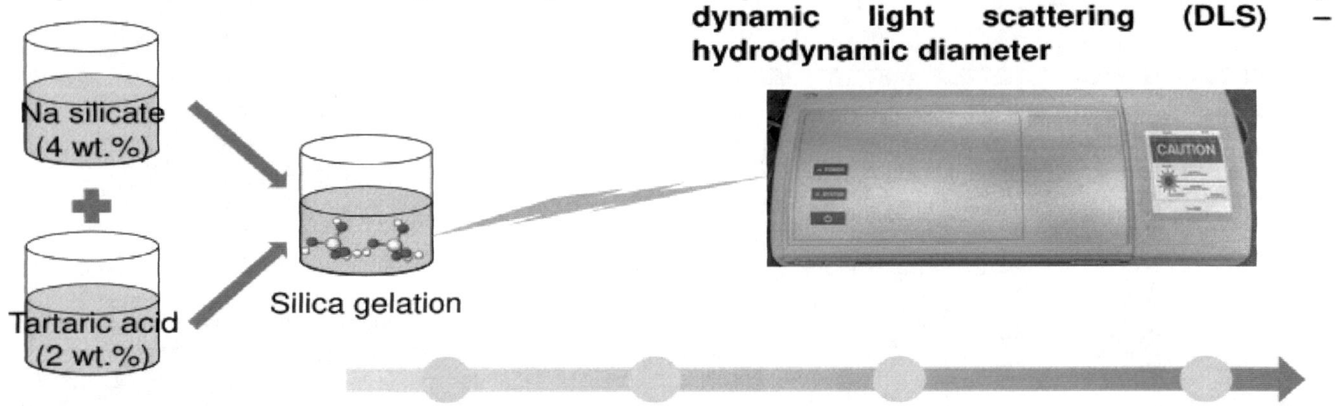

As-prepared *(t = 0)* | 1st aged point *(t = 30 min)* | 2nd aged point *(t = 60 min)* | … | Final aged point *(t = 180 min)*

Idea inspired from: *Sci. Technol. Adv. Mater. 9 (2008) 035006*

Methodology Development:
Etch Product Dispersant Screening (2/2)

Si etchants
$A+B+C+D+H_2O_2$
E (large/small) & E1

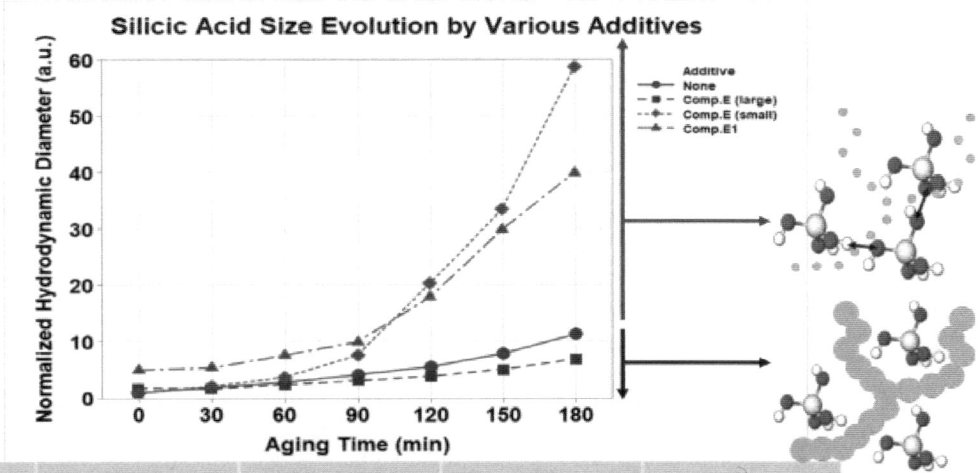

Additive	E (Large)	None	E (Small)	E1
Size	*Larger silicic acid size* →			
Uniformity	*Worse T2B Uniformity* →			

- Only Comp.E (large) can suppress silicic acid growth.

Si Chelator Loading Effect on Surface Roughness

Si etchants
$A+B+C+H_2O_2$

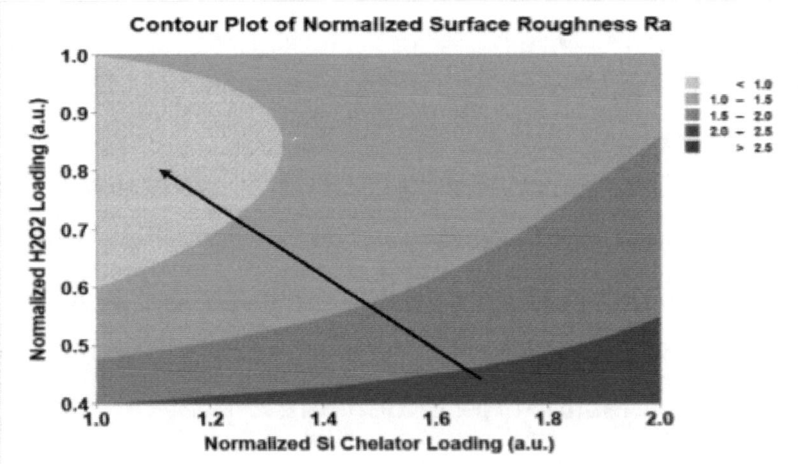

- Both Si chelator and H_2O_2 loadings are critical to surface roughness.
- Less Si chelator with more H_2O_2 resulted in smoother post-etched surface.

Si Chelator Effects on Etch Pits – Applying Si (111) for Evaluation

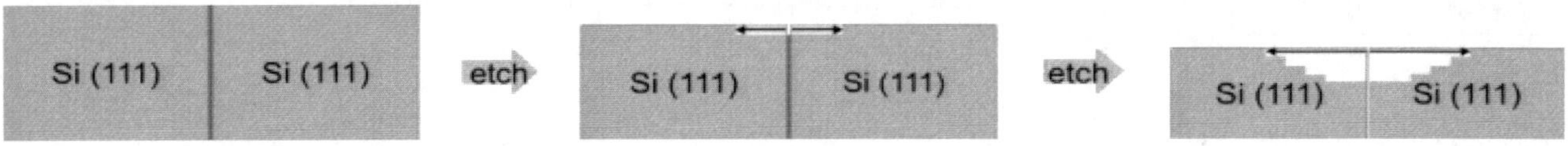

Component	Etchant-1	Etchant-2	Etchant-3	Etchant-4	Etchant-5	Etchant-6
Si chelator	A	A1	A2	A3	TMAH	NH_4OH
Comp.B	V					
Comp.C	V					
Etch Pits (by AFM)	X	X	V	X	V	V

- By changing Si chelator types, etch pit formation tendency can be different.
- The widely used bases (TMAH and NH_4OH) led to formation of undesired etch pits.

Conclusions

- ❑ DuPont has developed wet etchants for uniformly etching polySi channels from top to bottom (T2B).
  - By adding effective additives with different functions, T2B uniformity can be significantly improved.
  - Innovative eyes for etch component function have also been developed.
- ❑ Methodology for etch product dispersant screening has been developed – silicic acid etch product growth rate can correlate to T2B uniformity.
- ❑ Si chelator and H_2O_2 loadings are critical to lowering surface roughness.
- ❑ Etch pit formation is Si-chelator-type dependent.

Acknowledgements

Advices

Jean Yang, Katie Gramigna, Paul Bernatis, and Kyoung Sik Chin

Justin Kim

Experimental support

Carol Cheng, Lester Lee, and Owen Liu

Amy Chen, Annabelle Wang, and Vic Lin

All internal collaborators who supported

All external collaborators who supported

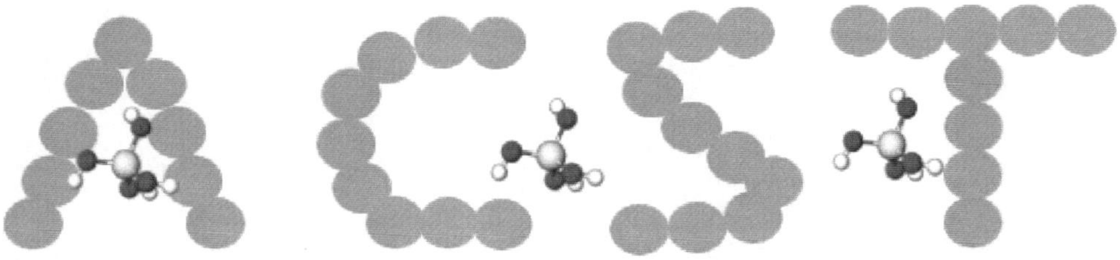

DUPONT™

THE SURFACE PREPARATION AND CLEANING CONFERENCE (SPCC)

SPCC 2024 | April 24-25, 2024

Investigation of Surface Characterization of 4H-SiC for Better Cleaning Performance

Yoon-Ji Ra[1], Juyeol Lee[1], Jooyoung Yang[2], Tae-Uk Kim[2], Tae-Gon Kim[1], and Jin-Goo Park[1*]

[1] Hanyang University ERICA, [2] SK Siltron

*jgpark@hanyang.ac.kr

The Engine of Korea

Contents

1. Introduction and Motivation
2. Objectives
3. Surface Characteristic of SiC
   - Surface chemistry changes with chemical treatment
4. Understanding of SiC Cleaning Mechanism
   - Evaluation of cleaning method applied with physical force
5. Summary

Introduction and Motivation

Why Silicon Carbide?

- **Material properties of 4H-SiC vs Si**

Properties	4H-SiC	Si
Crystal Structure	**Hexagonal**	Diamond
Bandgap energy (eV)	**3.26**	1.12
Saturation drift velocity (x 10^7cm/s)	**2.7**	1
Thermal conductivity (W/cm · K)	**4.9**	1.5
Breakdown field voltage (V/cm) X 10^6	**3**	0.3

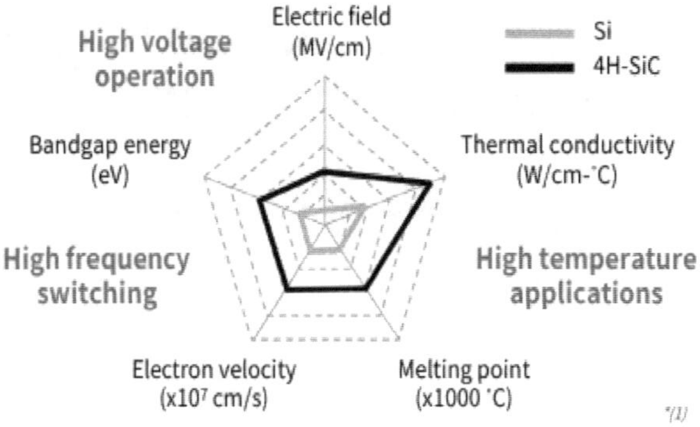

*(1)

- These exceptional attributes position SiC wafers as the preferred choice for power semiconductor applications.

- **Applications of SiC**

Automotive

Energy

Transportation

Industrial

1. *Syed Khaja Aarief. (2018). Diffusion Soldering for High-temperature Packaging of Power Electronics.*

Why Silicon Carbide?

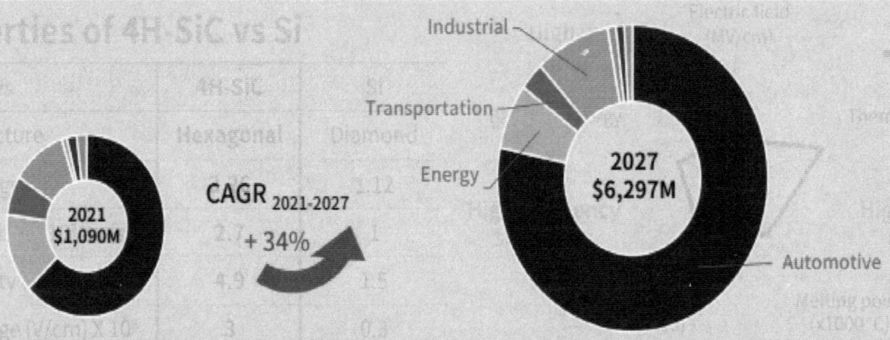

- With the escalating demand for SiC devices, there's a growing imperative to intensify research and development efforts focused on advancing SiC device technology.

- **Applications of SiC**

Automotive

Energy

Transportation

Industrial

1. *Power SiC Report. (2022). Yole Intelligence.*

Motivation

- **Manufacturing process of SiC wafer**
  - The research is underway on each of the manufacturing process of SiC wafer

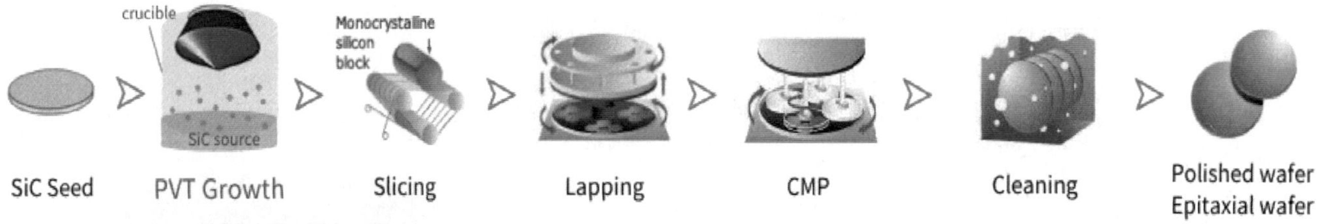

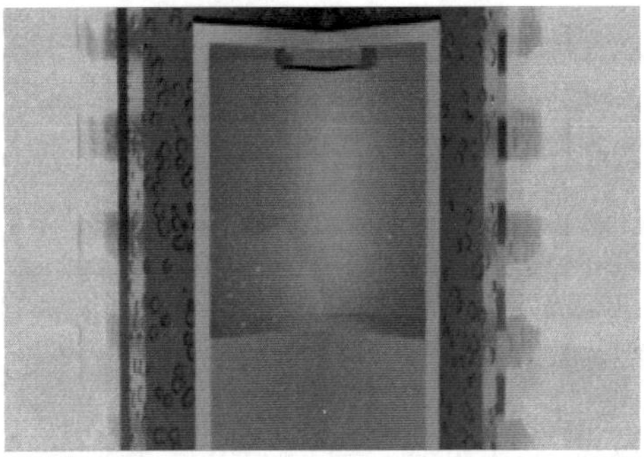
*(1)

- PVT method is notably slow, capable of growing only a few millimeters of SiC substrate in a day.
- Therefore, subsequent processes demand high technical precision to ensure the yield and reliability.

1. *ROHM Semiconductor Europe. (2021). Spotlight on SiC crystal growth[Video]. YouTube.*

Motivation

- **Importance of cleaning process for SiC wafer**

* Particle and damage free cleaning

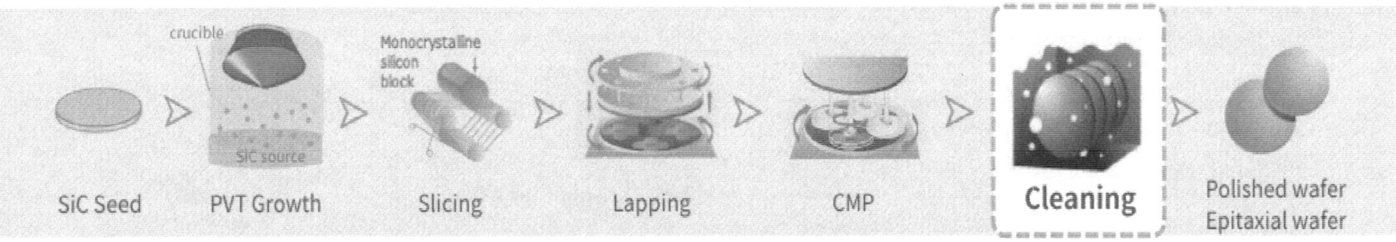

- Cleaning efficiency and defect level stand as the crucial final steps in the process, given their direct impact on the yield and reliability of the device.

Aspects of device

- Material quality
- Device performance
- Yield improvement

Aspects of process

- Enhanced efficiency
- Cost reduction
- Innovation & competitiveness

- In aspects of the process, the cleaning process significantly contributes to heightened efficiency and reduced costs.

Challenges in cleaning SiC wafer

- **Requirement of research for cleaning SiC wafer**

Si wafer

Conventional cleaning method
SC-1, SC-2, SPM, DHF

When using conventional methods to clean SiC,
a notable issue arises:
long time & contaminants remain

SiC wafer

Need to understanding cleaning method for SiC

- This is largely due to the intrinsic properties of SiC, which pose limitations on conventional cleaning techniques.

[Literature on surface preparation & cleaning of SiC]

- **Ozone Cleaning** (SUMITOMO ELECTRIC. (2011). Method of cleaning silicon carbide semiconductor. 5540919 B2.)
- **Plasma etching** (Artem A. Osipov. (2020). High-temperature etching of SiC in SF6/O2 inductively coupled plasma.)
- **Defect inspection** (Po-Chih Chen. (2022). Defect Inspection Techniques in SiC.)

- However, despite its necessity, for the cleaning process of SiC, there are very few studies regarding the cleaning mechanism and surface chemistry of SiC.

Mechanism of conventional cleaning

- **Conventional RCA cleaning**

Process	Chemicals (ratio)	Mechanism
SC-1	$NH_4OH : H_2O_2 : H_2O$ (1:2:10)	Oxidation & Etching
SC-2	$HCl : H_2O_2 : H_2O$ (1:1:10)	
SPM	$H_2SO_4 : H_2O$ (4:1)	Oxidation
DHF	$HF : H_2O$ (1:100)	Etching

- **Mechanism of particle removal: SC-1**

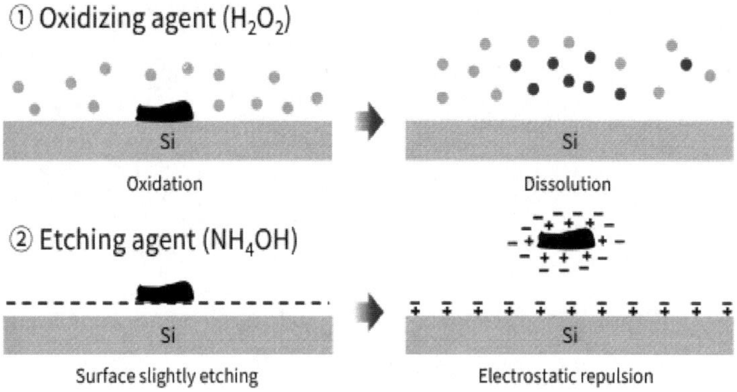

- The mechanism of SC-1, involves the oxidation of organic contaminants for dissolution from the surface and slightly etching enhances the contact area between the particles and the surface, promoting effective cleaning.
- Other cleaning methods operate through oxidation and etching mechanisms to remove the particles.
- Therefore, surface characterization of SiC with oxidation and etching is necessary to understand the cleaning mechanism for SiC.

Research Objectives

Research objectives

Investigation of Cleaning of SiC

01 Surface characterization

- Fundamentals of surface chemistry analysis, focusing on chemical reactions

02 SiC cleaning mechanism

- Understanding of cleaning mechanism for SiC
- Evaluation cleaning efficiency by cleaning method

Surface Characterization of SiC

- Surface chemistry changes with chemical treatment

Materials and analysis

Wafers

4H- SiC

Si

Chemicals

	Concentration	Temp.	Effects
SPM	$H_2SO_4 : H_2O_2 = 4 : 1$	100 ˚C	Heavy organic
SC-1	$NH_4OH : H_2O_2 : H_2O = 1 : 2 : 10$	70 ˚C	Particles
DHF	1 wt%	20 ˚C	Native oxide
H_2O_2	30 wt%	20 ˚C	Oxidizing agent
$KMnO_4$	1 wt%	20 ˚C	Strong oxidizing agent

Analysis tools

Surface roughness	Wettability	Oxide formation

AFM

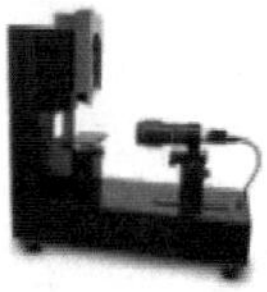

Contact Angle

XPS

Experimental procedure

Pretreatment → ✓Analysis → Oxidation → ✓Analysis → Etching → ✓Analysis

Si 4H- SiC

Dipping in DHF
(10 min)

DHF	20 ˚C

Si 4H- SiC

Dipping in oxidizing agent
(10 min)

SPM	100 ˚C
SC-1	70 ˚C
H_2O_2	20 ˚C
$KMnO_4$	20 ˚C

Si 4H- SiC

Dipping in etching agent
(10 min)

DHF	20 ˚C

Motivation

- **Importance of cleaning process for SiC wafer**

* Particle and damage free cleaning

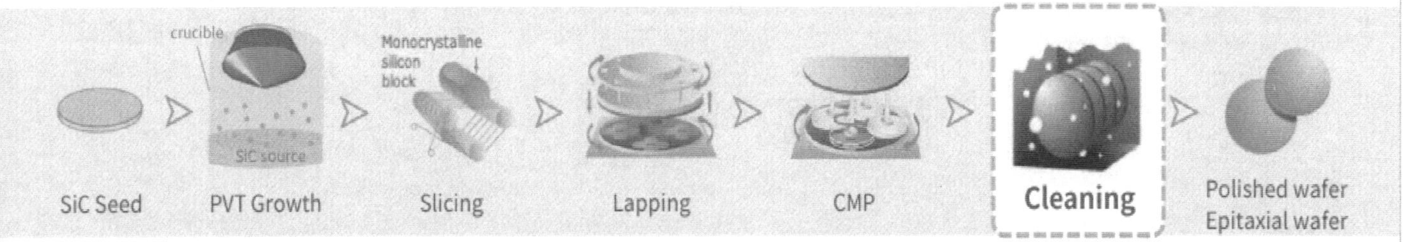

- Cleaning efficiency and defect level stand as the crucial final steps in the process, given their direct impact on the yield and reliability of the device.

Aspects of device

- Material quality
- Device performance
- Yield improvement

Aspects of process

- Enhanced efficiency
- Cost reduction
- Innovation & competitiveness

- In aspects of the process, the cleaning process significantly contributes to heightened efficiency and reduced costs.

Challenges in cleaning SiC wafer

- **Requirement of research for cleaning SiC wafer**

Si wafer
Conventional cleaning method SC-1, SC-2, SPM, DHF

When using conventional methods to clean SiC,
a notable issue arises:
long time & contaminants remain

SiC wafer
Need to understanding cleaning method for SiC

- This is largely due to the intrinsic properties of SiC, which pose limitations on conventional cleaning techniques.

[Literature on surface preparation & cleaning of SiC]

- **Ozone Cleaning** (SUMITOMO ELECTRIC. (2011). Method of cleaning silicon carbide semiconductor. 5540919 B2.)
- **Plasma etching** (Artem A. Osipov. (2020). High-temperature etching of SiC in SF6/O2 inductively coupled plasma.)
- **Defect inspection** (Po-Chih Chen. (2022). Defect Inspection Techniques in SiC.)

- However, despite its necessity, for the cleaning process of SiC, there are very few studies regarding the cleaning mechanism and surface chemistry of SiC.

XPS analysis

After chemical treatment

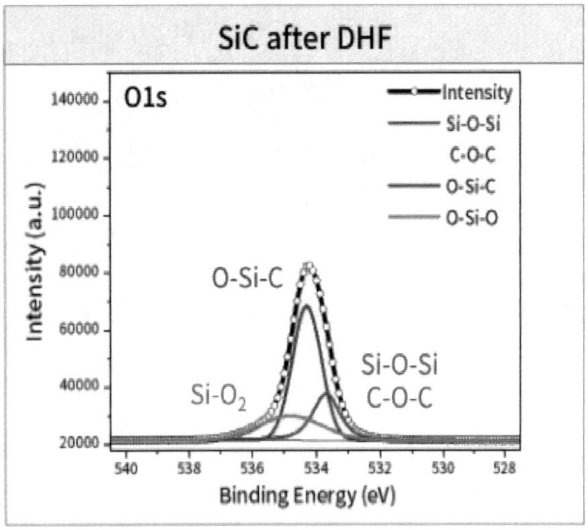

DHF Pretreatment	
Element	Atomic %
Si2p	45.9
C1s	45.9
O1s	8.1

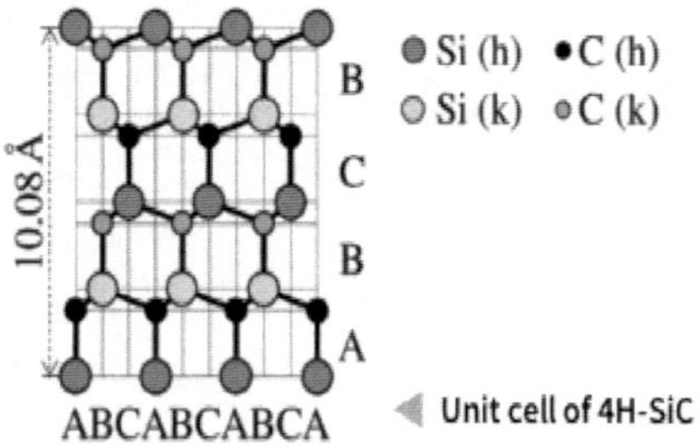

Unit cell of 4H-SiC

Element	Bond		Biding Energy
O1s	Si-O-Si	C-O-C	533.7
	O-Si-C		534.3
	Si-O$_2$		534.8

- The oxide formation can be confirmed by increasing or decreasing the intensity of the O1s peak.

1. *Rescher, G. (2018). Behavior of SiC-MOSFETs Under Temperature and Voltage Stress.*

Change of XPS O1s peak

After chemical treatment: *SPM → DHF*

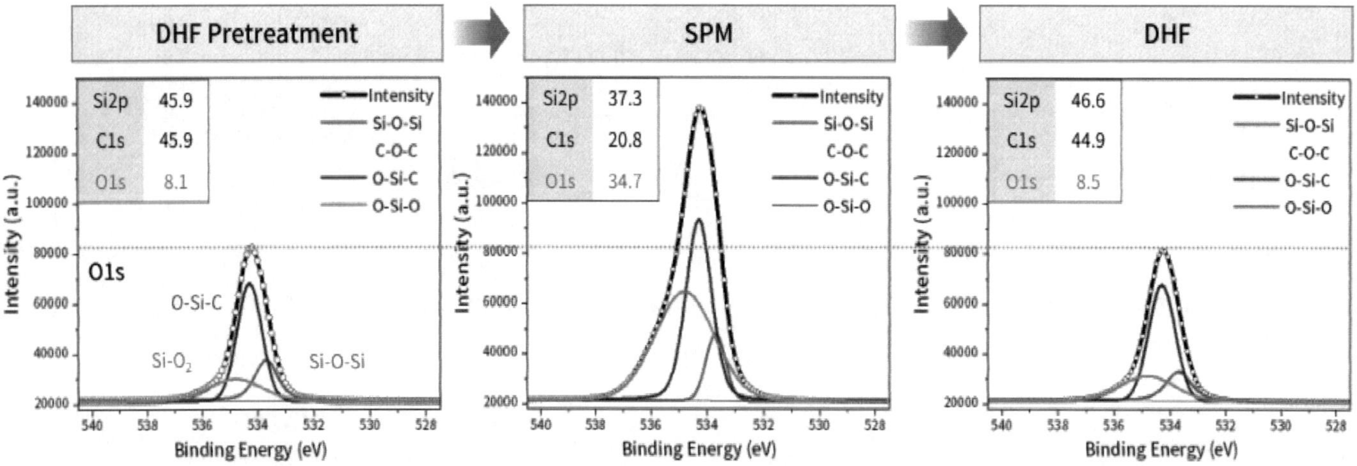

- After SPM treatment, it shows the contents and total intensity of O1s on the surface increased.
- After DHF again, it decreased to the same level as the initial state.
- SPM oxidize SiC surface, and the generated oxide is slightly removed by DHF.

Change of XPS O1s peak

After chemical treatment: *$KMnO_4$ vs SPM*

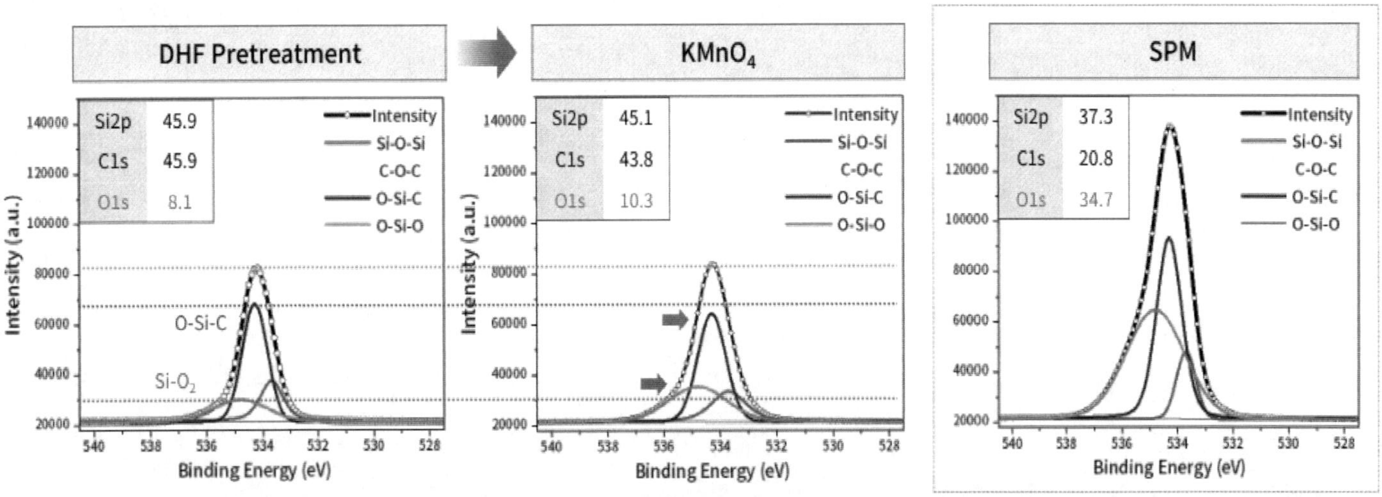

- However, when comparing $KMnO_4$ treatment used as a strong oxidizing agent, the O1s content slightly increased, and it can explain the non-hydrophilic behavior in the contact angle.
- The $KMnO_4$ and SPM reaction with the SiC surface, but they have different surface chemical properties because of the different oxidation behavior on the surface.

Oxidation of SiC surface

After chemical treatment: *$KMnO_4$ vs SPM*

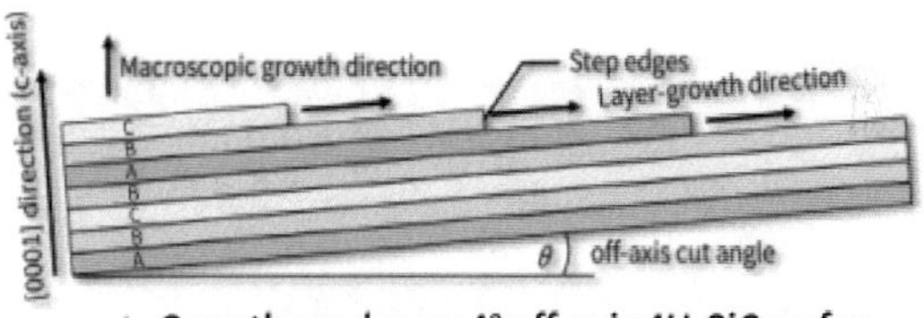

Growth modes on 4° off-axis 4H-SiC wafer

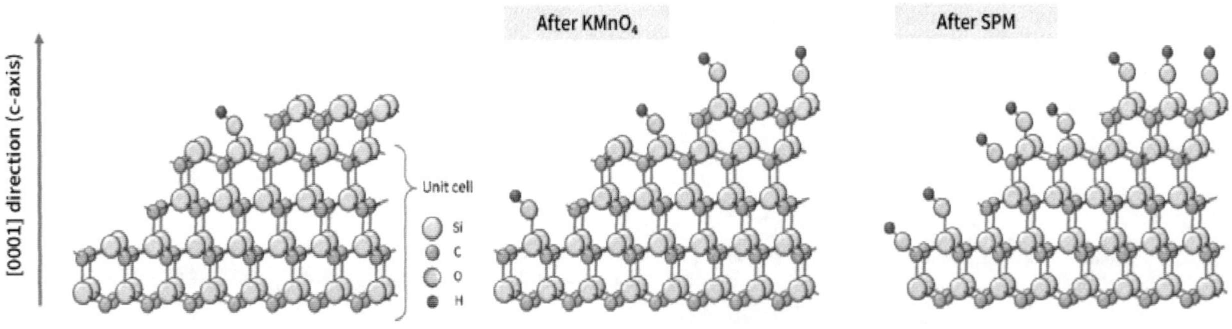

Initial SiC surface

Partially oxidized SiC surface

- The SiC surface reacts by adsorption of -OH groups rather than formation of the oxide layer.
- This oxidation behavior is expected by the difference in the density of -OH group adsorbed on the surface, which determine the surface chemical properties of SiC.
- The surface chemical properties are important for understanding the cleaning mechanism of SiC.

Understanding of SiC Cleaning Mechanism

- Evaluation of cleaning method applied with physical force

Different cleaning behavior of SiC

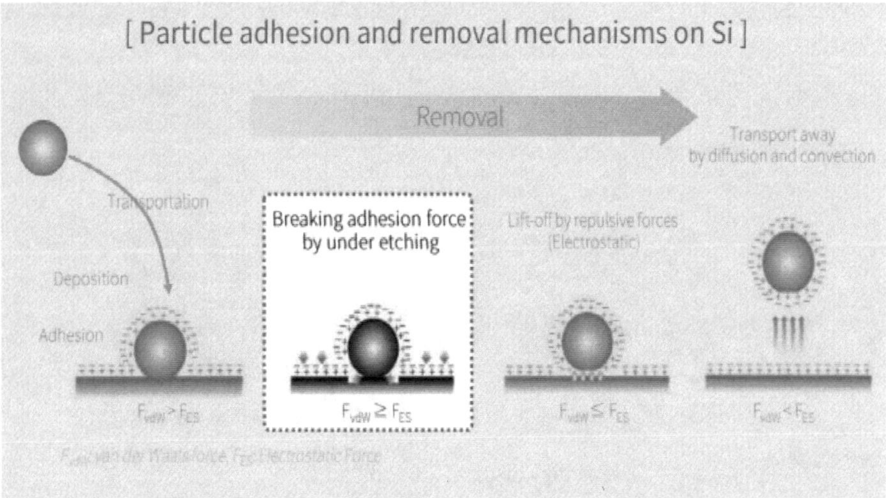

- For Si wafer cleaning, oxidation and etching reaction of more than 3 nm from the surface is required to ensure sufficient cleaning efficiency.[*(1)]
- Unlike Si, SiC does not undergo oxidation and etching, reacting mainly to Si and C atoms on SiC.

1. *T. Hattori, Ed. (1998). Ultraclean Surface Processing of Silicon Wafers, Springer.*

Different cleaning behavior of SiC

[Particle adhesion and removal mechanisms on Si]

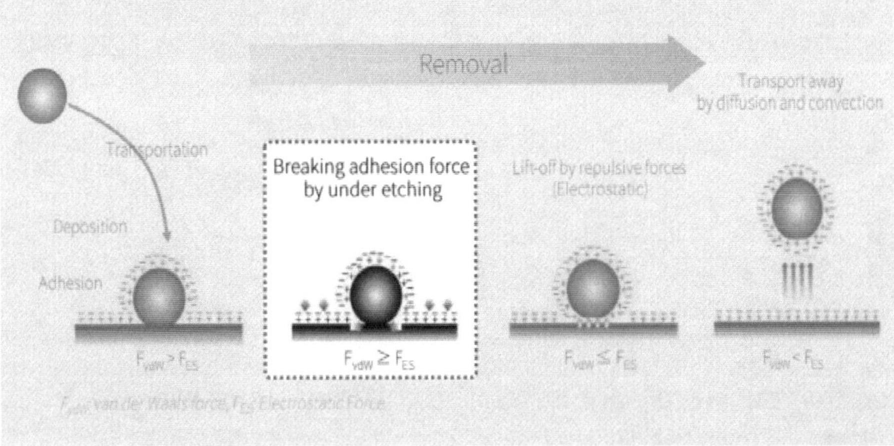

[Surface reaction on SiC]

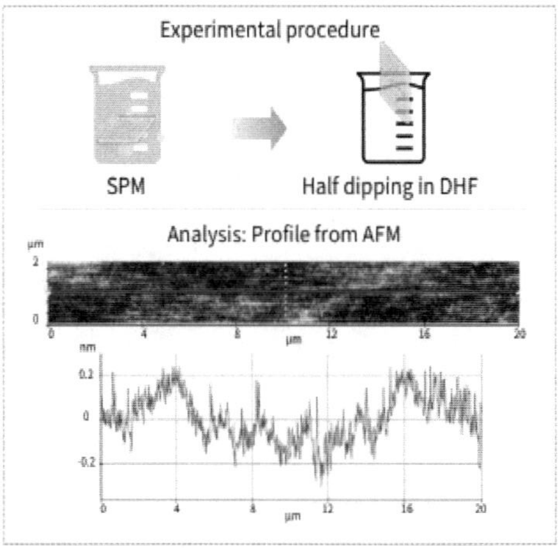

- After treating the sample with SPM and partially immersing only one half in DHF, we confirmed the profile boundary using AFM but no significant change was observed.

Approached a cleaning method applied with physical force for SiC.

Evaluation of cleaning method for SiC

Experimental procedure

1. Contamination

- Contamination with commercial SiC slurry (Al_2O_3 and $KMnO_4$ based, 300 nm)

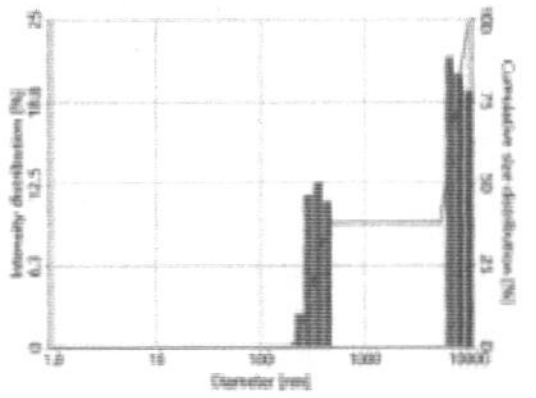

[Particle size distribution of slurry]

- Contamination sequence:

CMP for 1 min

↓

Drying

2. Cleaning

- Physical cleaning in DIW

	Type	Frequency	Solution	Time
Ultrasonic	Bath	40 kHz	DIW	1 min
Megasonic	Single	1 MHz	DIW	1 min
PVA Brush			DIW	1 min

- Chemical cleaning

	Concentration	Temp.	Time
SPM	$H_2SO_4 : H_2O_2 = 4 : 1$	100 ˚C	10 min
SC-1	$NH_4OH : H_2O_2 : H_2O = 1 : 2 : 10$	70 ˚C	10 min
DHF	1 wt%	20 ˚C	10 min

3. Analysis

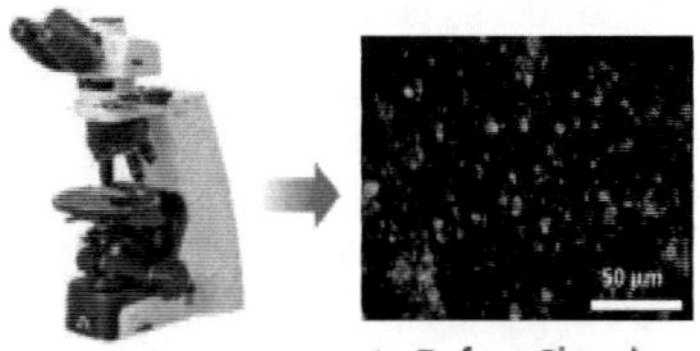

▲ Before Cleaning

Optical Microscope

- Surface image
- Particle removal efficiency

Evaluation of cleaning method for SiC

Sonication cleaning

▲ Single type sonication cleaner

[Principle of sonication cleaning]

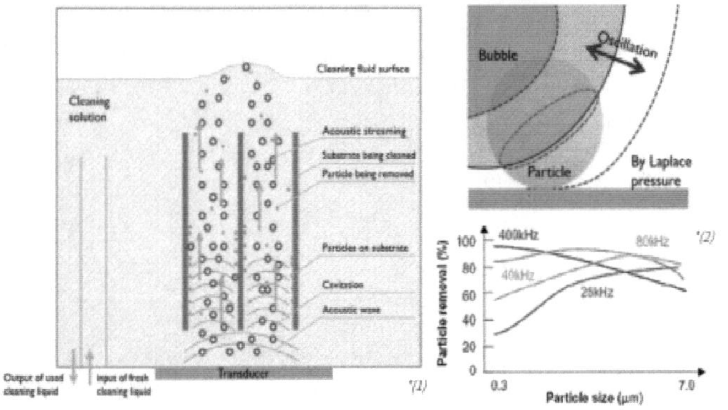

- (Cavitation) If the shock wave forces from bubble collapse are great enough to overcome particle adhesion forces, particles will be dislodged from the surface.
- (Acoustic Streaming) Efficient for particle removal because increasing the acceleration of a particle reduces the boundary layer.
- The size and efficiency of particle removal vary depending on the frequency.

1. *Megasonic Technology Innovator. ProSys. https://prosysmeg.com/technology-overview/*
2. *IC Knowledge LLC. Effectiveness of Particle Removal Relative to Frequency.*

Evaluation of cleaning method for SiC

PVA brush scrubber

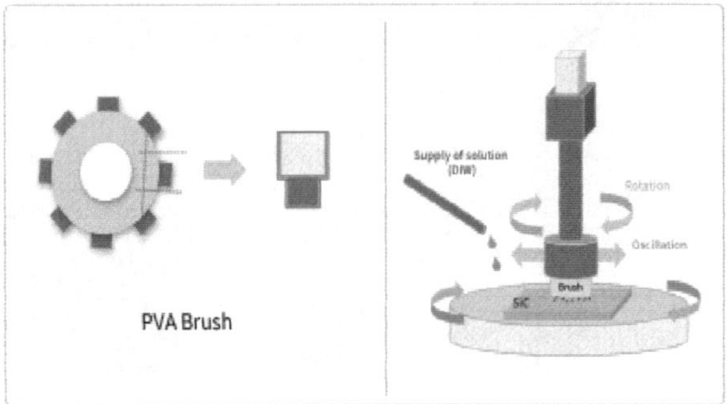

▲ Pencil-type scrubber

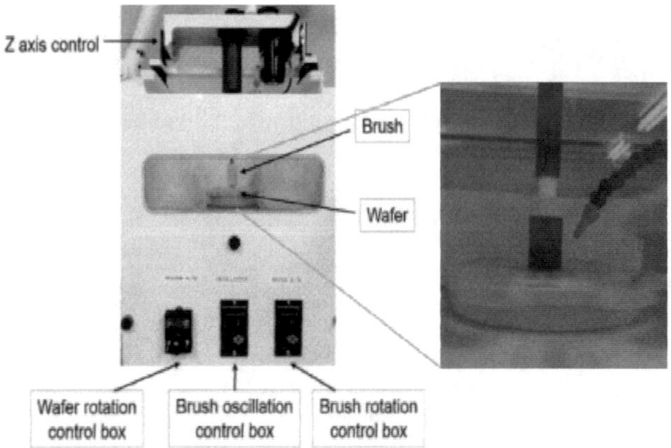

- PVA brush cleaning in post-CMP involves using soft and flexible PVA brushes with a cleaning solution to effectively remove particles and residues on the surface.
- Brush cleaning removes particles as the brush rotates while in contact with the wafer, and the wafer sample rotates.

Results of cleaning efficiency

Chemical cleanings vs Physical cleanings

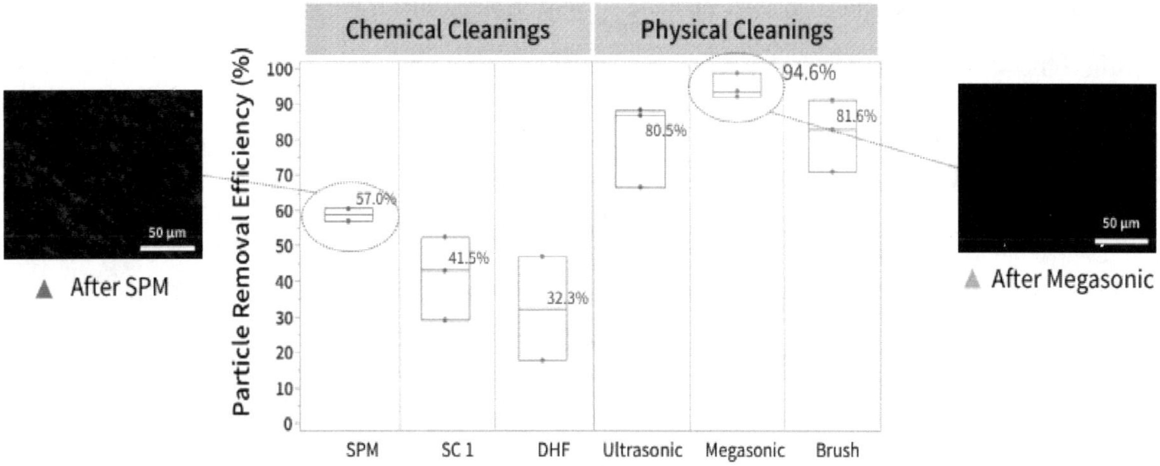

- Generally, physical cleanings showed higher cleaning efficiencies than chemical cleanings.
- Single type megasonic cleaning showed the highest PRE.

Physical force might be needed for SiC wafer cleaning.

Summary

Summary

1. Investigated surface chemistry of SiC with POR Si cleaning solutions.
2. Identified the surface reaction behavior on SiC surface.
    - No other reactions were found in POR chemicals except SPM.
    - Adsorption of –OH is the only reaction found with POR chemicals.
3. Evaluated cleaning methods with physical force for SiC.
    - The sonication cleaning method is the most efficient.
    - Propose hybrid chemical-physical cleaning method for SiC.

By recognizing the distinct properties of SiC and reevaluating cleaning methods, we can develop more effective strategies for better cleaning performance.

HANYANG UNIVERSITY

Use of Hybrid Metrology to Develop Holistic Filtration Solutions in Hydrogen Peroxide

Kusum Maharjan*, Nicole Williams, Sally Huang, Siddarth Sampath – Entegris

Brianna Dufek, Suhas Ketkar, Austin Schultz – Elemental Scientific Inc.

*Primary Author: Kusum Maharjan
978-436-6846
129 Concord Rd, Billerica, MA, 01821
Kusum.Maharjan@entegris.com

Introduction

All semiconductor manufacturers are currently driving to advance their processes to deliver improved performance. To achieve such improvements with each generation while maintaining high reliability and yield, strict contamination control must be established, especially for process chemicals.

The diversity of particles, gels, and dissolved species (ions, organics, and others) challenge semiconductor manufacturing process performance.

Development of contamination control solutions traditionally have focused on using standard particles as a measure of contamination retention properties, made using materials such as gold (Au) or polystyrene latex (PSL) beads.

To implement the correct filtration solutions, we need to understand step by step as below:

- **Part 1:** The native contaminants present in the semiconductor grade 30% H_2O_2
- **Part 2:** The effect of traditional filtration solutions and implementation on contamination control

In this poster presentation, we focus on 30% hydrogen peroxide due to its use in critical wet etch and clean (WEC) and chemical mechanical polishing (CMP) unit processes

Experiment

Part 1: Contamination profiling of 30% native H_2O_2

To understand the quality variations that are a result of containers, we looked at 30% H_2O_2 from the same lot of chemical but packaged in different HDPE containers

We utilized the following techniques to profile the peroxide quality:

1. **Particles >30 nm using Rion KS-19F**
   - Direct analysis of 30% H_2O_2 sampled from HDPE bottles resulted in coincidence errors reported by the LPC sensor
   - The coincidence errors were due to extremely high particle counts (>100,000) and resulted in an under-estimation of particles when directly injected
   - To overcome this challenge, we developed a dilution model to enable estimation of the >30 nm particles present in the 30% H_2O_2 containers. The method involved diluting the H_2O_2 with UPW at different aliquots that did not result in coincidence errors on the LPC. The measured counts from these aliquots were used to create a curve that was extrapolated to estimate the counts of particles present in the original 30% solution.
2. **Metals using ICPMS (Agilent 8900)**
3. **Scanning mobility particle sizer (SMPS) using TSI SMPS with nanoDMA**
4. **Organics using TOF-MS in collaboration with Elemental Scientific**

Entegris

Results and Discussion

Part 1: Contamination profiling of 30% H_2O_2 – Liquid particle counter

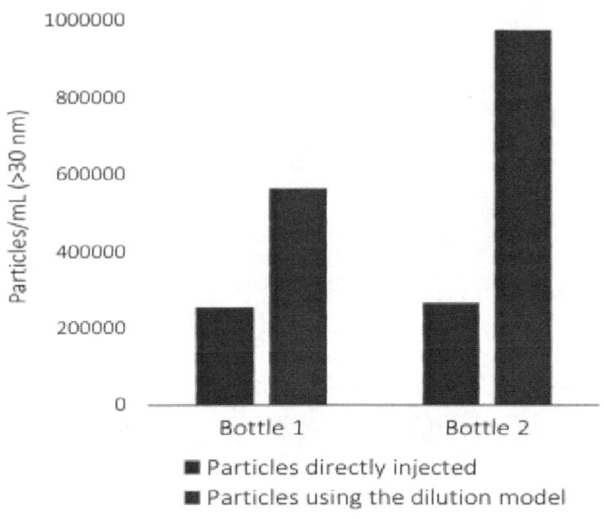

Figure 1: Impact of coincidence errors for estimation of real particle counts in 30% H_2O_2

- 30% H_2O_2 from two different bottles of the same lot were analyzed using the dilution method
- Using the dilution model, we were able to detect particle differences between Bottle 1 and Bottle 2, which was missed with direct injection
- Even though the H_2O_2 was from the same lot/batch, we observed big differences in the >30 nm particles from each bottle, highlighting the impact packaging material has on final chemical quality

Entegris

Results and Discussion

Part 1: Contamination profiling of 30% H_2O_2 – Total metals by ICPMS

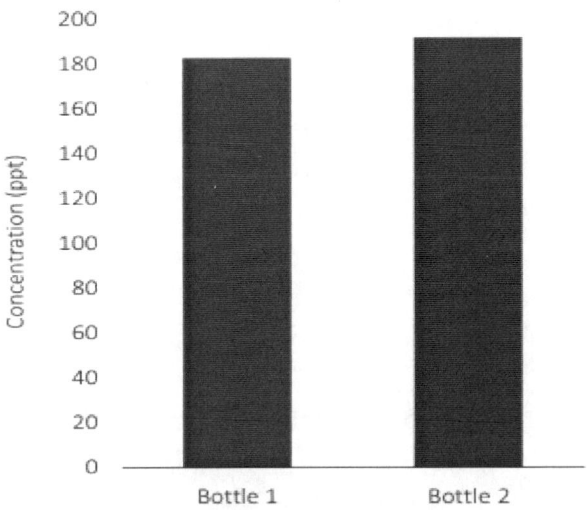

Figure 2: Total metals in two different grades of 30% H_2O_2

- 30% H_2O_2 from Bottle 1 and Bottle 2 were analyzed for 21 metals using ICPMS 8900
- ICPMS analysis showed that the ionic profiles for each bottle are similar

Entegris

Results and Discussion – SMPS with nanoDMA

Part 1: Contamination profiling of 30% H_2O_2 – SMPS

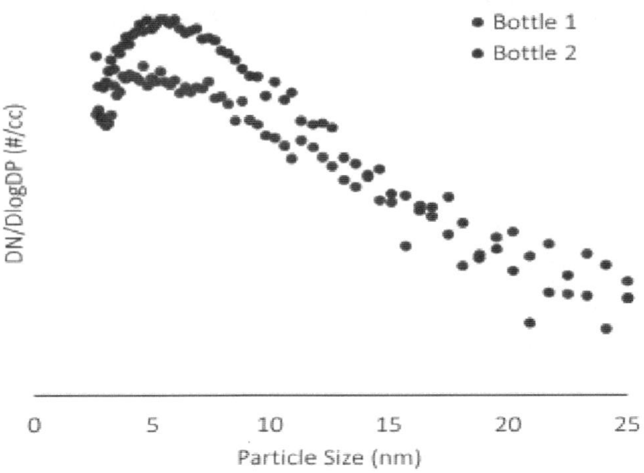

Figure 3. Difference in particles <15 nm between Bottle 1 and 2 using SMPS with a nano-DMA

- We observe a difference between the different bottles of peroxide (matching results observed from the 30 nm LPC analysis with the dilution model)
- The SMPS data shows that there are differences in the quality of the H_2O_2 <15 nm, similar to the trend seen with the >30 nm LPC

Entegris

Results and Discussion

Part 1: Contamination profiling of 30% H_2O_2 – prepFAST Carbon

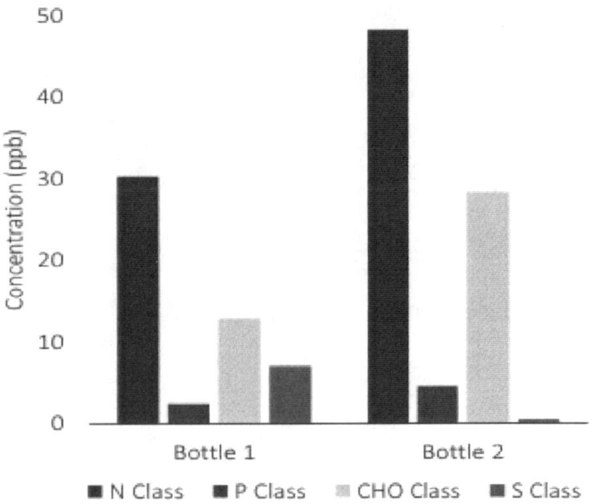

Figure 4: prepFAST Carbon organics analysis of native H_2O_2

Organics analysis of the 30% H_2O_2 samples was done using a SCOUT Carbon system in collaboration with ESI that utilized an electrospray ionization + TOF-MS system for chemical analysis

- The system provided a semi-quantitative report for different classes of compounds present in the peroxide
- Classification of the compounds shows a higher signal of N, P, and CHO class compounds in Bottle 2 while Bottle 1 has a higher signal for S class compounds

Entegris

Summary and Conclusions

Part 1: Contamination profiling of 30% H_2O_2

Test	Bottle 1	Bottle 2
Liquid particle counter	●	●
Metals	●	●
SMPS	●	●
prepFAST carbon	●	●

Table 1. Summary of test results comparing native H_2O_2 from same batch with different metrology techniques

- Liquid particle counter using the dilution model is effective at highlighting >30 nm particle contaminants present in the 30% H_2O_2
- Bottle 2 shows a higher particle concentration than Bottle 1 in terms of particles <15 nm, metals as well as higher concentrations of several organic class compounds

Entegris

Experiment

Part 2: Retention with different filter media in native peroxide

- Native 30% peroxide was dynamically passed (single pass) through a 90 mm pre-cleaned coupon holder with a filter membrane
- We selected two different filter media with similar surface properties but with different bubble points to understand the effect of pore size on contamination control

Tests used:

- Inline analysis
  - 30 nm LPC
- Offline analysis
  - Ionic ICPMS
  - SMPS

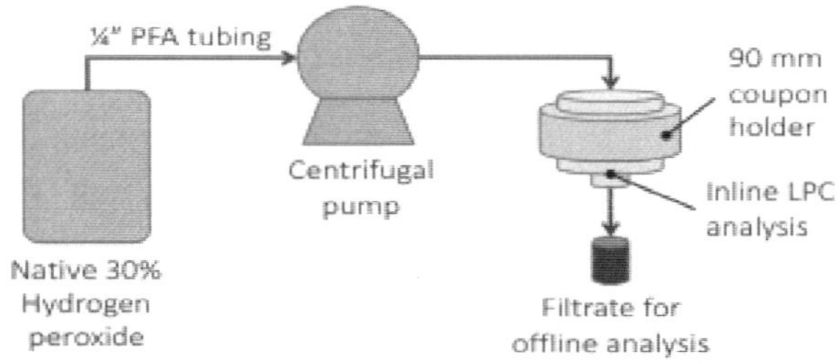

Figure 5: Experimental setup used for single pass filtration

Entegris

Results and Discussion

Part 2: Retention with different filter media in native peroxide – Standard particle retention

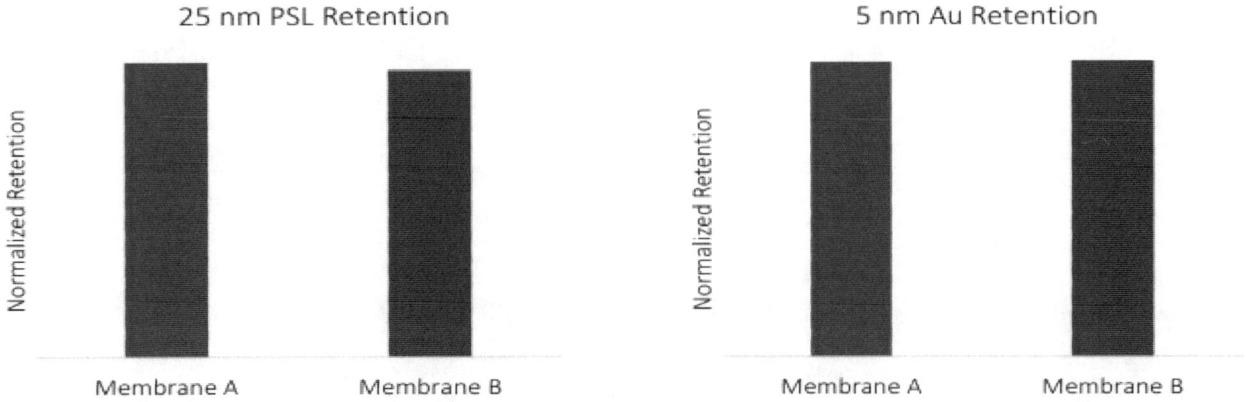

Figure 6: Standard particle retention between 2 different sets of membrane

- Standard particle retention testing shows similar performance between the two PTFE membranes with PSL and slightly better retention from Membrane B with Au

Entegris

Results and Discussion

Part 2: Retention with different filter media in native peroxide – LPC (inline analysis)

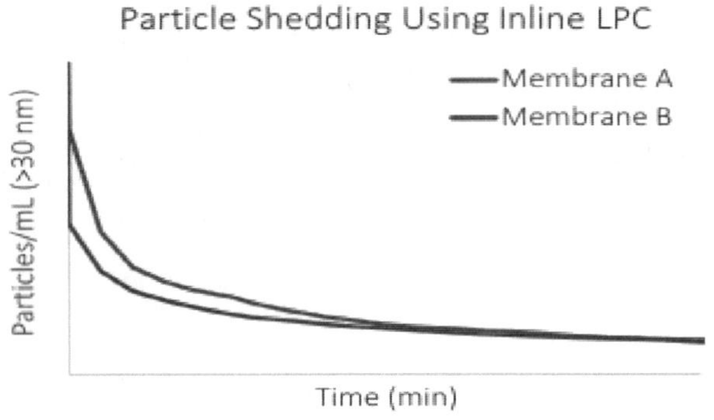

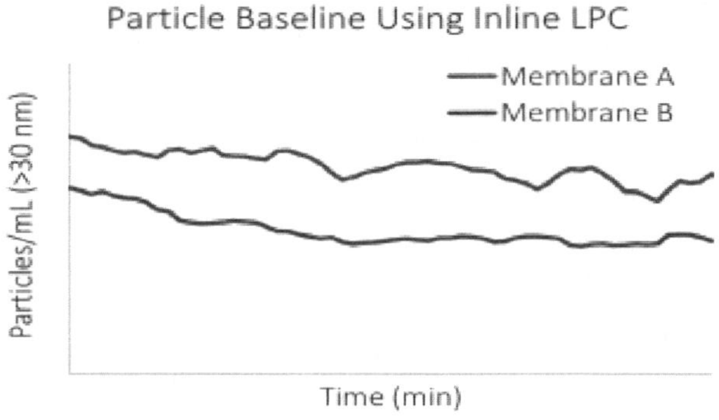

Figure 7: Particle shedding/baseline using 30 nm LPC

- In-line particle testing using a 30 nm particle counter showed that Membrane B was cleaner in start-up than Membrane A
- The results also highlight a significant retention performance difference between the two membranes, something that was not as clearly discernable with standard particle testing

Entegris

Results and Discussion

Part 2: Retention with different filter media in native peroxide – Metal removal by ICPMS

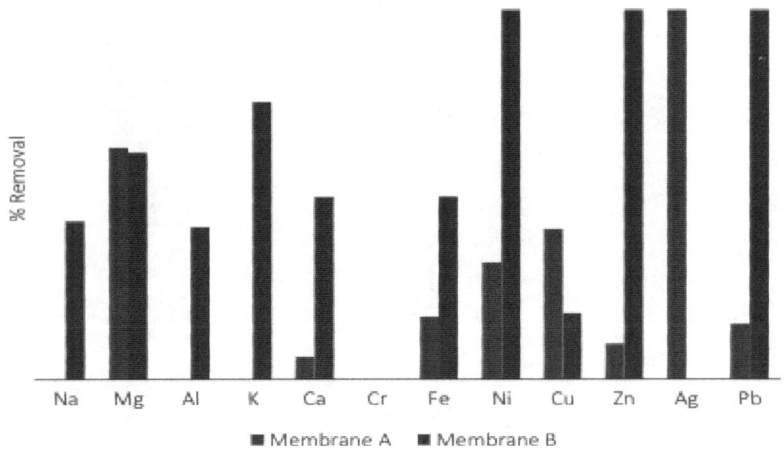

Figure 8: Metal removal efficiency in 30% H_2O_2 – 12 common elements of interest

- Membrane B shows better metal removal performance in comparison to Membrane A in native 30% peroxide

Entegris

Results and Discussion

Part 2: Retention with different filter media in native peroxide – SMPS

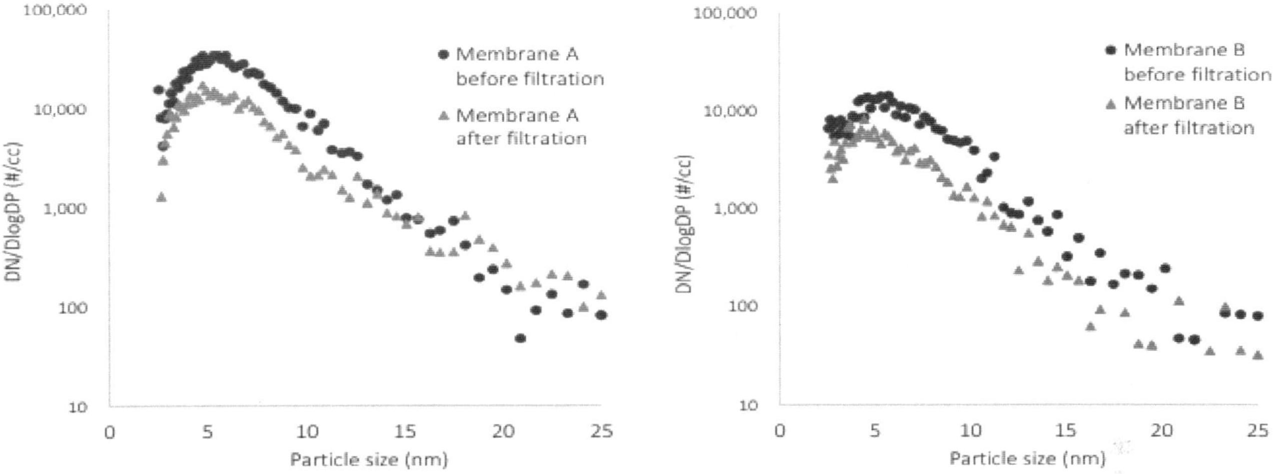

Figure 9: Size distribution before and after filtration for membrane A and membrane B

- Both filter membranes show reduction in SMPS signal post filtration, confirming that the media is helping clean up the 30% H_2O_2

Entegris

Summary

Part 2: Retention with different filter media in native peroxide

Test	Membrane A	Membrane B
Standard particle retention	● (grey)	● (grey)
Liquid particle counter (>30 nm)	●	●
Metal removal by ICPMS	● (grey)	●
SMPS	● (grey)	● (grey)

Table 2. Summary of analysis of 30% H_2O_2 after filtration through two PTFE membranes

- The filtrate through both PTFE membranes showed comparable results for standard particle retention and SMPS
- Membrane B showed better shedding and retentive properties in 30% H_2O_2 compared to Membrane A for LPC and metal removal
- By using the holistic approach, we were able to characterize the impact of filtration with Membrane B beyond standard particle retention

Entegris

Conclusion

- Hydrogen peroxide has diverse contamination profile
  - Techniques such as liquid particle counter only show a portion of the signal present
  - With use of complementary analytical techniques, we're able to see differences in the chemical quality that are not seen with LPC or ICPMS analysis
- Utilizing a hybrid metrology approach with inline analysis and grab sampling, we were able to develop a powerful method to speciate the contaminants
- Speciation of the chemical enables us to better understand the contaminants present and develop targeted solutions to address these specific contamination challenges

Entegris

2024 SPCC

Quantitation of Potential Critical Organic Contaminants from Polymer Components

Larry Zazzera[1], Allen Chan[2], Gary Van Schooneveld[1], Ali Ö. Altun[3], Bob McIntosh[4]

[1]CT Associates, Inc., 7121 Shady Oak Road, Eden Prairie, MN 55344
[2] Balazs™ NanoAnalysis, 46409 Landing Parkway, Fremont, CA 94538
[3] UNISERS AG, Baslerstrasse 60, 8048 Zürich, Switzerland
[4] Enviro-Energy Solutions, 13023 NE Highway 99, Vancouver, WA 98686

CT Associates, Inc.

2024 SPCC

This type of study can not be done alone!

IRDS Critical Components Task Force

Charter – define and understand material properties needed to support the IRDS roadmap for UPW and bulk chemical systems.

Ali Altun	Unisers
Allen Chan	Balazs NanoAnalysis
Archita Sengupta	Intel
Darren Conner	Evantic
David Stiles	Asahi
Gary Van Schooneveld	CT Associates
James Henry	Arkema
Jenell Mccall	Chemours
Jim Tu	Entegris
John Leys	Entegris
Kamila Faryna	Georg Fischer
Katrin Wallheinke	Georg Fischer
Kimberly Farnsworth	Chemours
Larry Zazzera	CT Associates
Marie Tripp	UNISERS
Mark Litchy	CT Associates
Matt Orr	Daikin
Paolo Toniolo	Syensqo
Per Nelson	Daikin
Quynh Le	Balazs NanoAnalysis
Robert McIntosh	Enviro E
Stephane Domy	St Gobain
Thomas Labour	Arkema
Tommaso Crisenza	Syensqo

2

2024 SPCC

IRDS Critical Components Team

Background and Objectives

- Organic contamination as film or particle is posing an increasing risk to semiconductor yield and reliability.
  - Evolving chip design and yield constraints.
  - Limited effectiveness of filtration and mitigation options.
  - Challenging analytical requirements.
- Critical organic contamination as currently defined is of highest concern
- Objectives of tests:
  1. *Determine if organics extracted from "critical components" can deposit on a silicon wafer under spin processing conditions.*
  2. *Determine what portion of the extract remains following various elevated drying conditions.*

3

2024 SPCC

Particle Precursors and Critical Organics are a Problem

Particle Precursors: Dissolved matter < 10 nm that can result in wafer defects after drying and thermal processing.

Critical Organics: Adhere to the wafer and are more likely to result from non-volatile and semi-volatile organics.

- Critical organics definitions in the IRDS Yield tables (YE3)
  - UPW - > 500,000 Dalton and > 200°C BP
  - Chemicals - > 10 carbons
  - Critical components - > 150mw and > 150°C BP

We did not know if the organics from polymer components would meet the definition of critical organics.

We know now that polymer components are potential sources of critical organics.

A problem of unquantified risks

- More data needed to fill fundamental knowledge gap
- Must identify particle precursors and understand pathways to potential "critical organics" on wafers.

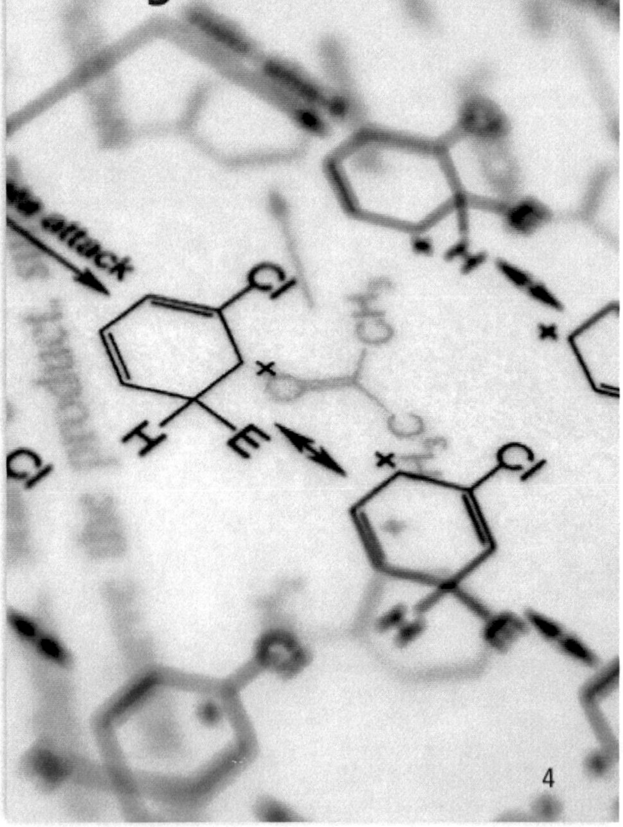

4

2024 SPCC Polymer Components and Potential Particle Precursors

Source: G. Van Schooneveld et. al., SPCC 2022

Source: L. Zazzera et. al., UPM 2023

Hot UPW Fluoropolymer Piping Extract

100 and sub-10nm particles

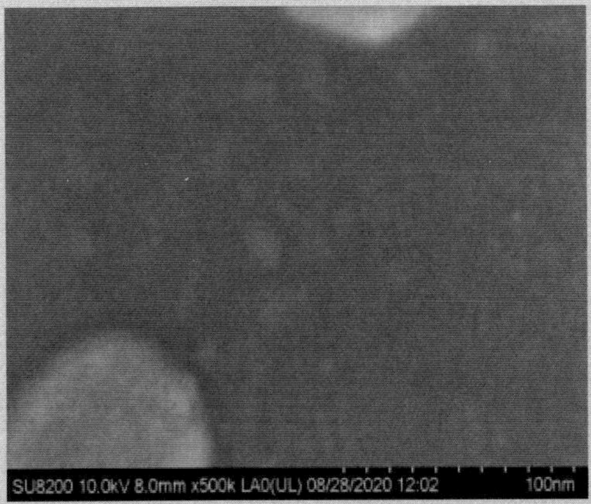

High MW polymers: an obvious potential source of particles

Precursors from Hot UPW IX Resin Extract

Small reactive molecules

PSS

Triton B

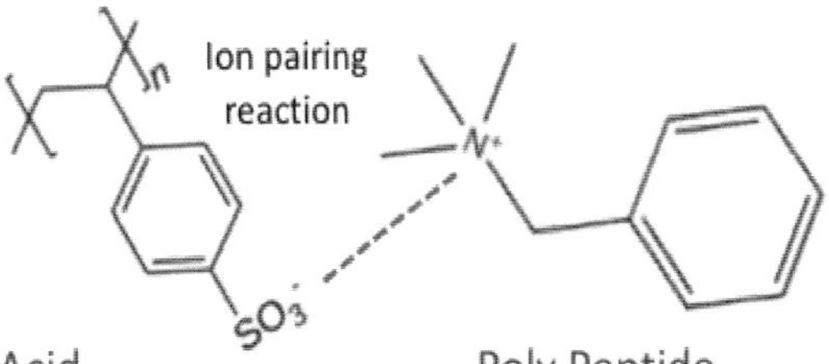

Amino Acid

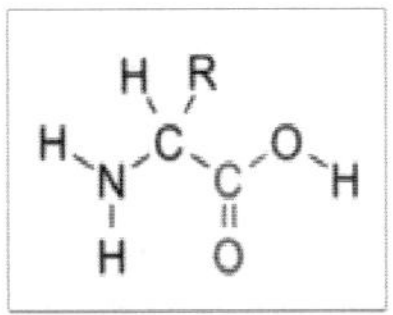

condensation reaction

Poly Peptide

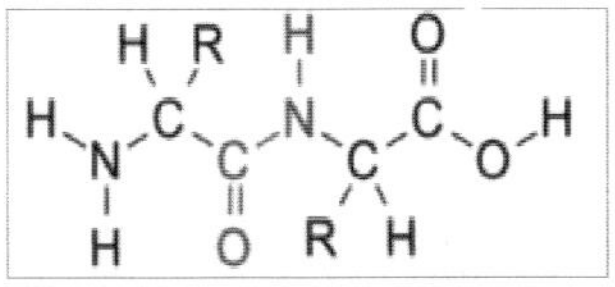

Low MW molecules: a less obvious source of particles

What are the compositions of particle precursors from polymer components and what are the pathways to potential particle formation?

5

6

2024 SPCC Sample and Organic Analysis Process Flow

Extract Preparation

CTA (USA)

"Generic" polymer extracts

Modified SEMI-F40 , 14 days, 85°C

PFA tube and PVDF pipe

Surface area normalized data

Extract Analysis

CTA (USA)

Total Organic Carbon (TOC)

ng/cm² of wetted polymer

Particle and particle precursor counts (4nm -65nm) by Liquid Nanoparticle Sizing (LNS)

Wafer Defect Analysis

CTA (USA), UNISERS (Switzerland)

Defects on silicon wafers

Sequential Spin Coating (SSC)

Surface Enhanced Particle Sizing (SEPS)

How much sticks

Residue Analysis

CTA (USA)

Non-volatiles by Attenuated Total Reflection Infrared (ATR-IR)

(25°C, 50°C, 90°C @ 2 torr)

How much stays (remains)

Balazs Nanoanalysis (USA)

Semi-volatiles by Thermal Desorption Gas Chromatography Mass Spectrometry (TD GC/MS)

(50°C, 100°C, 160°C)

How much goes

6

2024 SPCC

Extract Analysis for Particles and Particle Precursors by LNS

Dilution Normalized Dilution Series

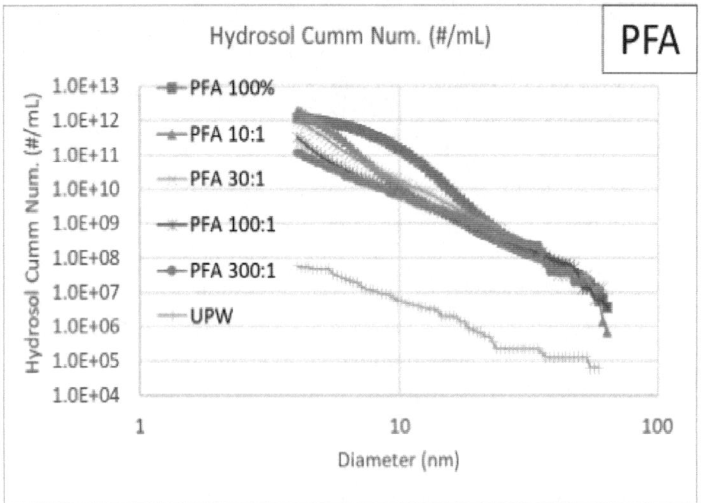

Dilution Normalized Dilution Series

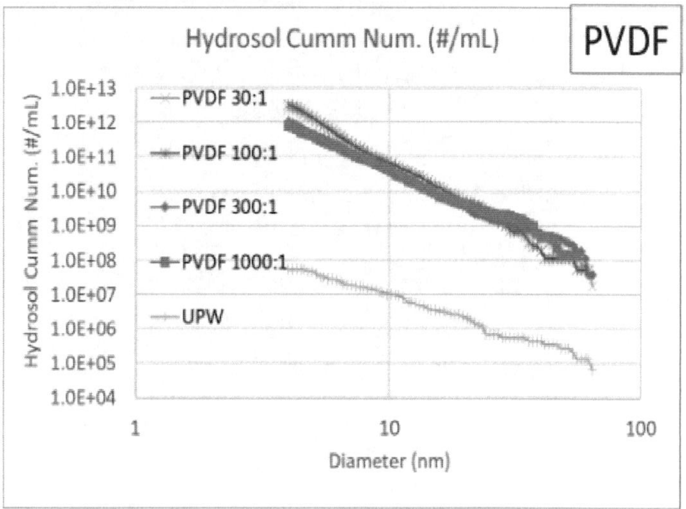

More particles and particle precursors in PVDF extract on an absolute basis, but a greater fraction of PFA extract is particle precursors

7

2024 SPCC

Non-volatile Organic Residues by ATR-IR

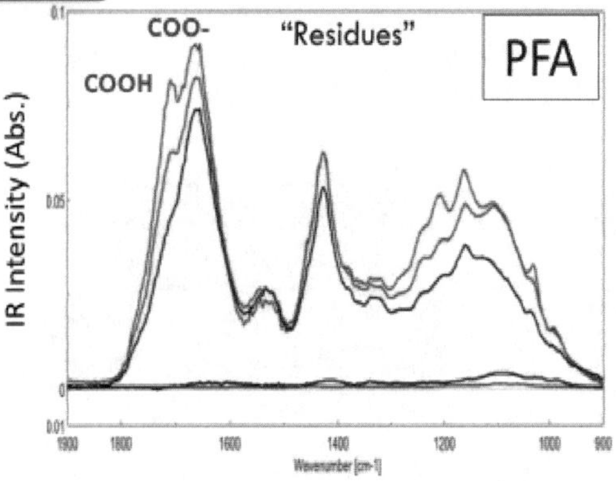

How much remains

- Extract 25°C Vac.
- Extract 50°C Vac.
- Extract 90°C Vac.
- Method Blank 25°C Vac.
- No Deposition 25°C Vac.

IR Intensity Scale 4x →

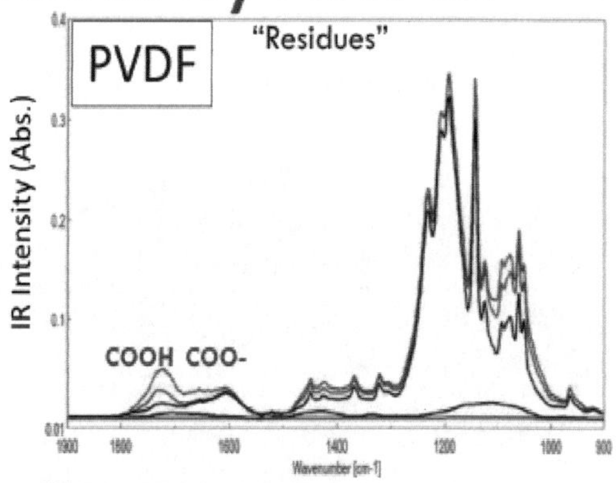

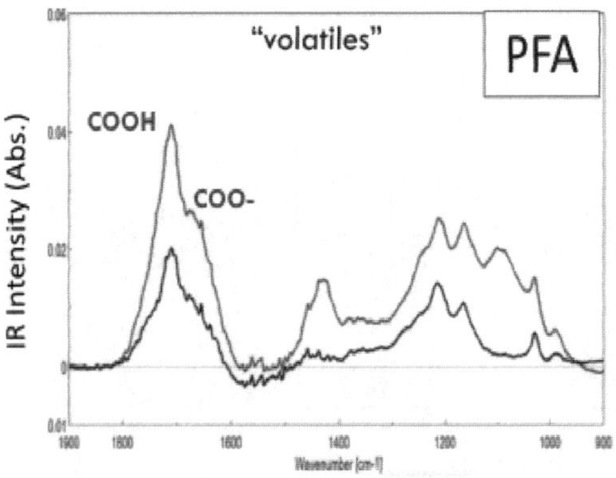

- Extract 90°C Vac.
- Extract 50°C Vac.

IR Intensity Scale 1.2x →

Physical properties for reference

Molecule	Pvap@22 Deg. C (torr)	B.P. (Deg. C.)
H2O	19	100
PFBA	5	121
Cyclohexanone	5	156
PGMEA	4	146
Cyclohexanol	1	162

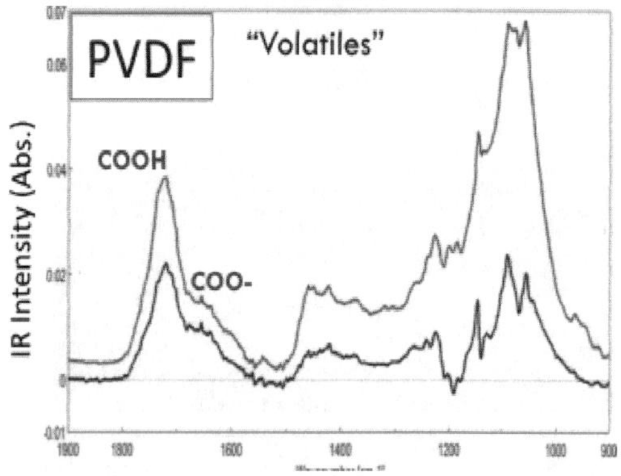

More residue from PVDF extract but a greater fraction of PFA extract is volatile 8

2024 SPCC

Semi-volatile Organic Residues by TD-GC/MS

How much goes

Compare absolute amounts of different semi-volatiles from each PFA and PVDF sample at 50°C, 100°C and 160°C
Plot semi-volatiles from each BP range as mass per unit surface area by temperature

Normalized for different surface areas of polymers

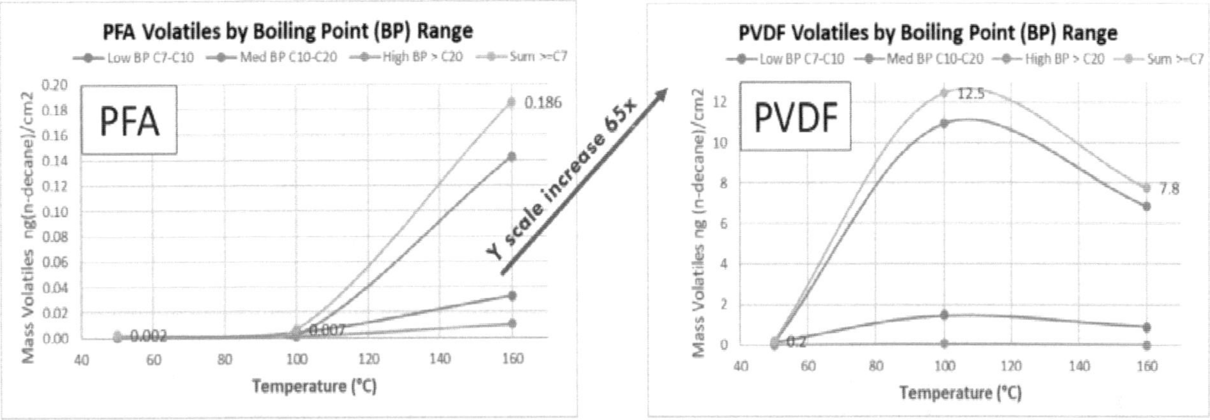

More semi-volatiles from PVDF vs PFA for same part area but the PFA extract seems "stickier" (desorbs at higher temp)

9

2024 SPCC

Wafer Defect Analysis by SSC and SEPS

How much sticks

On-wafer Particle Concentration Added[1] On-wafer Particle Concentration[2]

Equivalent particle concentration in liquids during sequential spin coating

1. From 2022 IRDS CC study

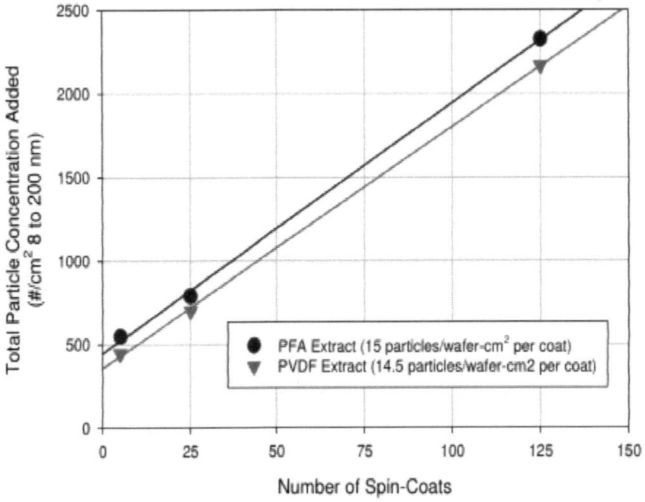

2. From 2023 IRDS CC study

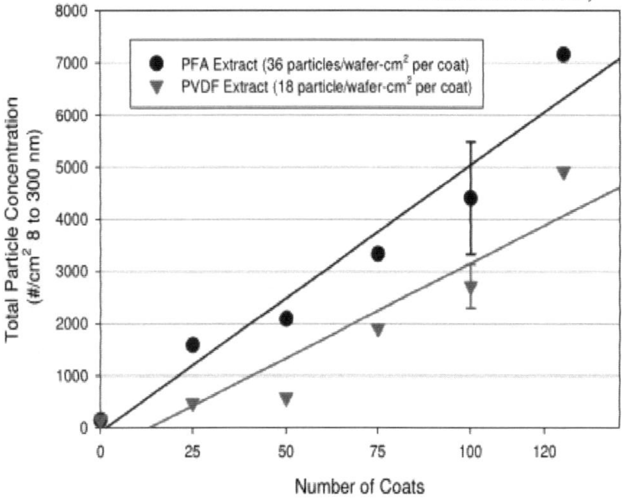

Are particle precursors from PFA extract more reactive or "stickier"?

10

2024 SPCC

Summary and Take-aways

- Polymer components are a potential source of particle precursors.
  - Leachates or dissolved organic matter less than 10 nm in diameter
- Different types of particle precursors can arrive at the wafer
  - Higher molecular weight organics as native particles
  - Lower molecular weight organics as films or aggregates with native particles
- Critical organics are more likely to stick to the wafer and are of highest concern
- We observed potential critical organics from polymer components
  - Non-volatile and semi-volatile portions of the PFA and PVDF extracts match the definition of critical organics
- Unknown risks from critical organics at specific steps in the manufacturing process
  - Propose specific extraction limits for specific polymer components in specific applications

Continue to address the fundamental knowledge gap on critical organics to quantify and manage risks to yield

11

THE SURFACE PREPARATION AND CLEANING CONFERENCE (SPCC)

SPCC 2024 | April 24-25, 2024

Effect of SDS Surfactant in H_2-DIW on Cavitation Bubble Dynamics in Ultrasonic and Megasonic Fields

Mir Jalal Khan[1], Jenasree Hazarika[1], YunHee-Cho[2], Wonseob Cho[2], Andreas Klipp[3],

and

Jin-Goo Park[1*]

[1]Department of Materials Science and Chemical Engineering, Hanyang University, Ansan, 15588, Republic of Korea

[2]Electronic Material R&D Center Asia, BASF, Suwon, 16419, Republic of Korea

[3]BASF SE, Ludwigshafen, 67056, Germany

*jgpark@hanyang.ac.kr

The Engine of Korea

Contents

I. Introduction and motivation

II. Objectives

III. Experimental materials and methods

III. Results and discussion

1. Bubble interaction studies
2. PRE and pattern damage analysis

IV. Summary

Introduction and motivation

Introduction to megasonic cleaning

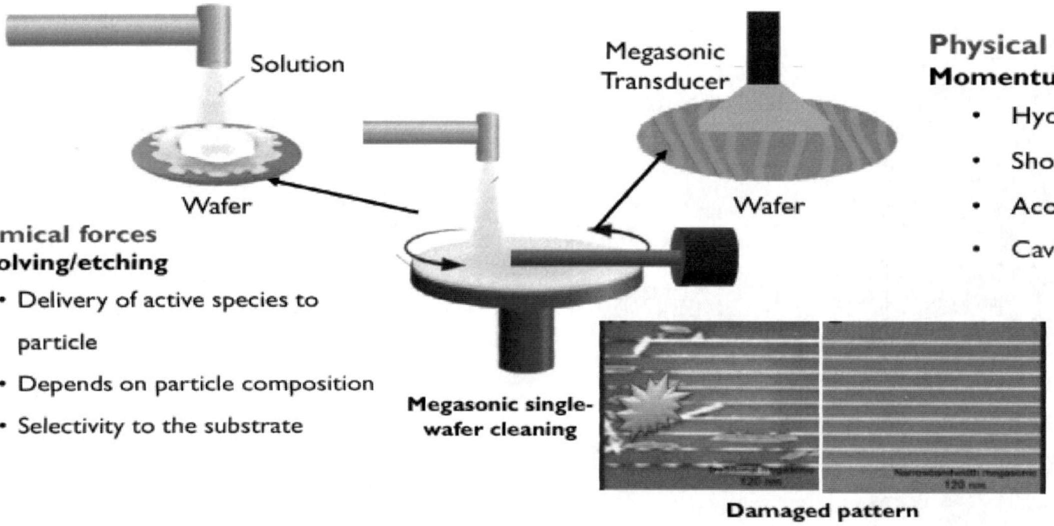

Chemical forces
Dissolving/etching

- Delivery of active species to particle
- Depends on particle composition
- Selectivity to the substrate

Physical forces
Momentum transfer

- Hydrodynamic flow
- Shockwave
- Acoustic streaming
- Cavitation

Pattern Damage

- Structural
- CD change

- Most effective wet cleaning process but, causes severe and uncontrolled pattern damages due to cavitation
- Megasonic cleaning is believed to be a more effective if the cavitation could be controlled

Megasonic cleanings of patterned wafers

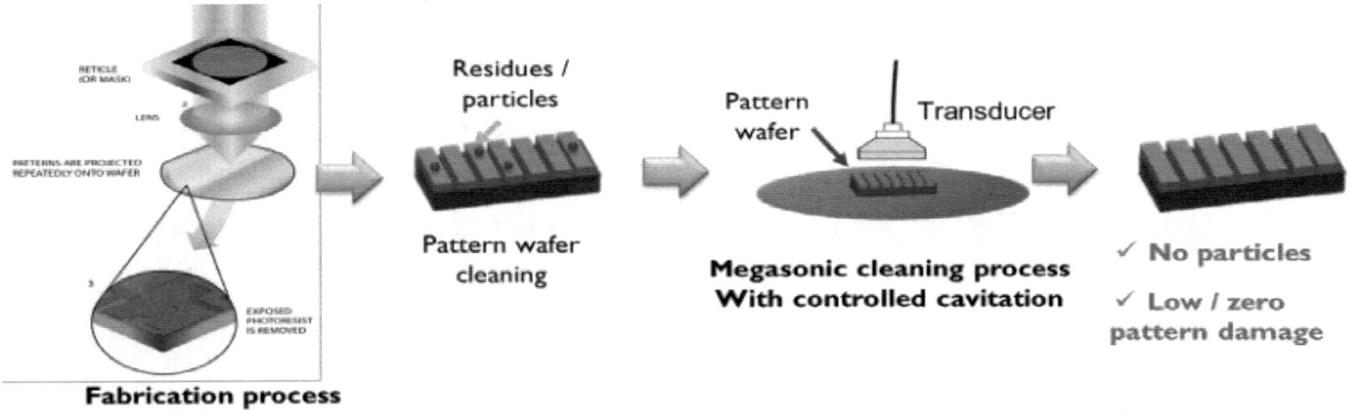

- Pattern wafer cleaning is a critical step in the semiconductor fabrication process
- Megasonic cleaning is an effective method to remove particles and contaminants from the surface of the wafer
- Megasonic conditions and selection of surfactant could lead to higher PRE and lower PD

Megasonic cleaning mechanism

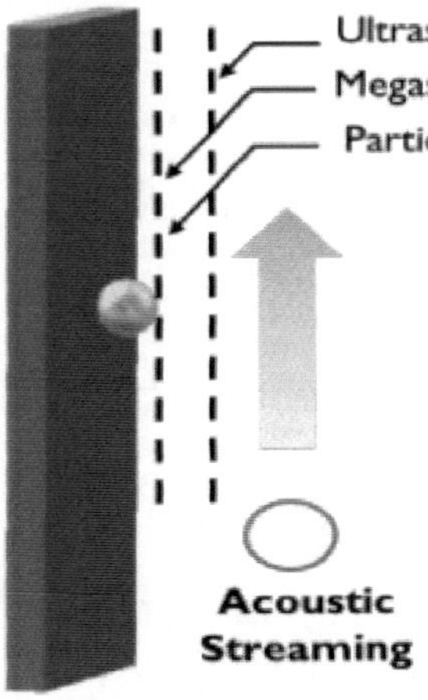

1. Acoustic streaming
2. Boundary layer thickness reduction
3. Cavitation

- Decrease in boundary layer thickness can remove particles faster
- Controlling bubble cavitation could improve particle removal and lower pattern damage

Rajiv Kohli and K.L. Mittal. Developments in Surface Contamination and cleaning. Vol3 (2011) p38

Pattern damages Issues due to cavitation

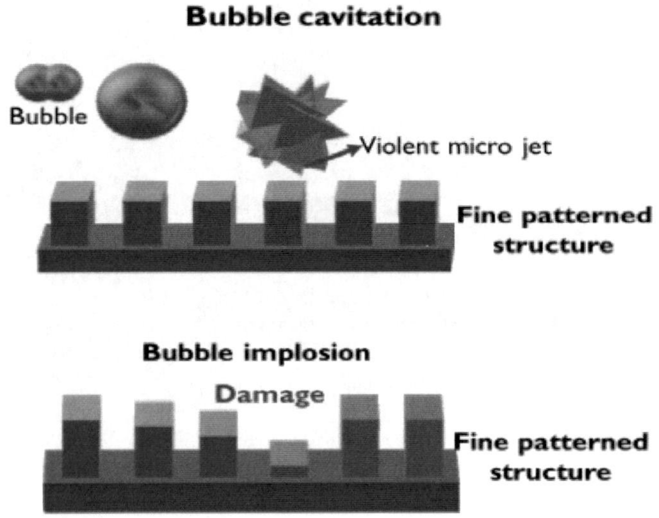

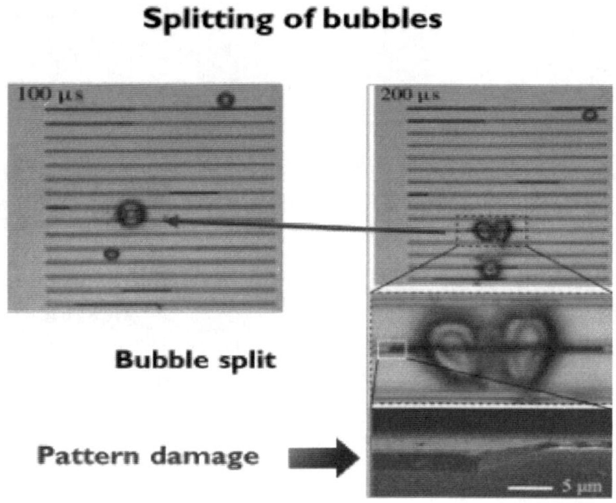

- Transient cavitation should be controlled to reduce damages
- Bubble split and cavitation cause severe pattern damages due to shock waves

D. H. Wang et al, "Damage-Free Cleaning of Advanced Structure Using Timely Energized Bubble Oscillation Megasonic Technology", Solid State Phenomena. 282, 64-72, (2018)
T.H. Kim and H.Y. Kim, "Disruptive bubble behavior leading to microstructure damage in an ultrasonic field", J. Fluid Mech. 750, 355–371, (2014)

Effect of dissolved gas on bubble behavior during megasonic cleaning

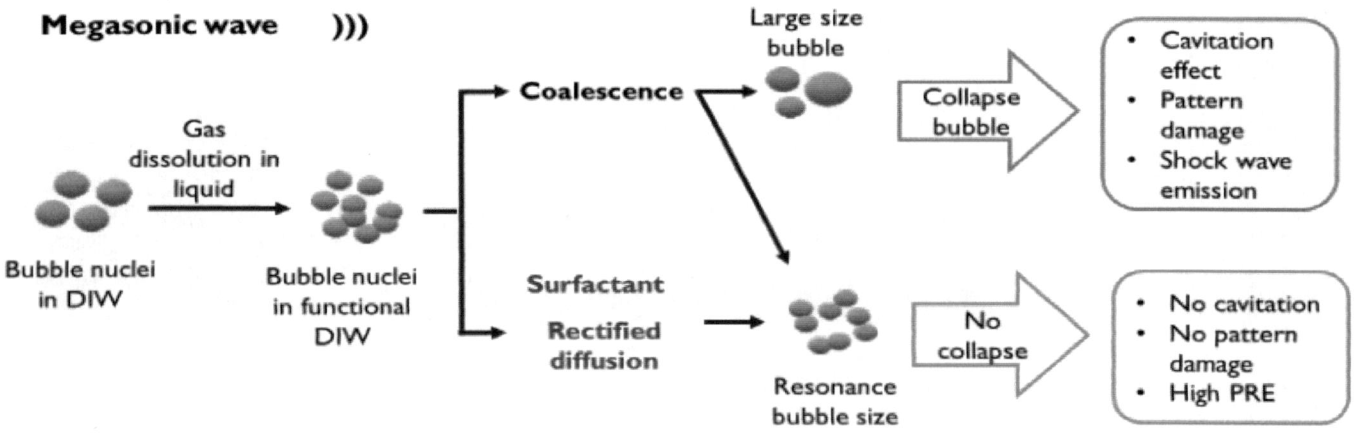

- Presence of dissolved gasses can increase the bubble population during megasonic cleaning
- Bubble size increases due to bubble coalescence
- Rectification of diffused gasses can increase in resonance size bubbles
- Bubble collapse causes cavitation effect, pattern damage, and generate shock waves

Kang, B. K., Kim, M. S., Lee, S. H., Sohn, H., & Park, J. G. (2011). Effect of acoustic cavitation on dissolved gases and their characterization during megasonic cleaning. ECS Transactions, 41(5), 101.

Dissolve gases effect on PRE and PD in DIW during megasonic

Pattern damage due to shock waves

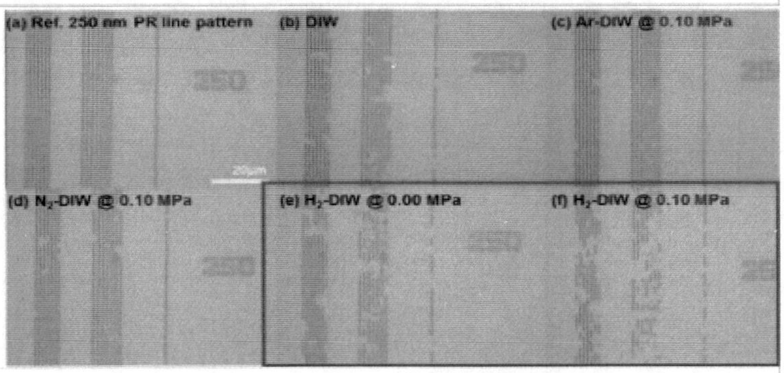

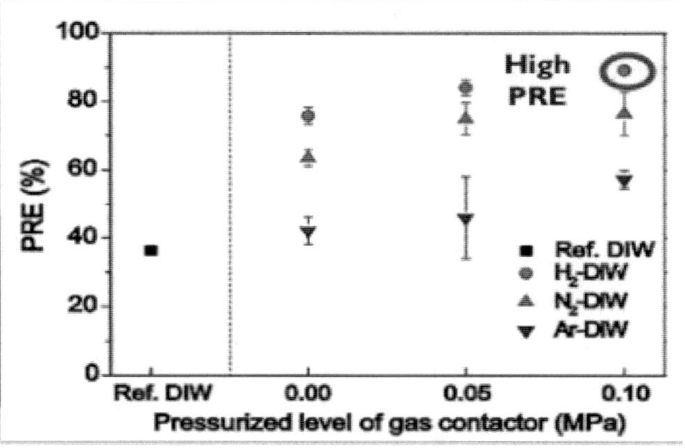

High Pattern damage

- Dissolved gases in DIW can cause severe pattern damage due to high cavitation
- High PRE can be attained by dissolved gases in DIW during megasonic cleaning
- No pattern damage and high PRE can achieved by controlling cavitation using surfactants

Kang, B. K., Kim, M. S., & Park, J. G. (2014). Effect of dissolved gases in water on acoustic cavitation and bubble growth rate in 0.83 MHz megasonic of interest to wafer cleaning. Ultrasonics sonochemistry, 21(4), 1496-1503.

Effect of surfactant to control bubble cavitation and coalescence

DIW

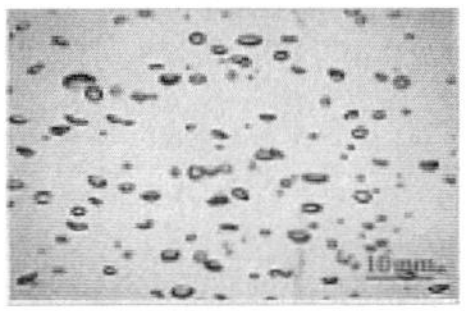

Larger bubbles size
Coalescence

DIW + Surfactant

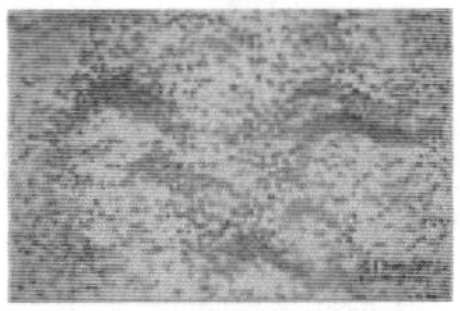

Small bubbles size
No coalescence

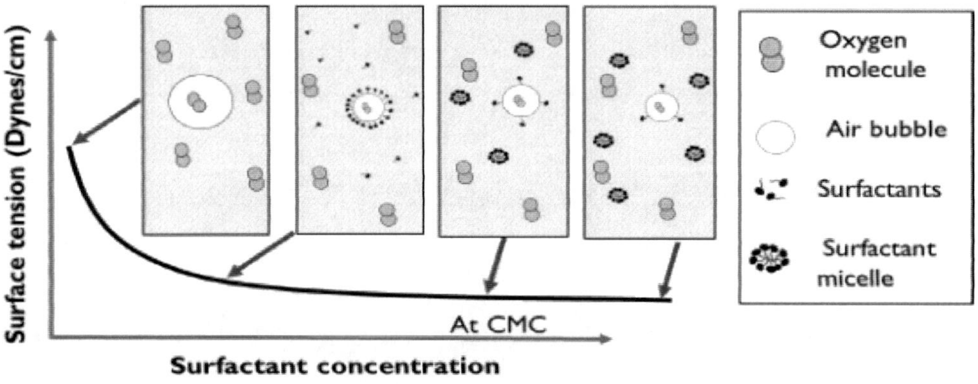

- Surfactants significantly control cavitation and prevent bubble coalescence
- Bubble size decreases by increasing surfactant concentration due to an increase in buoyancy forces

- Mateus, Marcos Vinicius, et al. "Molecular interactions and modeling of anionic surfactant effect on oxygen transfer in a cylindrical reactor." Environmental Engineering Science 36.2 (2019): 180-185.
- Takagi, Shu, et al. "The effects of surfactant on the multiscale structure of bubbly flows." Philosophical Transactions of the Royal Society A: Mathematical, Physical and Engineering Sciences 366.1873 (2008): 2117-2129.

Objectives

Research objectives

Evaluation of SDS surfactant in DIW and H_2-DIW during ultrasonic and megasonic fields

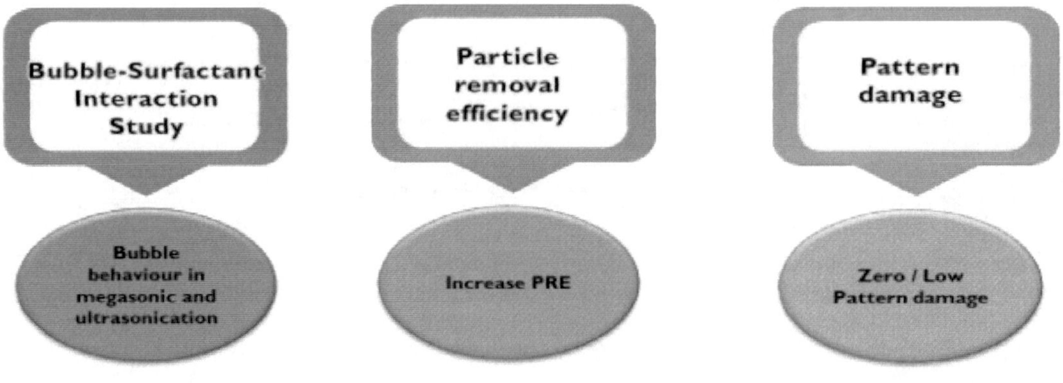

Experimental materials and methods

Experiment materials and methods

Materials and methods

- **Wafer:** Bare Silicon, PR pattern (coupon size)
- **Slurry:** 100nm Si_3N_4
- **Surfactant:** SDS (32 dyne/cm at 2000ppm)
- **Particle deposition method:** Drop deposition
- **Air infuse generator:** Harvard PHD2000
- **Solutions:** DIW and H_2-DIW

Analysis methods

- **High speed camera:** Phantom VEO710
- **PD analysis:** FE-SEM
- **PRE analysis:** Optical microscope
- **Bubble interaction studies:** ultrasonic and megasonic bath

Equipment

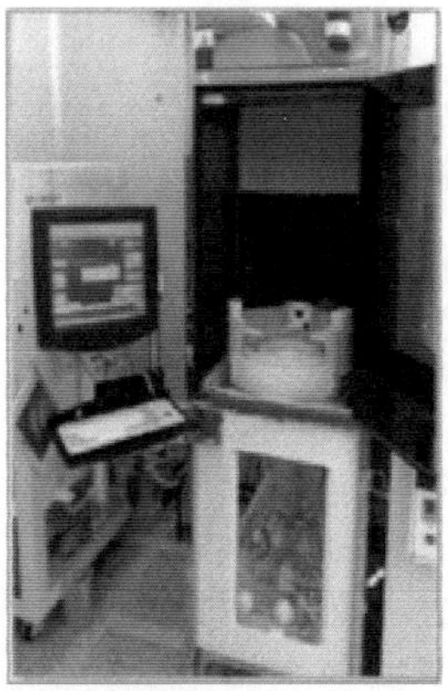

Akrion GF-1,USA (PRE and PD)

Experimental Conditions	
Megasonic frequency	0.83 MHz
Power for PRE	Low
Power for PD	High
Process time	10 seconds
Brush gap distance	0.74mm
H_2 concentration	0-94~1.15 ppm

Single bubble analysis during ultrasonic and megasonic condition

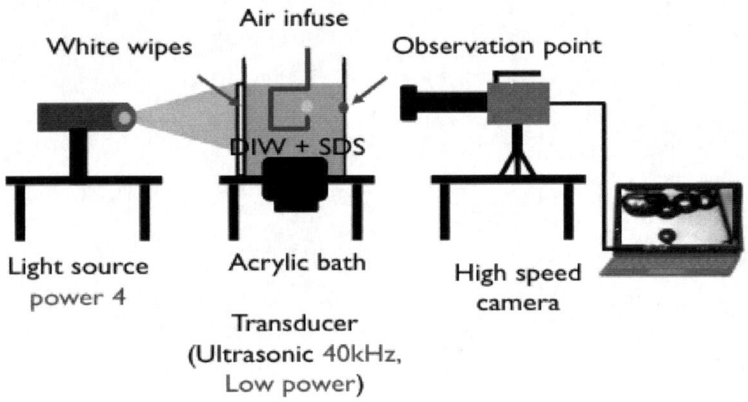

Ultrasonic bath parameters

S.No.	Parameter	Magnitude
1	Resolution	2048 x 1952
2	Ultrasonication	40 KHz
3	Power	low
4	Temperature	~22 °C
5	Air infuse rate	2 (ml/min)
6	Surfactant concentration (SDS)	Zero, low, high

- **Analysis to study:**
  - **Single bubble behavior with and without ultrasonication in DIW and H_2-DIW**
  - **Effect of surfactant SDS concentration on bubble behavior**
- Single bubble behavior was evaluated at 40KHz with low power

Experimental procedure: Effect of megasonic on bubble dynamics

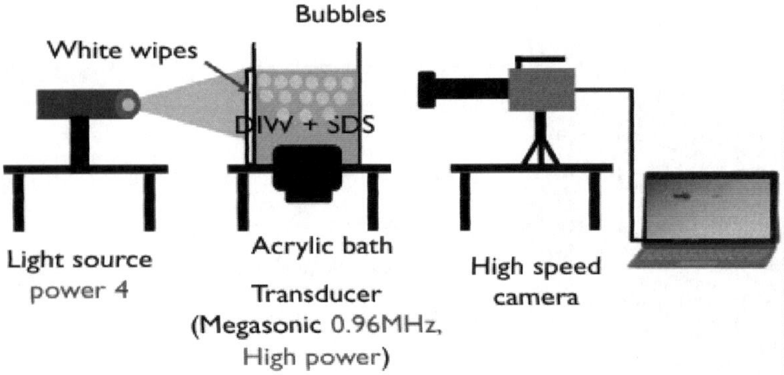

Megasonic waves (0.96 MHz, Power: high, λ= 1.56mm)

Megasonic bath parameters

S.No.	Parameter	Magnitude
1	Resolution	2048 x 1952
2	Ultrasonication	0.96 MHz
3	Power	High
4	Temperature	~22 °C
6	Surfactant conc. (SDS)	Zero, low, high

- Bubble behavior (population and size) under megasonic condition
- Effect of surfactant SDS concentration on bubble behavior
- Single bubble behavior was evaluated at 0.96MHz with high power

Particle removal efficiency (PRE) and pattern damage (PD) evaluation

1. Particles and patterns

- 100nm Si_3N_4 particle on Si wafer
- Poly-Si 70nm pattern

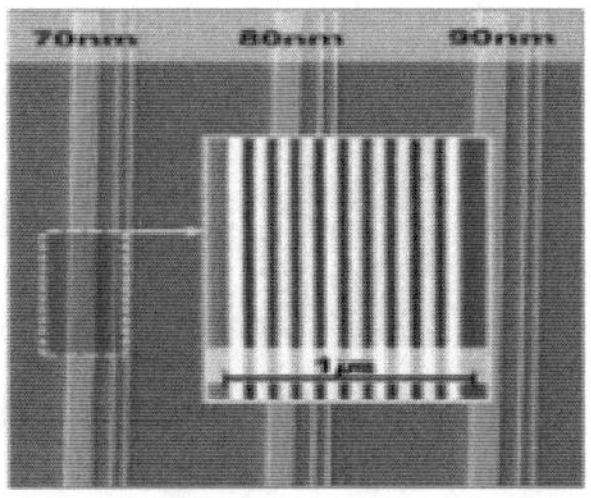

Width : 70nm
Height : 250nm
Aspect ratio : 3.6

2. Cleaning

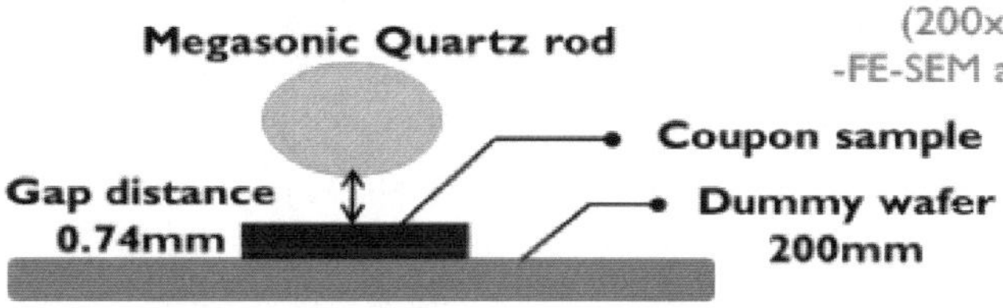

- **Cleaning conditions**
  - Frequency : 0.83MHz,
  - Power : low (**PRE**) and high (**PD**)
  - Process time: 10sec
  - H_2 concentration: 0.98~1.15ppm

- ***High power, 95~100% PRE***
- ***Low power, no/zero pattern damage***

3. PRE and PD analysis

- Optical microscopy dark field mode (200x, 800ms)
- FE-SEM analysis

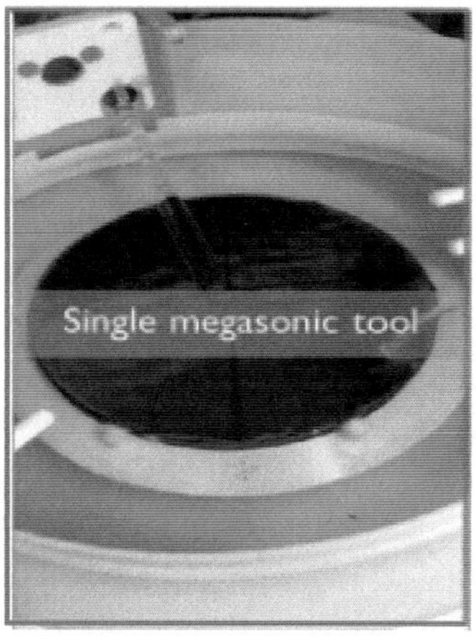

Results and discussion

1. Bubble interaction studies
2. PRE and pattern damage analysis

Bubbles coalescence analysis in DIW and H_2-DIW without ultrasonication

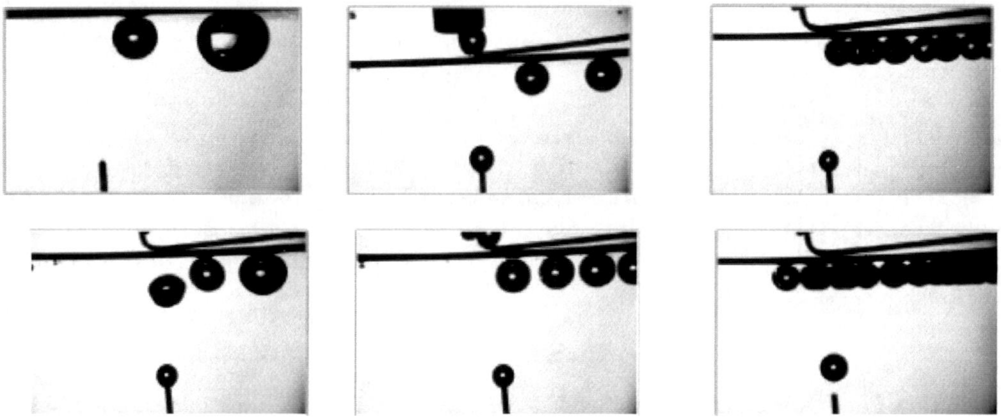

- No bubble coalescence was observed in the presence of SDS surfactant in DIW
- Decrease in bubble size was observed in H_2-DIW as compared to DIW
- Same bubble behavior was observed with the addition of dissolved hydrogen gas compared to DIW

Bubbles with ultrasonication in DIW/H_2-DIW

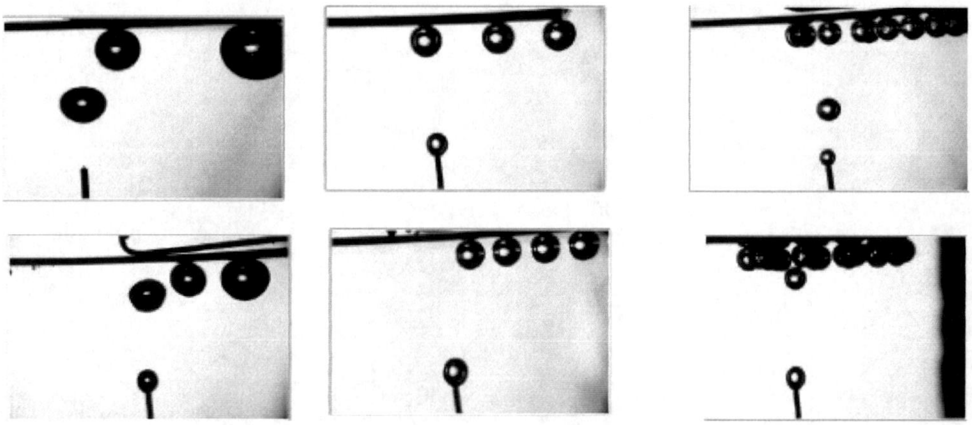

- No bubble coalescence was observed in the presence of SDS surfactant
- Increase in surfactant concentration increases the bubble population
- Similar bubble behavior was observed in H_2-DIW as compared to DIW

Effect of SDS surfactant on bubble size in DIW – Megasonic

High-Speed Camera images at different concentrations SDS under megasonic cleaning

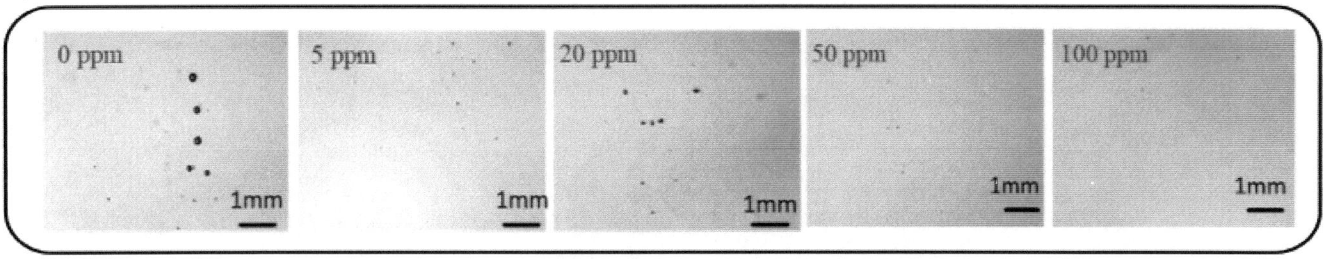

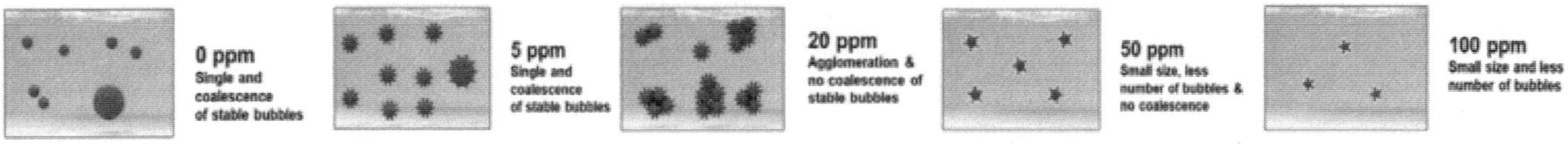

- Increase SDS concentration up to 20 ppm the number of bubbles increased, and clusters were observed
- Resonance-sized bubbles increased at lower concentrations of surfactant
- Less and smaller sized bubbles were observed at 50ppm and 100 ppm of SDS
- Coalescence of stable bubbles was observed in the absence of surfactant in DIW

Effect of SDS surfactant on bubble size in H_2-DIW – Megasonic

High-Speed Camera images at different concentrations SDS under megasonic cleaning

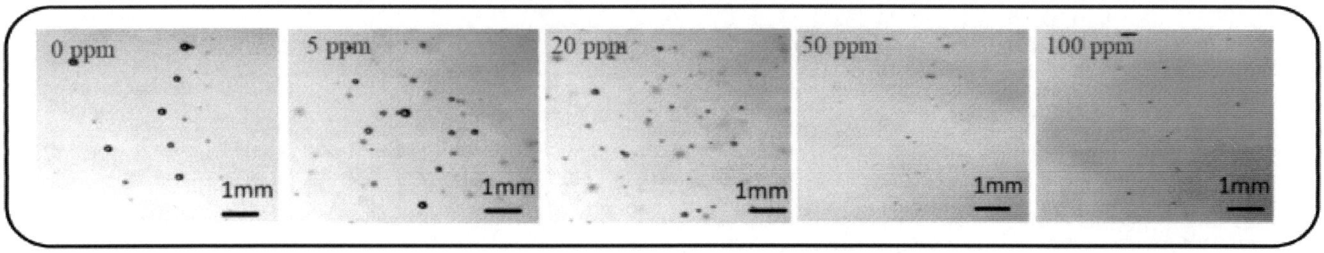

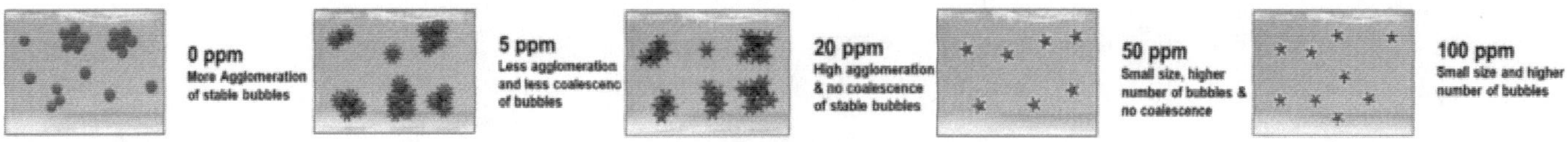

- The number of bubbles increased significantly in H_2-DIW compared to DIW
- Dissolved hydrogen gas facilitates the bubble acoustic cavitation
- At 20 ppm, a higher number of resonance bubbles were observed
- Higher bubble population was observed at 100ppm in H_2-DIW as compared to DIW

PRE and PD evaluation of SDS (anionic surfactant) in DIW & H_2-DIW

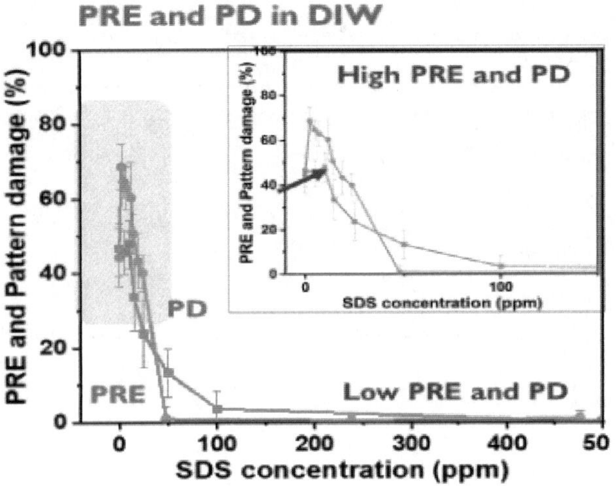

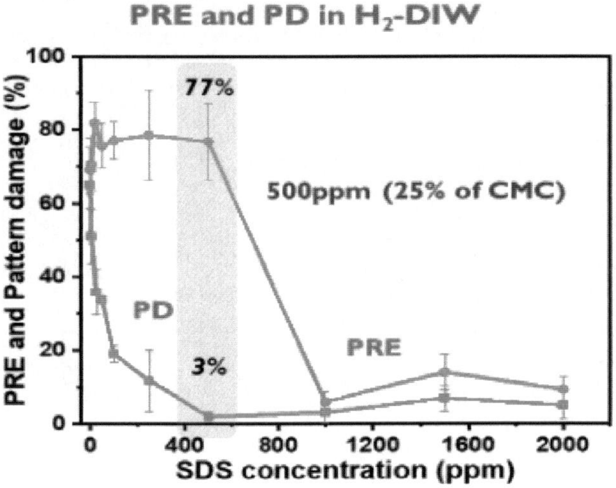

- H_2-DIW significantly improves the PRE as compared to the DIW condition
- Higher PRE (up to 82%) was observed in H_2-DIW + SDS surfactant
- Decrease in PRE was observed by the increase in surfactant concentration in DIW

High PRE and low PD mechanism in surfactant added H_2-DIW

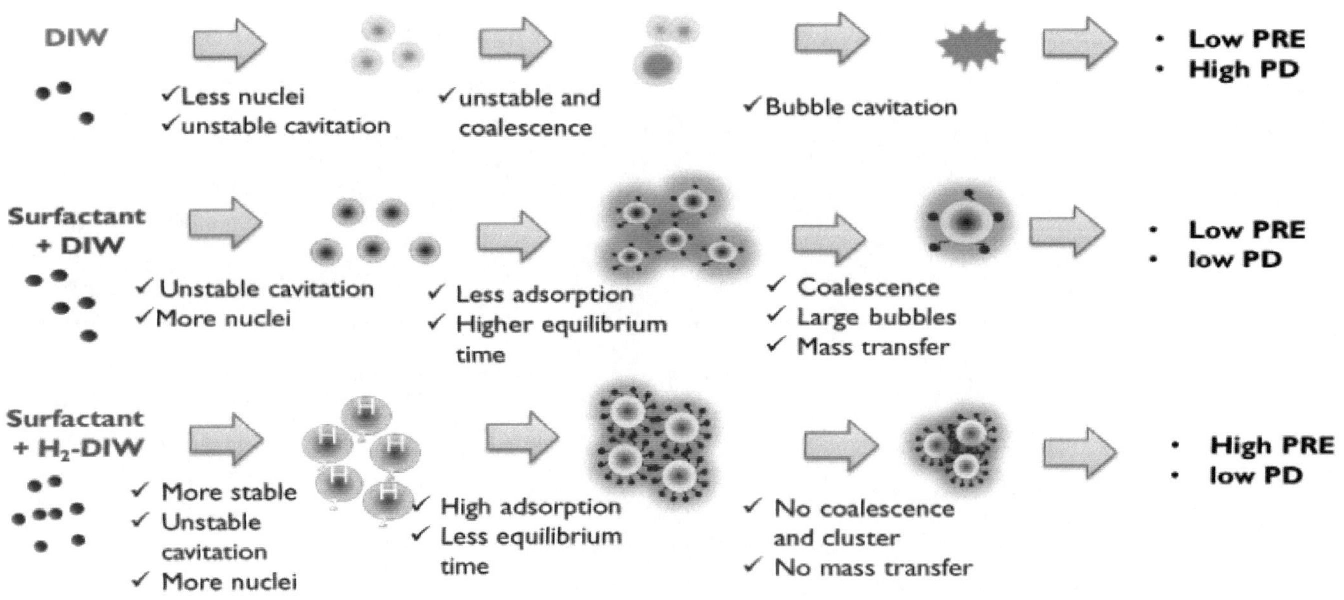

- Presence of dissolved H_2 in DIW and SDS surfactant can improve PRE and PD by preventing bubble coalescence

Change of bubble characteristics with surfactant concentration

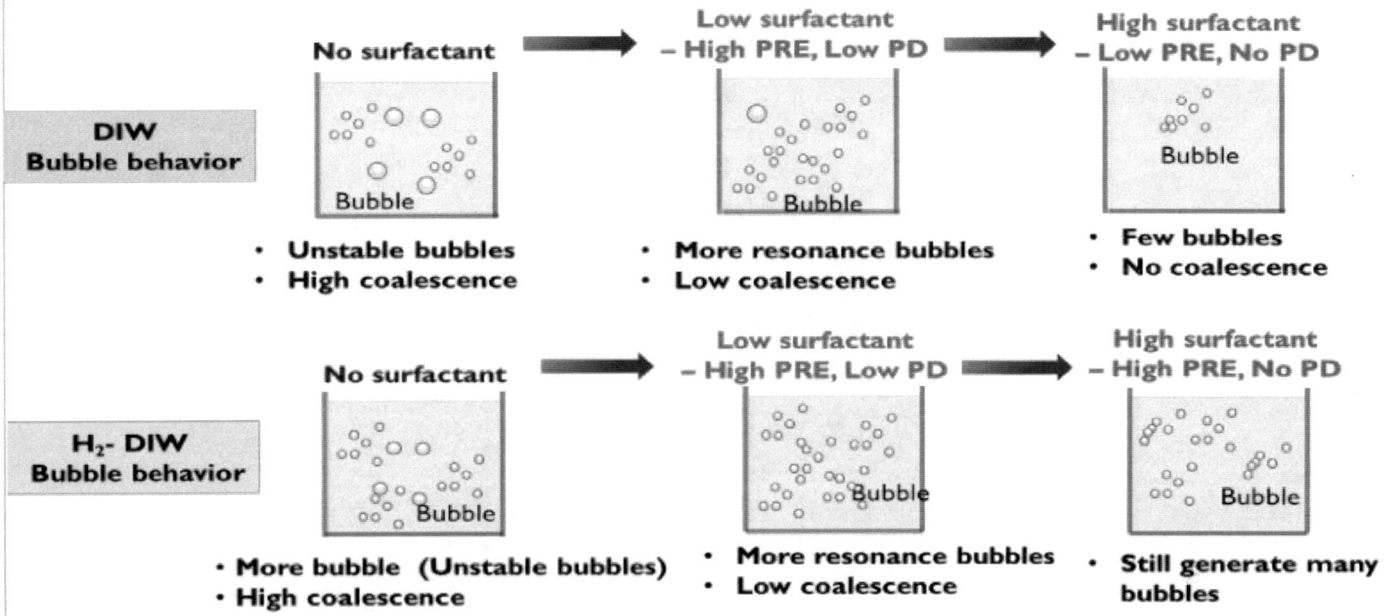

- It was observed that with an increase in surfactant concentration, the bubble resonance reduce significantly

Summary

- H_2-DIW can significantly increase PRE and decrease pattern damage due to increased aquatic streaming in the presence of surfactant SDS
- High bubble coalescence was observed in DIW and H_2-DIW in the absence of surfactant
- No bubble coalescence was observed in the presence of surfactant
- Bubble population increased significantly with H_2-DIW compared to DIW
- Increasing SDS concentration the bubble size decreased up to the CMC value in DIW and H_2-DIW

HANYANG
UNIVERSITY

한양대학교 ERICA 캠퍼스

SAMSUNG

Plasma-based Dry Clean Process Monitoring by Exhaust Gas Analysis with ToF-MS

Minhyoung Kim
Mechatronics Research
Samsung Semiconductor

E-Mail address : minhyoung.kim@samsung.com

SAMSUNG

Outline

01 | Monitoring for Plasma-based Dry Clean

Advantages & Limitations of Plasma-based Dry Clean

Process Monitoring via Exhaust Gas Analysis w/ ToF-MS

02 | *in-situ* Process Monitoring w/ ToF-MS

Exhaust Gas Intensity Chromatogram

Verification of Etch Component

Verification of Input Gas Influence

03 | Correlation Analysis w/ Process Data

Correlation between ToF-MS and Etch/Depo. Process Data

Process Dependency on Etch/Depo. Component Ratio

04 | Summary

2

01| Monitoring for Plasma-based Dry Clean

SAMSUNG

Advantages of Plasma-based Dry Clean

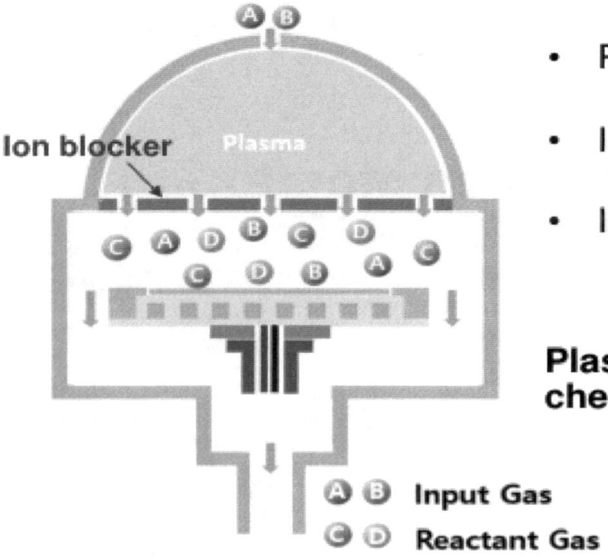

- Plasma creates reactant chemistry
- Ions are blocked, chemistry passes through
- Isotropic & highly selective etch

Plasma-based dry clean can creates tailored chemistry for specific target process

3

01| Monitoring for Plasma-based Dry Clean

SAMSUNG

Limitations of Plasma-based Dry Clean

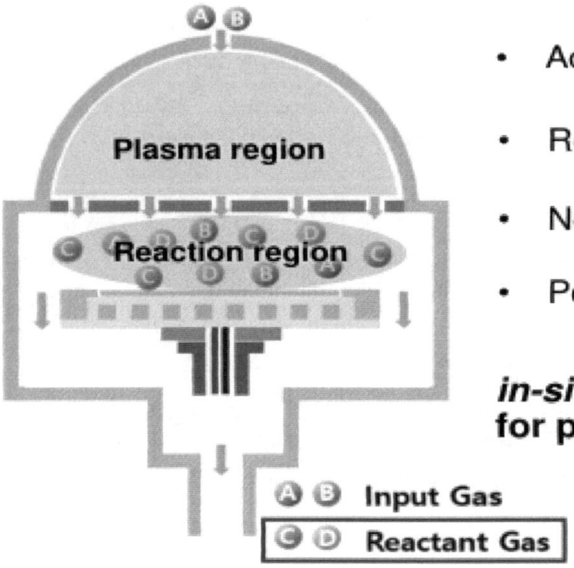

- Actual reactants are different from input gases
- Reaction and plasma regions are separated
- No monitoring tool for reaction region
- Post-process analysis with *ex-situ* inspection

***in-situ* process monitoring tool development for plasma dry clean chamber is needed**

4

01| Monitoring for Plasma-based Dry Clean

SAMSUNG

Process Monitoring via Exhaust Gas Analysis w/ ToF-MS

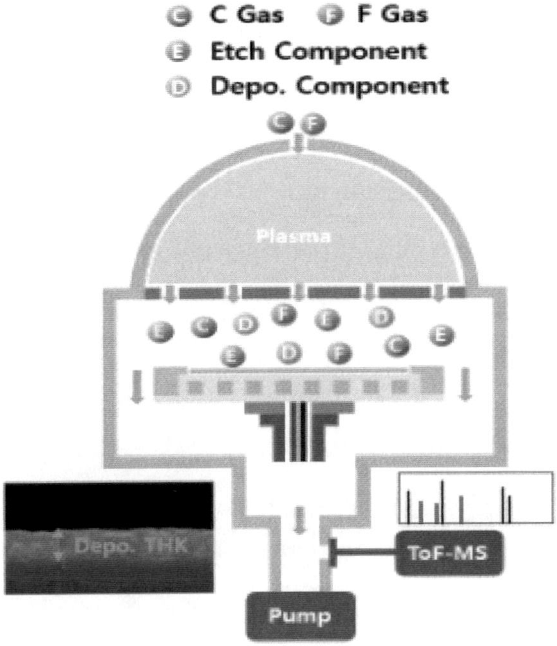

- ToF-MS (Time-of-Flight Mass Spectrometry)
  High data recording speed & High resolution
- Exhaust gas analysis
  Both reactant and byproduct can be detected
- Experimental
  Input : Fluorine-containing gas for etch component
  Carbon-containing gas for depo. component

  Monitoring the exhaust gas in real time

  Confirming the etch and depo. component

  Matching with the NPW process data

5

02| *in-situ* Process Monitoring w/ ToF-MS

SAMSUNG

Exhaust Gas Intensity Chromatogram @ Main Etch

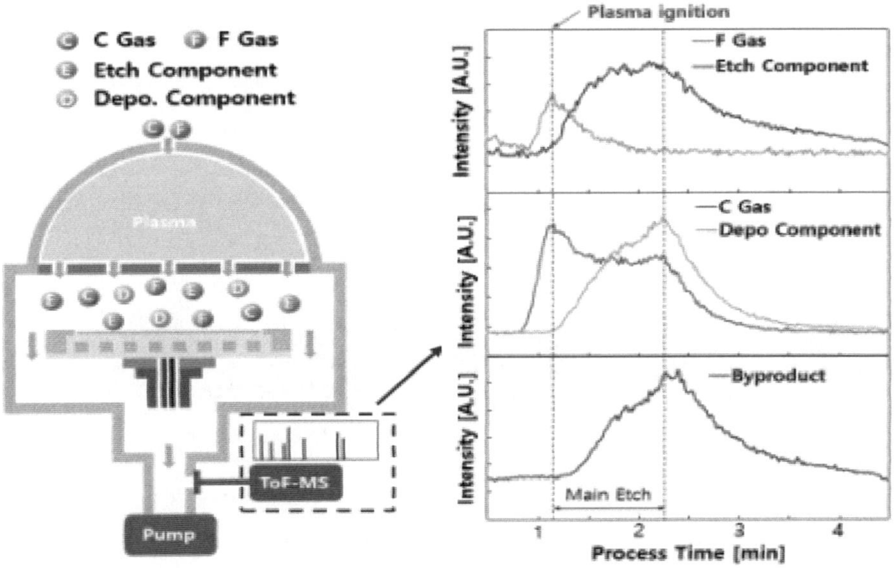

After Plasma Ignition

F Gas ↓

Etch Component ↑

C Gas ↓

Depo. Component ↑

Byproduct ↑

F Gas/C Gas are converted to Etch/Depo. component after plasma ignition

6

02| *in-situ* Process Monitoring w/ ToF-MS

SAMSUNG

Verification of Etch Component @ Main Etch

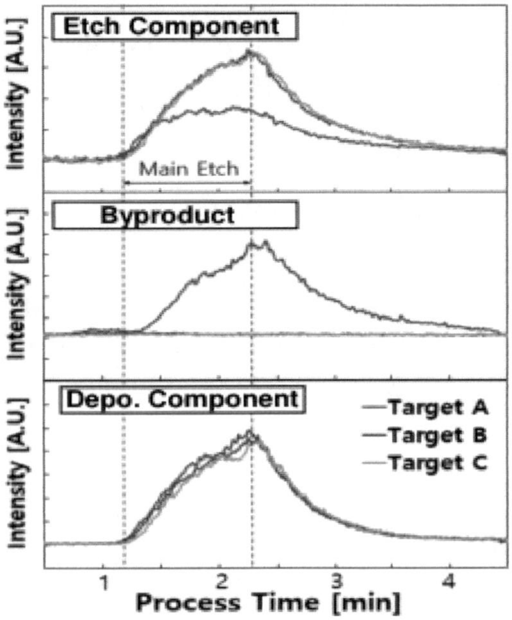

Target **A** : **Etchable**

Target **B**/C : Non-etchable

During the etch processing,

Etch Component ↓ (consumed)

Byproduct ↑ (produced)

Depo. Component ~

Actually, the etch component reacts with Target A to produce the byproduct

7

02| *in-situ* Process Monitoring w/ ToF-MS

SAMSUNG

Verification of Input Gas Influence @ Main Etch

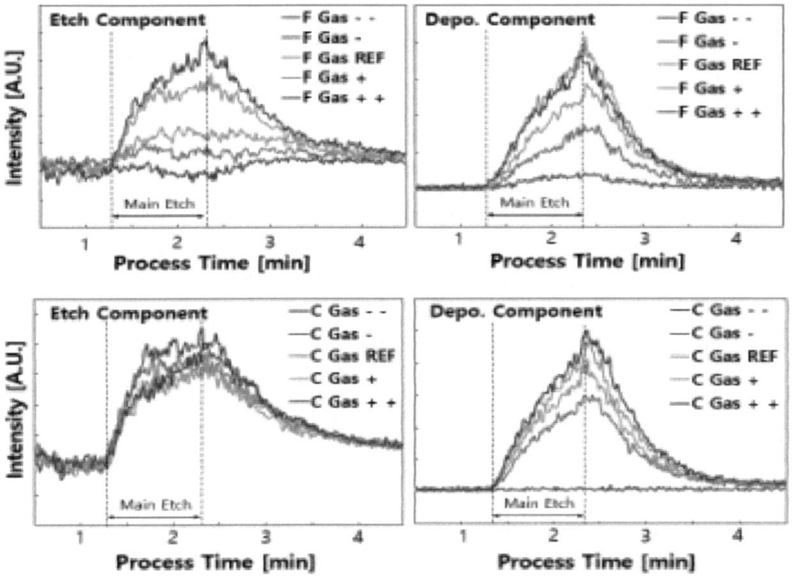

Input F Gas increase

Etch Component ↑

Depo Component ↑

F Gas was converted to Etch/Depo. Component

Input C Gas increase

Etch Component ~

Depo Component ↑

C Gas was converted only to Depo. Component

Etch Component : Only F gas is needed

Depo. Component : Both C gas and F gas are needed

8

03| Correlation Analysis w/ Process Data

SAMSUNG

Correlation between ToF-MS and Etch/Depo. Process Data

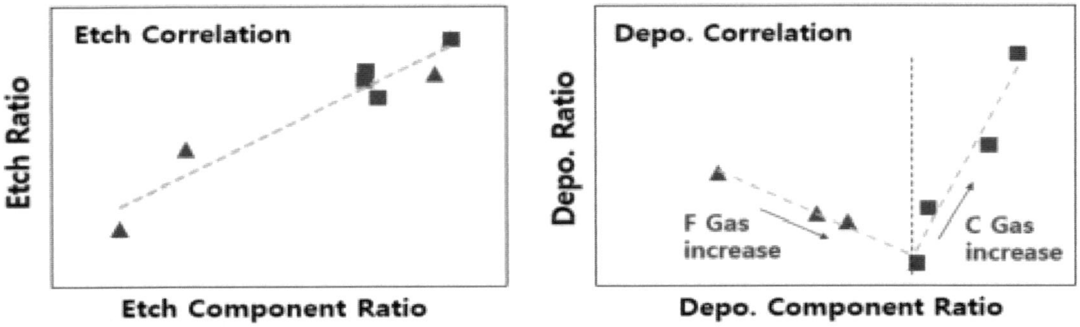

X-axis : Relative Signal of Etch/Depo. Component from ToF-MS

Y-axis : Relative Etch/Depo. Amount from Reflectometer

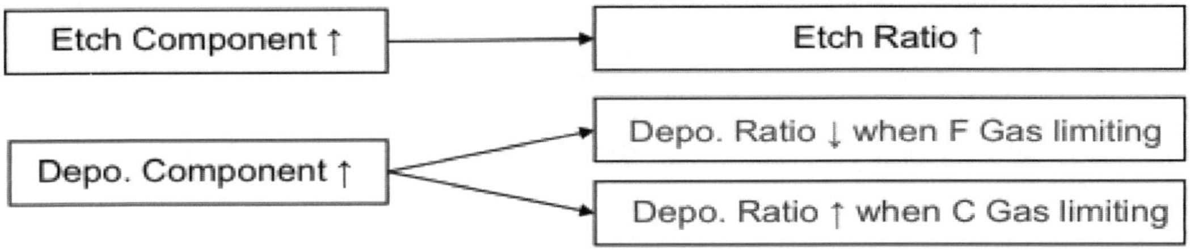

9

03| Correlation Analysis w/ Process Data

SAMSUNG

Process Dependency on Etch/Depo. Component Ratio

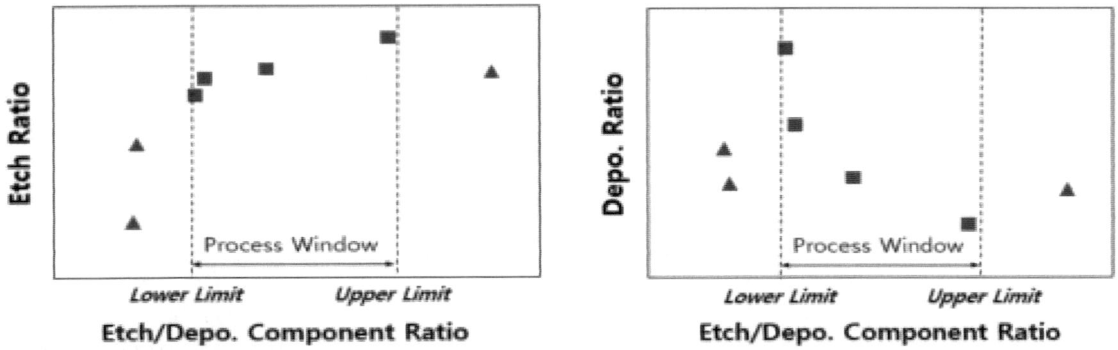

When F Gas limiting (red triangle),

Process data does not match with ToF-MS data → **outside the stable process window**

When C Gas limiting (blue square),

Etch ratio vs. Etch/Depo. Component ratio : proportional

Depo ratio vs. Etch/Depo. Component ratio : inversely proportional

Process data match well with ToF-MS data → **within the stable process window**

10

04| Summary

SAMSUNG

- New process monitoring technology for plasma-based dry clean has been developed to enable detect reactant gases in real time.
- With our monitoring tool, the actual reactants of process condition could be confirmed.
- The process data well matched with the detected gas data, verifying our monitoring technology is reliable.

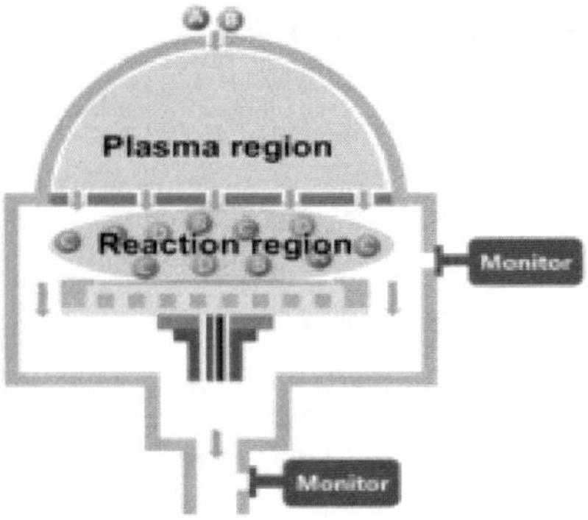

- With these monitoring technologies, real-time process condition diagnosis is possible without the conventional post-analysis-based verifications.
- Process monitoring tools will be essential for next-generation plasma-based dry clean chambers.

11

SAMSUNG

Thank you

12

TECHNIC

CLEAN

« Chimies Liquides Efficaces pour l'Amélioration des Nanotechnologies »

« Ce projet a été financé par le gouvernement dans le cadre de France 2030 »
« Financé par l'Union européenne - Next Generation EU dans le cadre du plan France Relance »

Backside metallic contamination removal focused on noble metals using an halogen-based solution

Marie-Claire Pignié, Laurence Gabette, Virginie Loup,
Laura Magnin, Jérôme Daviot, Marine Audouin

SPCC conference

24 - 26 April 2024

Speaker : Marie-Claire Pignié PhD

Technic France

CLEAN

bpifrance

TECHNIC

Metallic contamination removal processes

Where?

Metal contamination removal on backside & bevel

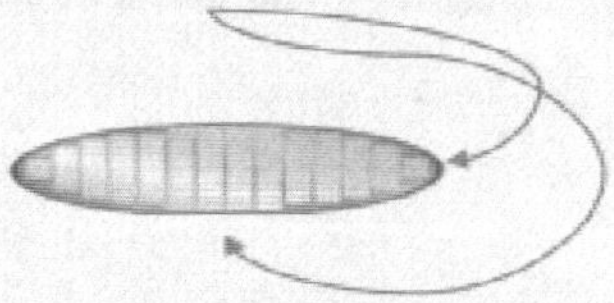

When ?

After potential contamination steps
Before subsequent thermal steps

How ?

Wet chemistry processes
With **backside injection** SW
(front side protected by N_2)

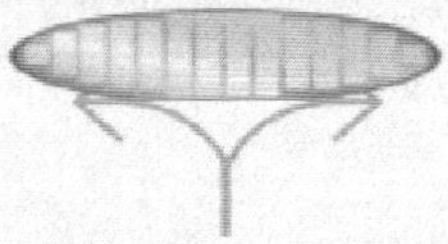

Why?

External sources

- Demos
- Transport
- ...

Internal sources

- Characterization tools
- Technologies cohabitation
- ...

In foresight of backside integration ?

CLEAN

Besson, P.; Backside and bevel contamination removal *Solid State Phenomena* **2015**
Broussous, L; Single backside cleaning on silicon, silicon nitride and silicon oxide *Solid State Phenomena* **2005**

bpifrance

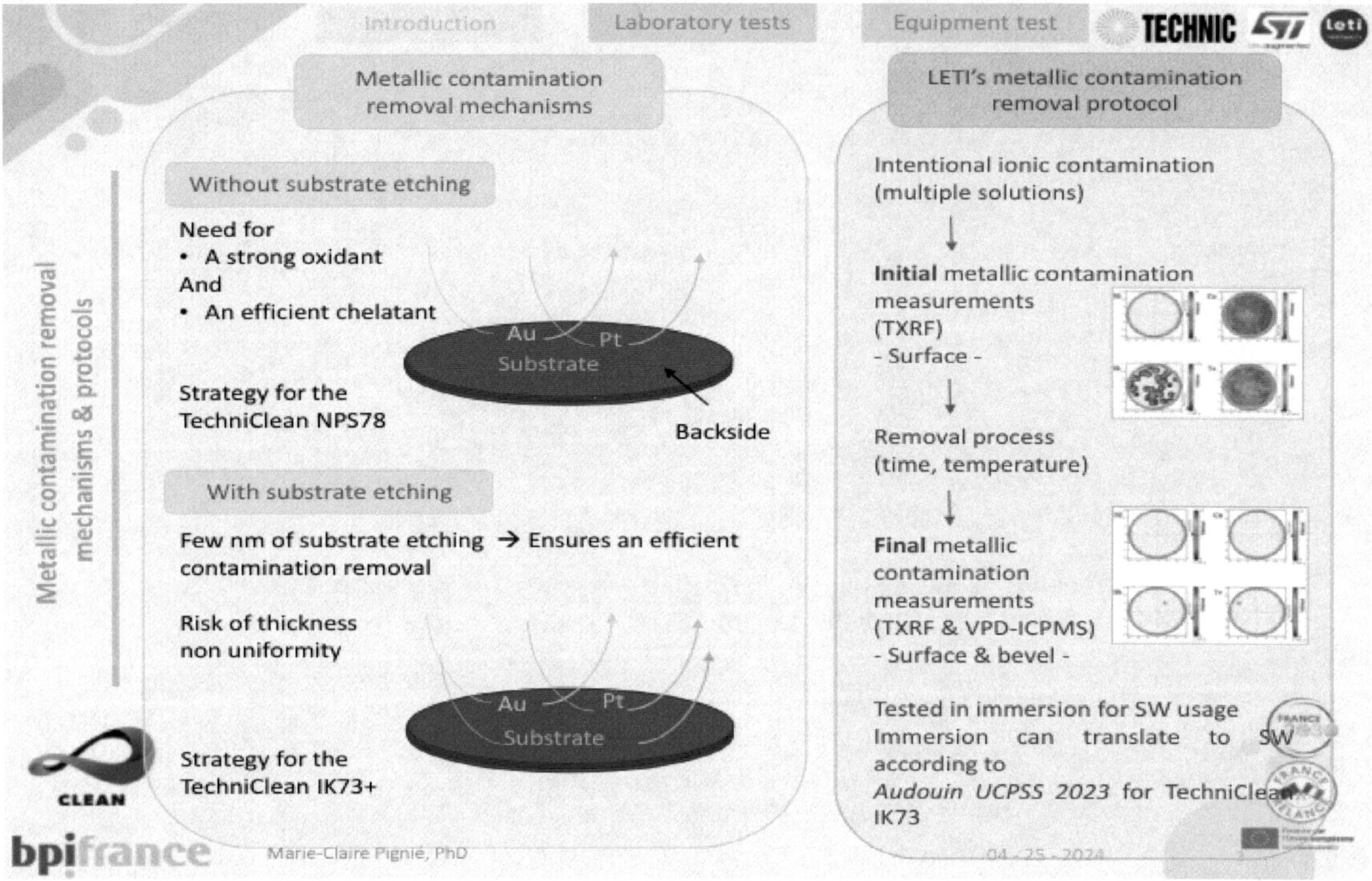
Introduction
Laboratory tests
Equipment test
TECHNIC
Metallic contamination removal mechanisms & protocols
Metallic contamination removal mechanisms
Without substrate etching
Need for
• A strong oxidant
And
• An efficient chelatant
Au
Pt
Substrate
Backside
Strategy for the TechniClean NPS78
With substrate etching
Few nm of substrate etching → Ensures an efficient contamination removal
Risk of thickness non uniformity
Au
Pt
Substrate
Strategy for the TechniClean IK73+
LETI's metallic contamination removal protocol
Intentional ionic contamination (multiple solutions)
Initial metallic contamination measurements (TXRF) - Surface -
Removal process (time, temperature)
Final metallic contamination measurements (TXRF & VPD-ICPMS) - Surface & bevel -
Tested in immersion for SW usage
Immersion can translate to SW according to
Audouin UCPSS 2023 for TechniClean IK73
CLEAN
bpifrance
Marie-Claire Pignié, PhD
04 - 25 - 2024

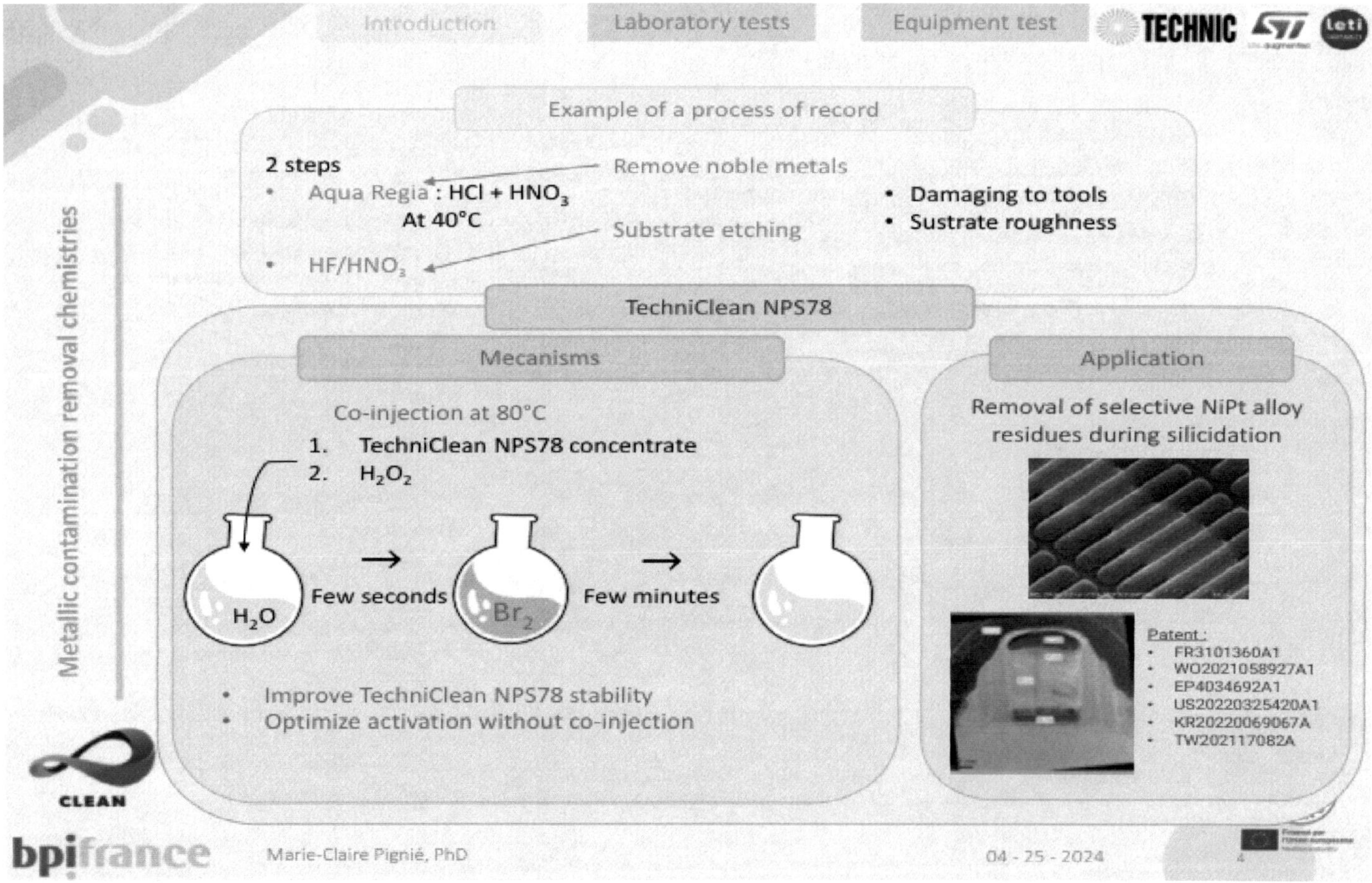
Introduction
Laboratory tests
Equipment test
TECHNIC
Metallic contamination removal chemistries
Example of a process of record
2 steps
• Aqua Regia : HCl + HNO3
At 40°C
• HF/HNO3
Remove noble metals
Substrate etching
• Damaging to tools
• Sustrate roughness
TechniClean NPS78
Mecanisms
Co-injection at 80°C
1. TechniClean NPS78 concentrate
2. H2O2
H2O
Few seconds
Br2
Few minutes
• Improve TechniClean NPS78 stability
• Optimize activation without co-injection
Application
Removal of selective NiPt alloy residues during silicidation
Patent :
• FR3101360A1
• WO2021058927A1
• EP4034692A1
• US20220325420A1
• KR20220069067A
• TW202117082A
CLEAN
bpifrance
Marie-Claire Pignié, PhD
04 - 25 - 2024
4

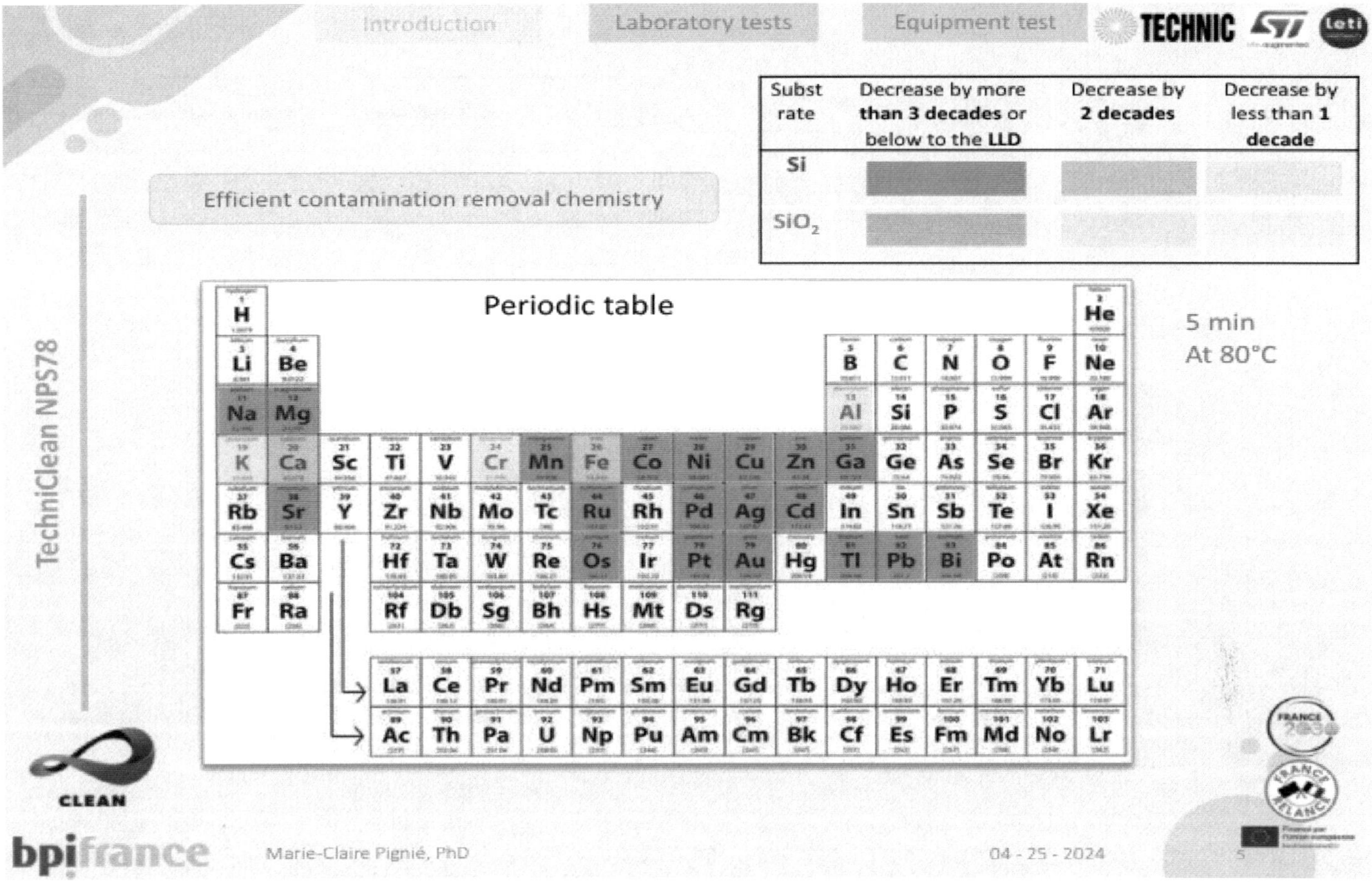

Introduction
Laboratory tests
Equipment test
TECHNIC
Subst rate
Decrease by more than 3 decades or below to the LLD
Decrease by 2 decades
Decrease by less than 1 decade
Si
SiO2
Efficient contamination removal chemistry
TechniClean NPS78
Periodic table
5 min
At 80°C
CLEAN
bpifrance
Marie-Claire Pignié, PhD
04 - 25 - 2024
5

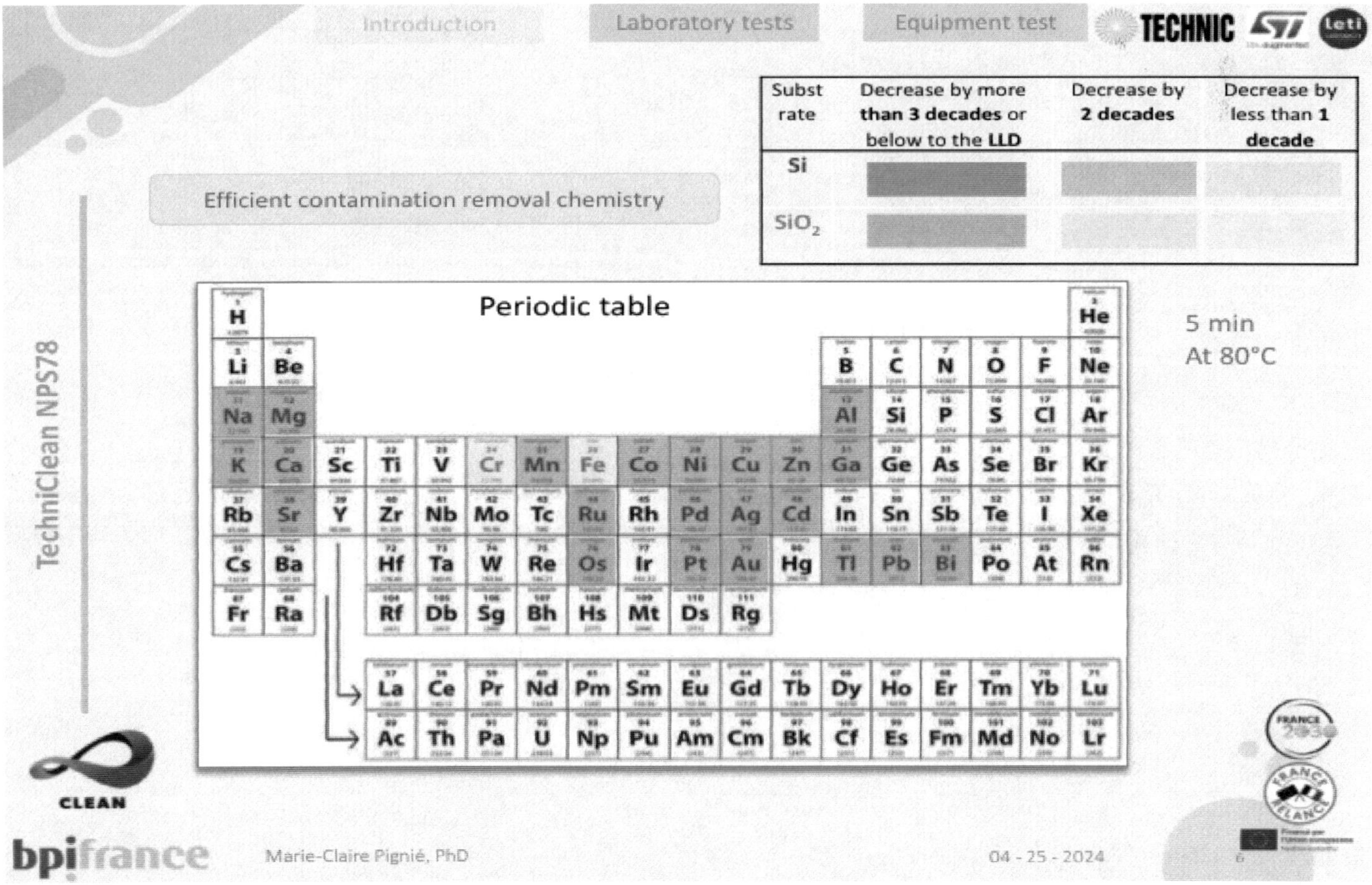

Introduction
Laboratory tests
Equipment test
TECHNIC
Subst rate
Decrease by more than 3 decades or below to the LLD
Decrease by 2 decades
Decrease by less than 1 decade
Si
SiO2
Efficient contamination removal chemistry
TechniClean NPS78
Periodic table
5 min
At 80°C
CLEAN
bpifrance
Marie-Claire Pignié, PhD
04 - 25 - 2024
6

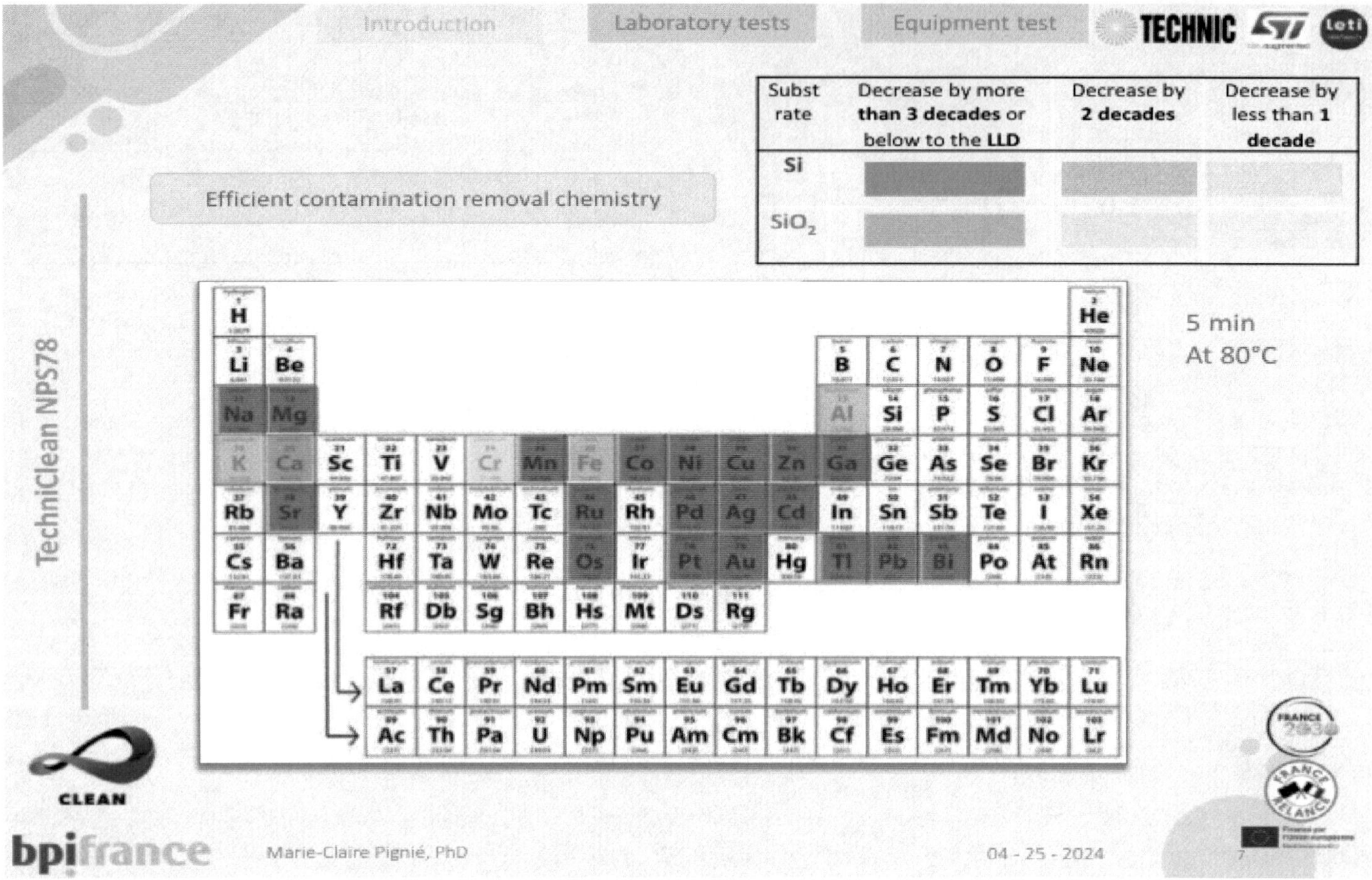
Introduction
Laboratory tests
Equipment test
TECHNIC
leti
Subst rate
Decrease by more than 3 decades or below to the LLD
Decrease by 2 decades
Decrease by less than 1 decade
Si
SiO2
Efficient contamination removal chemistry
TechniClean NPS78
H He
Li Be B C N O F Ne
Na Mg Al Si P S Cl Ar
K Ca Sc Ti V Cr Mn Fe Co Ni Cu Zn Ga Ge As Se Br Kr
Rb Sr Y Zr Nb Mo Tc Ru Rh Pd Ag Cd In Sn Sb Te I Xe
Cs Ba Hf Ta W Re Os Ir Pt Au Hg Tl Pb Bi Po At Rn
Fr Ra Rf Db Sg Bh Hs Mt Ds Rg
La Ce Pr Nd Pm Sm Eu Gd Tb Dy Ho Er Tm Yb Lu
Ac Th Pa U Np Pu Am Cm Bk Cf Es Fm Md No Lr
5 min
At 80°C
CLEAN
bpifrance
Marie-Claire Pignié, PhD
04 - 25 - 2024
7

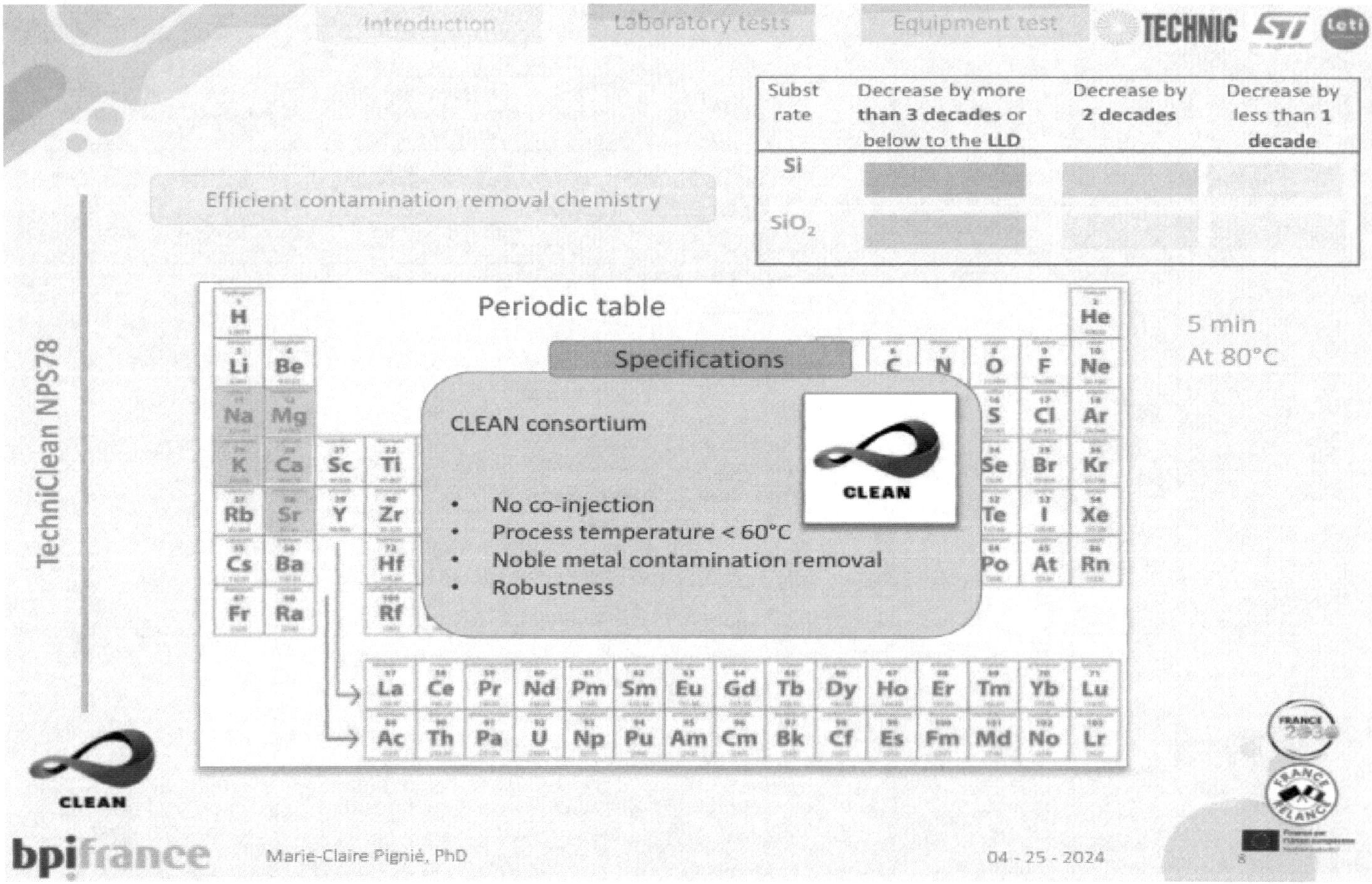
Introduction
Laboratory tests
Equipment test
TECHNIC
leti
Subst rate
Decrease by more than 3 decades or below to the LLD
Decrease by 2 decades
Decrease by less than 1 decade
Si
SiO2
Efficient contamination removal chemistry
TechniClean NPS78
Periodic table
Specifications
CLEAN consortium
No co-injection
Process temperature < 60°C
Noble metal contamination removal
Robustness
CLEAN
5 min
At 80°C
bpifrance
Marie-Claire Pignié, PhD
04 - 25 - 2024
8

Introduction | Laboratory tests | Equipment test

Etching behaviour

At room temperature

Br_2

- Present in the TechniClean NPS78 + H_2O_2
- Identified as the main etchant species
- Quantifiable by spectroscopy

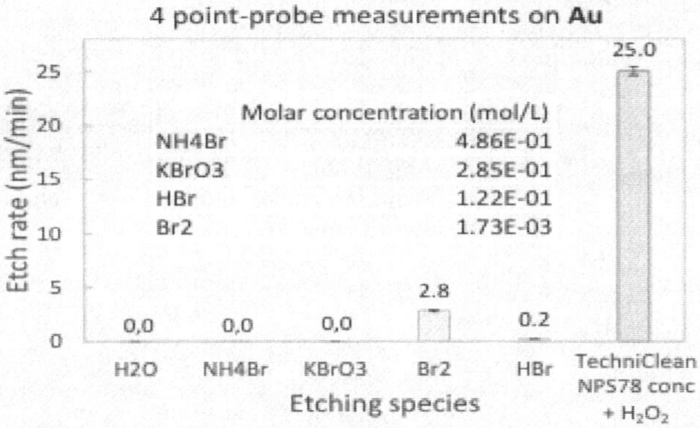

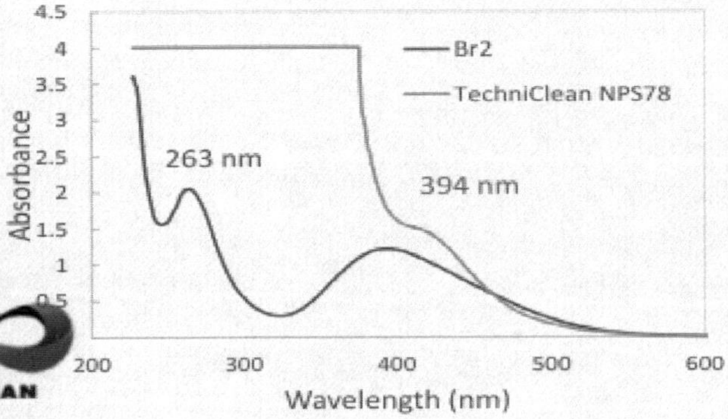

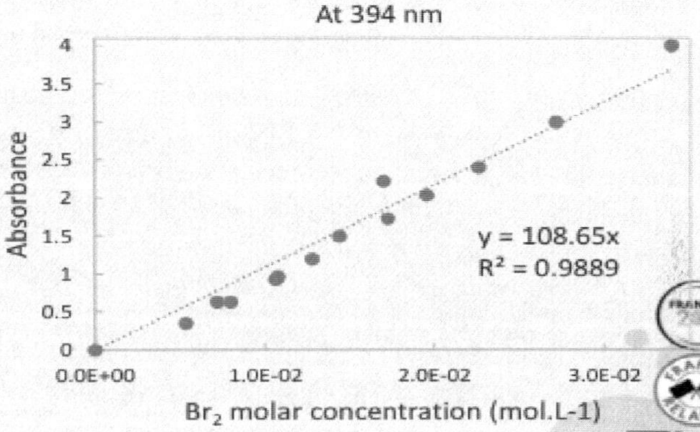

TechniClean NPS78

CLEAN

bpifrance

Marie-Claire Pignié, PhD

04 - 25 - 2024

9

Introduction | Laboratory tests | Equipment test

Specifications:

- ER(Au)=[10 - 20] nm.min^{-1}
- Stable bath for more than a week

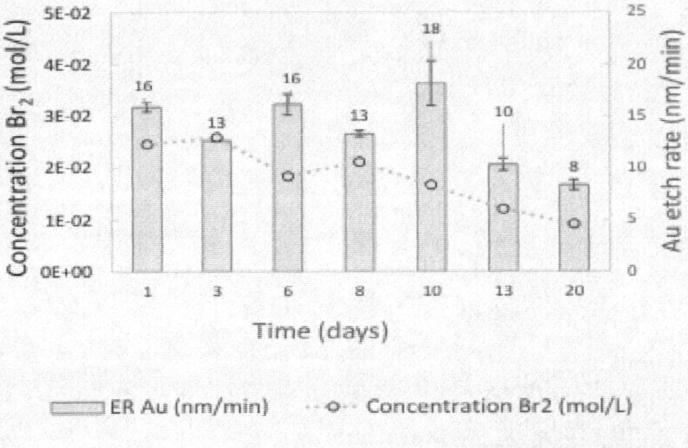

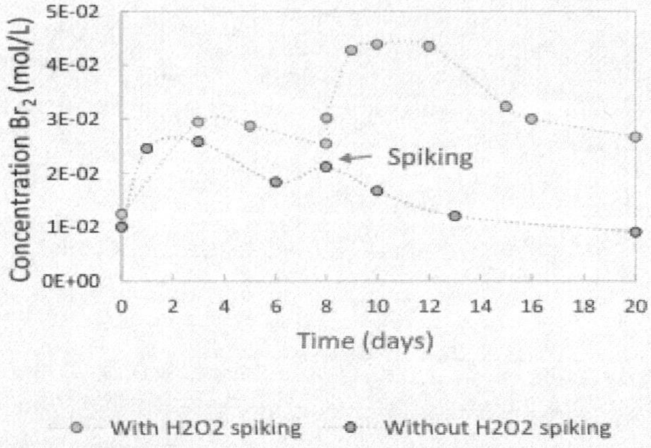

30% dilution TechniClean NPS78

- Stable ER(Au) between [10 ; 20] nm/min
- Stable [Br_2] i.e. stable solution for 14 days
- Reactivable product by H_2O_2 spiking at 7 days for a 18-20 days stability with [1/100; 1/150] as best volume ratio

Diluted TechniClean NPS78

CLEAN

bpifrance

Marie-Claire Pignié, PhD

04 - 25 - 2024

10

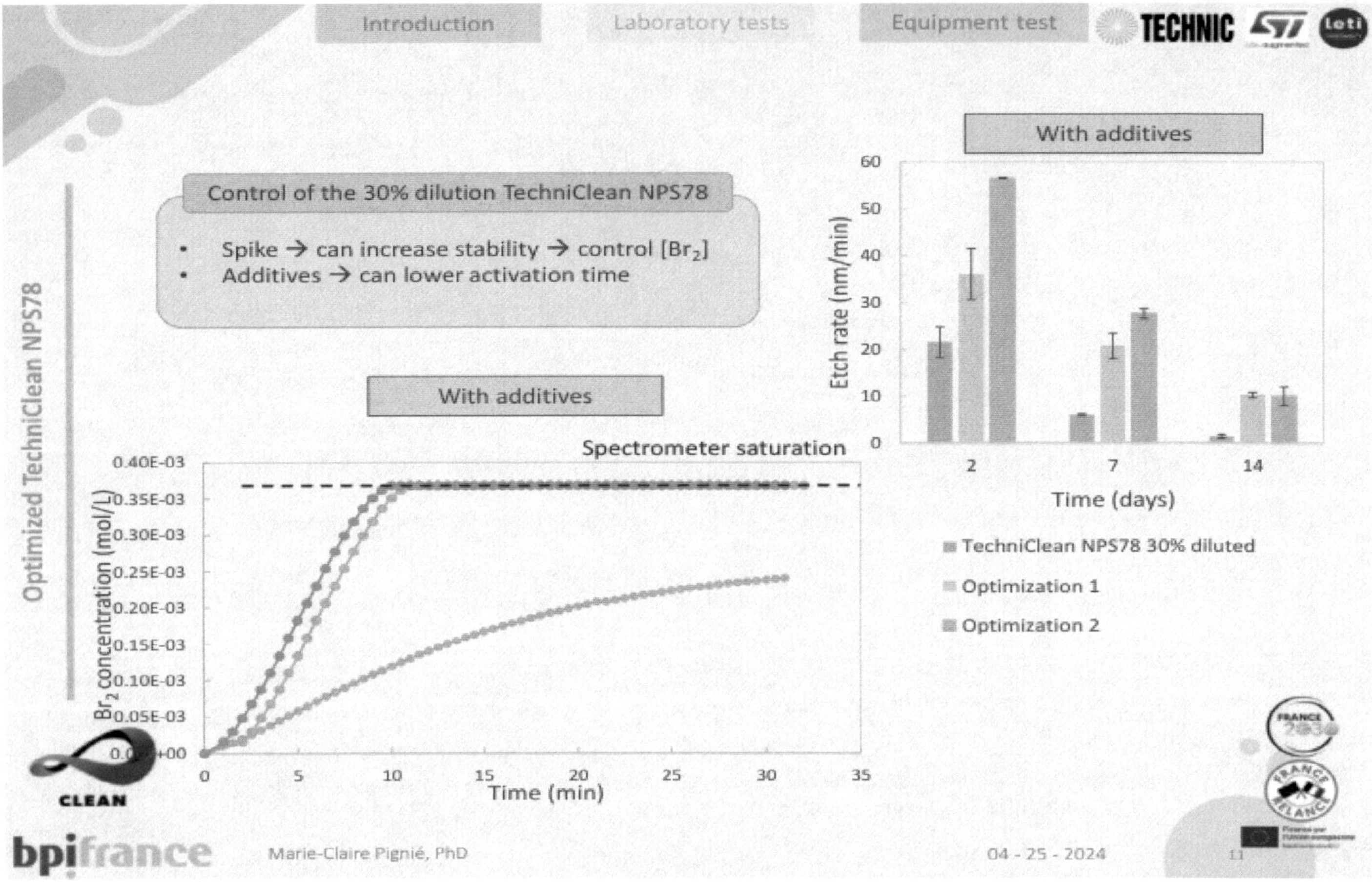
Introduction
Laboratory tests
Equipment test
TECHNIC
Optimized TechniClean NPS78
Control of the 30% dilution TechniClean NPS78
• Spike → can increase stability → control [Br2]
• Additives → can lower activation time
With additives
Spectrometer saturation
Br2 concentration (mol/L)
Time (min)
With additives
Etch rate (nm/min)
Time (days)
TechniClean NPS78 30% diluted
Optimization 1
Optimization 2
CLEAN
bpifrance
Marie-Claire Pignié, PhD
04 - 25 - 2024
11

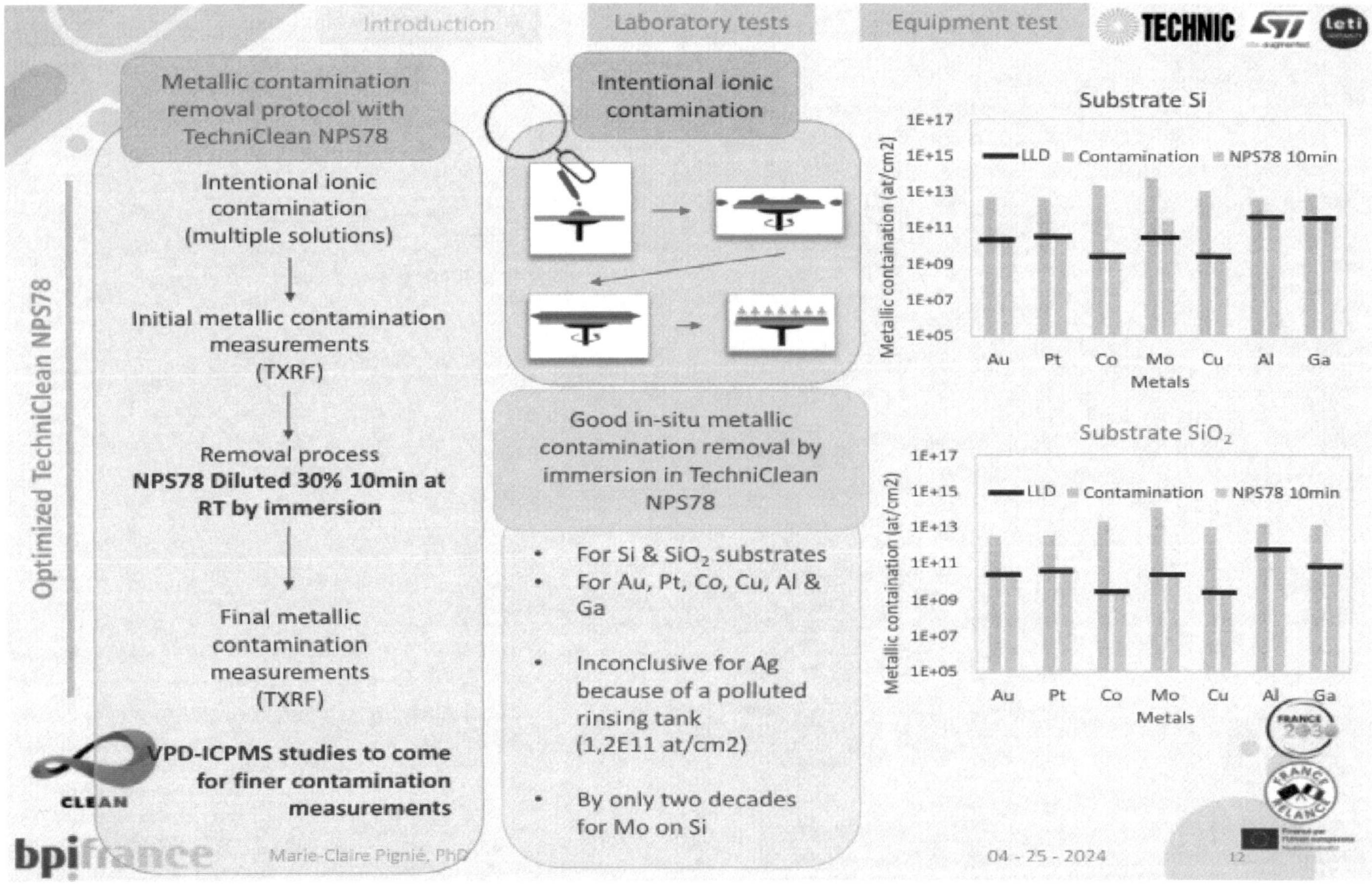
Introduction
Laboratory tests
Equipment test
TECHNIC
Optimized TechniClean NPS78
Metallic contamination removal protocol with TechniClean NPS78
Intentional ionic contamination (multiple solutions)
Initial metallic contamination measurements (TXRF)
Removal process
NPS78 Diluted 30% 10min at RT by immersion
Final metallic contamination measurements (TXRF)
VPD-ICPMS studies to come for finer contamination measurements
Intentional ionic contamination
Good in-situ metallic contamination removal by immersion in TechniClean NPS78
• For Si & SiO2 substrates
• For Au, Pt, Co, Cu, Al & Ga
• Inconclusive for Ag because of a polluted rinsing tank (1,2E11 at/cm2)
• By only two decades for Mo on Si
Substrate Si
LLD
Contamination
NPS78 10min
Metallic containation (at/cm2)
Metals
Substrate SiO2
CLEAN
bpifrance
Marie-Claire Pignié, PhD
04 - 25 - 2024
12

TECHNIC

Bring home message

Respect of LETI's specs

- No co-injection
- Stable for more than a week
- Able to etch gold [10 – 20] nm.min^{-1}
- Good metallic contamination removal for Au, Pt, Co, Cu, Al & Ga

No POR fits necessary requirements :

- Premix needed
- Too aggressive to tools (AR)
- Substrate roughness issues (HF/HNO_3)

TechniClean NPS78 : An industrial process for a current need

Metallic contamination removal is a current need for most foundries

Validated by immersion

Designed for **backside injection** SW (front side protected by N_2) for backside & bevel cleaning

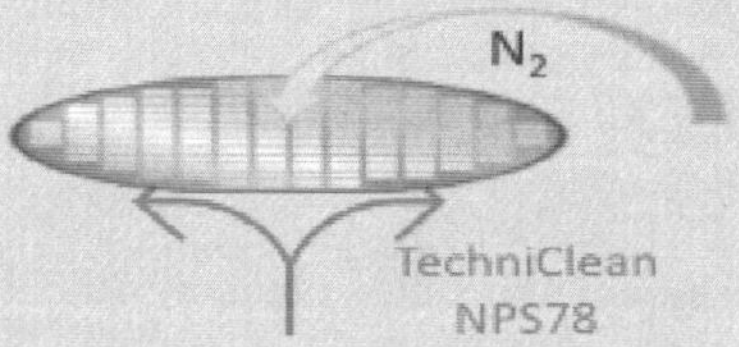

Industrial advantages

- Good bath control
- Room temperature process
- Tunable solution

Bath control

- More than 7 days stability
- Possible H_2O_2 spike for an added 7 days stability
- Efficiency weakly impacted by loading
- Possible monitoring

CLEAN

bpifrance

TECHNIC

Acknowledgments

CLEAN consortium team

Thank you for your attention
Feel free to share any questions

Pignié Marie-Claire, PhD
marie-claire.pignie@technic.fr
+33 149465137

CLEAN

bpifrance

Dummy Poly Gate Etch Clean Challenges in Advanced Nanosheet Devices

M. Sankarapandian (Sankar), Wei-Shang Lo, Alma Vela Ramirez, Sirish Rimal, Stephanie Reynoso, Manasa Medikonda, Chung Ju Yang, Nash Flores, Xuan Liu;
IBM Research, Albany, USA

Taiki Hinode, Henan Zhang;
SCREEN SPE USA, LLC

The Surface Preparation and Cleaning Conference (SPCC), 2024

IBM SCREEN Semiconductor Solutions Co., Ltd.

IBM -SCREEN Presentation

Outline

- **Background**
- **Gate RIE clean defects**
  - **Condensation defects**
    - **Mitigation pathways**
  - **Line Flop-over (LF) defects from pattern collapse**
    - **Defect reduction**
- **Conclusion**

3D Architecture – Nanosheet (NS) Device

- **The Gate length (Lg) shrinks with scaling of MOSFETs**
- **Improving electrostatics is needed for continued scaling**
- **New NS Gate-All-Around device architecture**
  - **Gate wraps around the channel and provides better electrostatic control and leakage**

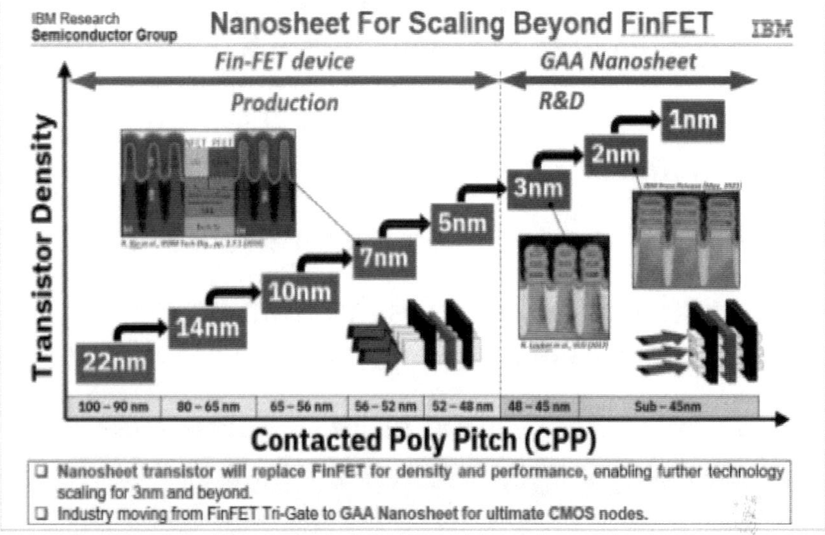

S. Mochizuki IBM NIMS 2023

3

Stacked GAA Nanosheet Process Sequence

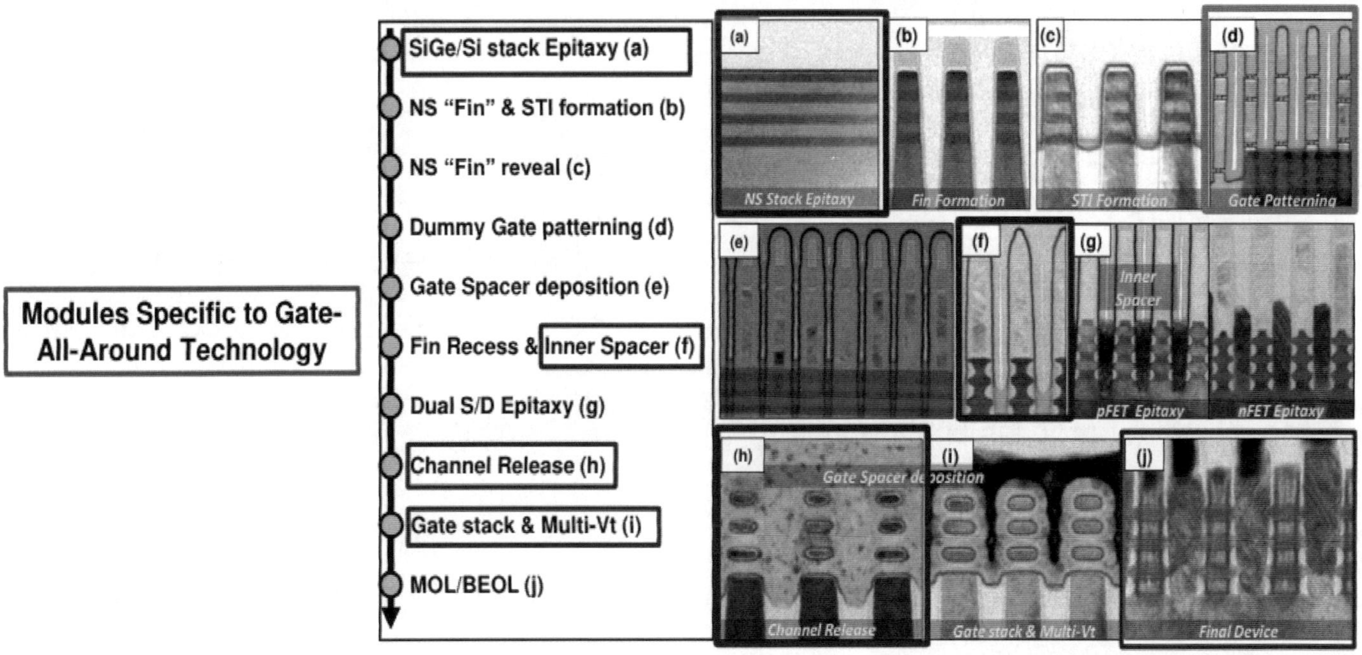

N. Loubet et al., VLSI (2017)

Gate RIE process and Post RIE Cleaning

- **Gate RIE provides a high aspect ratio vertical gate profile with high selectivity to gate oxide**
- **Bromine based etch chemistry used for selectivity**
- **Wet cleaning needed for residue removal**
- **Two major defectivity issues from gate etch residue removal cleaning:**
  1. **Queue time residue build up (condensation defects)**
  2. **Pattern collapse (due to high aspect ratio structures)**

Post Gate RIE Structure

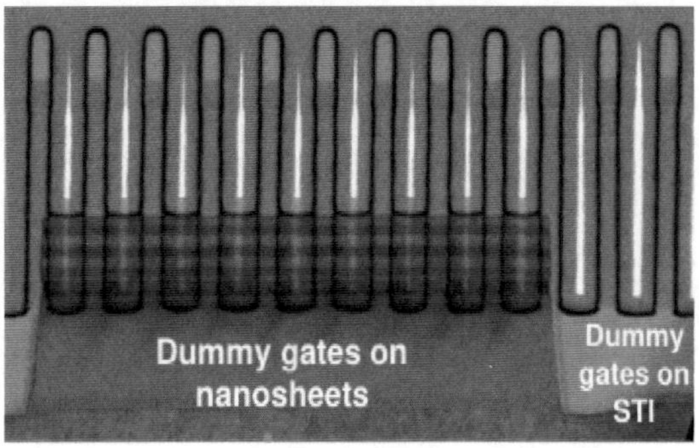

Defectivity from Gate Etch Cleaning – Condensation Defects

- HBr residue outgassing
- Chemisorption of bromine species and bromine terminated surface reaction with atmosphere
- HBr hydrate crystal formation*
- Defect size increases with queue time
- Monitored by process limiting yield (PLY) metrology

Si-O-Br + H2O → **HBr** + SiO_2

*AVS 49th International Symposium, Paper PS-MoM7; HBr Outgassing and Condensation from Silicon and Polysilicon Wafers after Plasma Etching
HBr Silicon etching process : US Patent US6358859B1
FOUP Contamination and Limitation of Cleaning Procedure IEEE ASMC 2021

Gate Structure Collapse During RIE Wet Cleaning and Drying

- Pattern collapse (Line flop-over (LF)) defects from drying
  - Surface tension induced capillary force causing the collapse
  - IPA Exchange with DIW for drying to reduce the surface tension
  - Additional drying methods to improve pattern collapse prevention
    - Surface modification drying
    - SCCO2 drying

Top-down SEM Image for LF defect

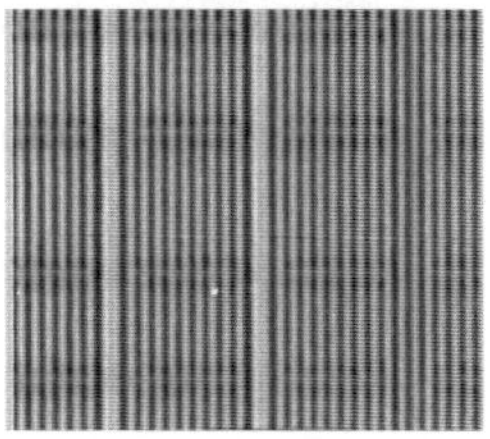

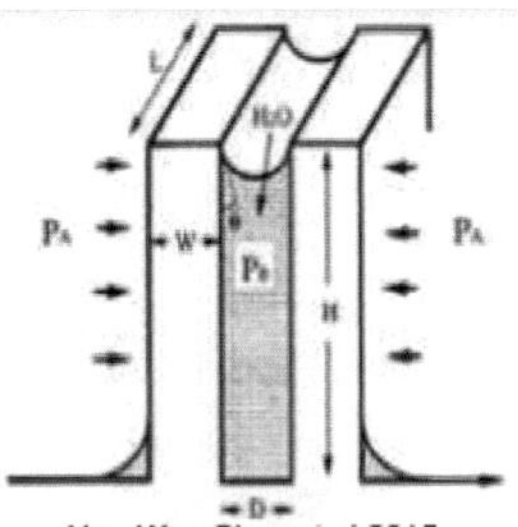

Han-Wen Chen et al 2015
ECS Trans. 69 119

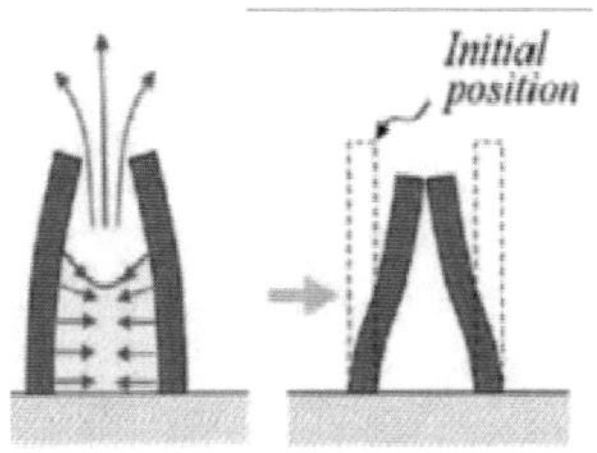

Yanming Wang et al; J. Vac. Sci. Technol. B **36**, 051602 (2018)

$$F = \frac{2\sigma H cos\theta}{w}$$

Condensation Defects Study – Queue time effect

- Condensation residues continue to grow with queue time even after wet clean
- XPS studies indicated intense Br signal post RIE
- Residual Br signal is modulated by RIE conditions, but it could not be completely eliminated
- Wet cleans showed no Br signal even with RIE-1 condition
- FOUP contamination and re-adsorption of HBr species on the wafer suspected

Br signal by XPS analysis of wafer surface

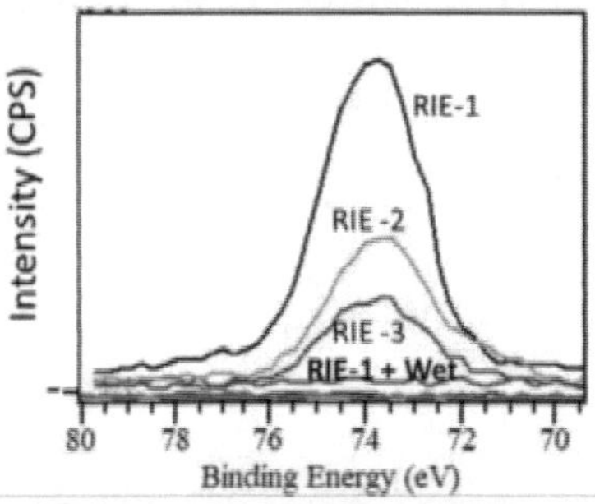

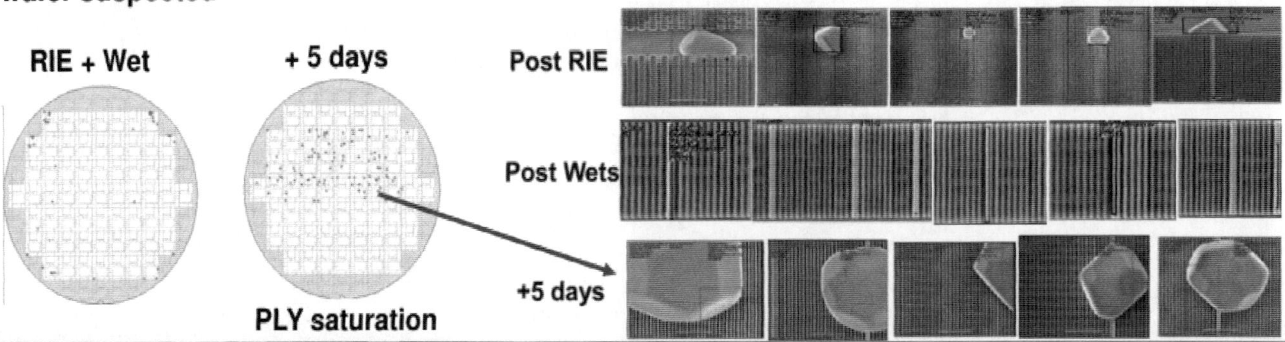

IBM SCREEN

N2 FOUP Use for Preventing Condensation Defects

- **Using N2 purged FOUPs prevented condensation residue growth**
- **No saturation of PLY defects / condensation observed after 1 week**
  - **No further increase in LF defects**

Wafers in N2 purge FOUP → RIE and N2 purge → Wets and N2 Purge → PLY

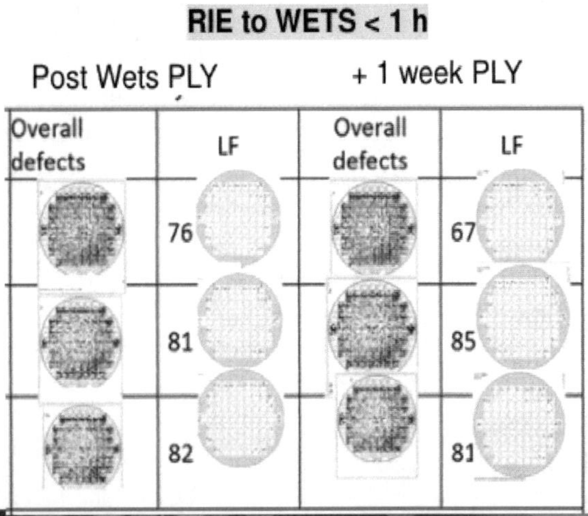

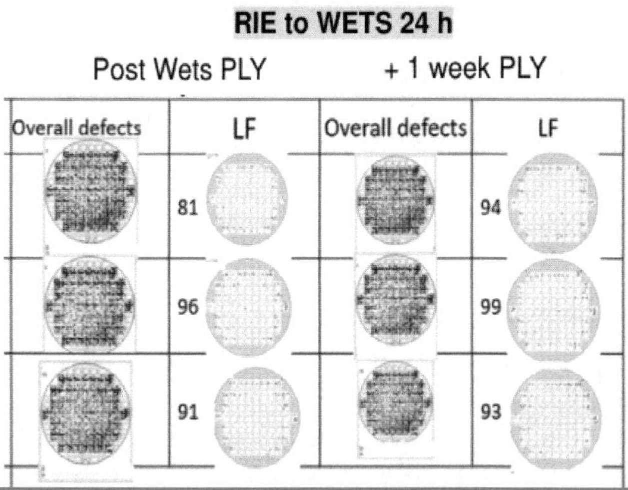

IBM SCREEN

FOUP Change Post Wets Processing

- **Post wets processing, wafers were loaded back to new FOUP**
- **No condensation defects and additional LF defects seen in new FOUP with queue time**

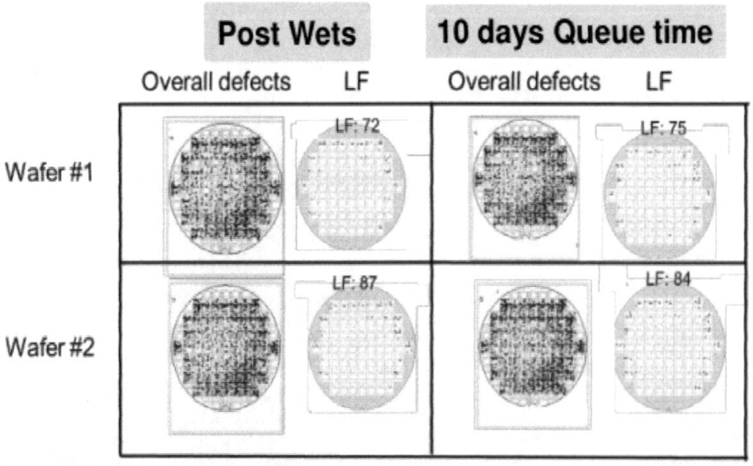

Line Flop-over Defects from Pattern Collapse

IBM SCREEN

LF Defects from Post Gate RIE Cleaning

- Litho shrink and aggressive etch were performed to obtain AR of 13-17:1
- With RT IPA as the drying solvent, the LF defects observed
  - High AR structures on STI (dummy gate structures)
  - STI erosion on either side of the gate structure due to HF based cleaning sequence
  - Hydrophobic last or hydrophilic last cleaning did not show a difference

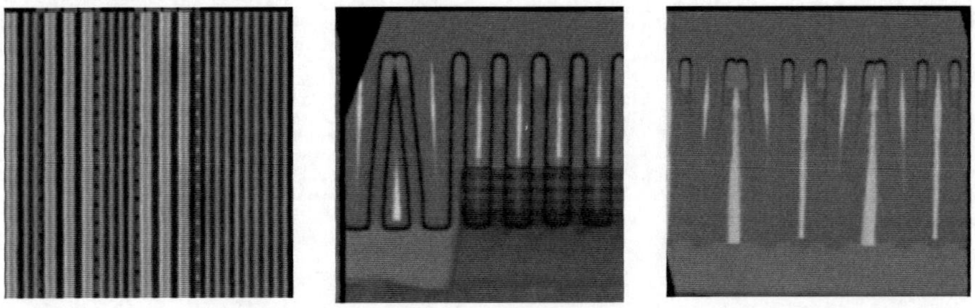

Line Flop-over Defects from Pattern Collapse

- Surface modification (SM) drying and Sublimation drying techniques were evaluated for preventing LF defects and results compared with IPA drying
- Surface modification was used to modify the surface for reducing the Laplace force
  - **Chemical sequence → DIW Rinse → IPA Rinse → Surface modification → IPA rinse**
- Novel sublimation drying was also employed to circumvent liquid surface tension effects
  - **Chemical sequence → DIW rinse → IPA rinse → sublimation chemistry → Dry**

$$F = \frac{2\sigma H \cos\theta}{w}$$

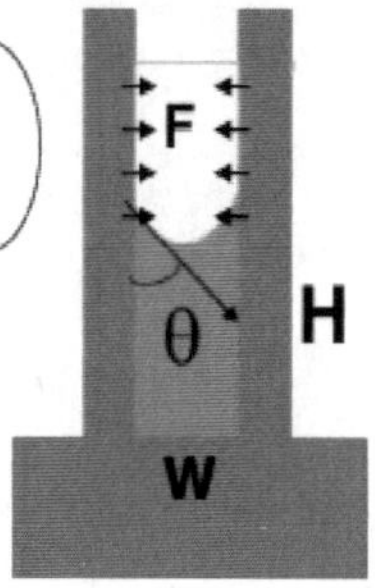

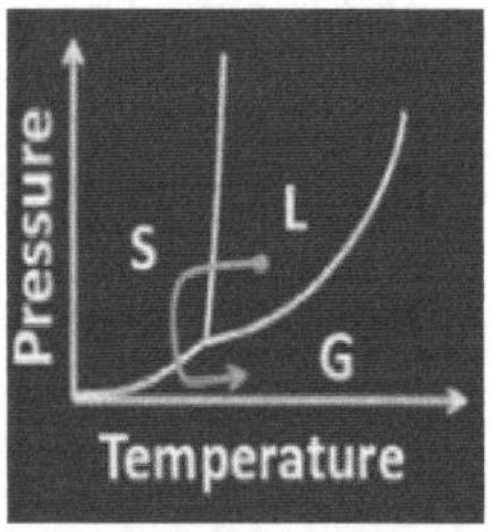

Advanced Drying Methods for LF Defects Reduction

- **Optimized SM drying provided improved performance over IPA drying**
- **Sublimation drying performed superior with very few LF defects**
- **Further optimization of sublimation drying could provide zero LF defects**

LF Defect count as a function of drying splits

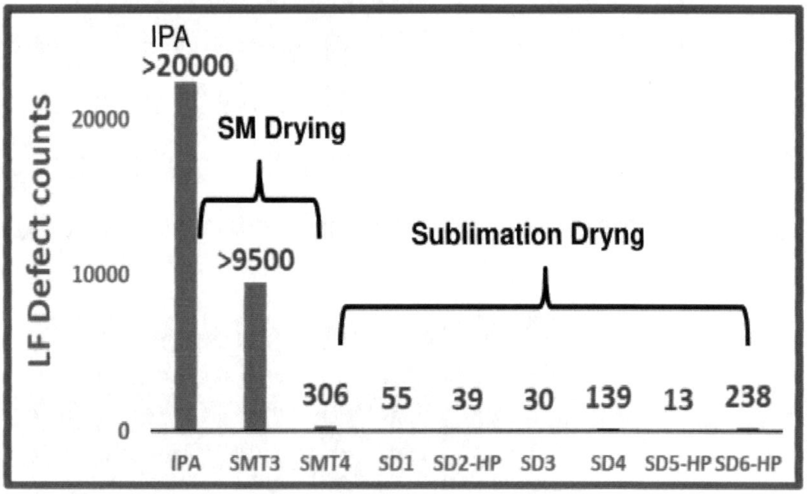

Wet Clean Processes Drying Roadmap

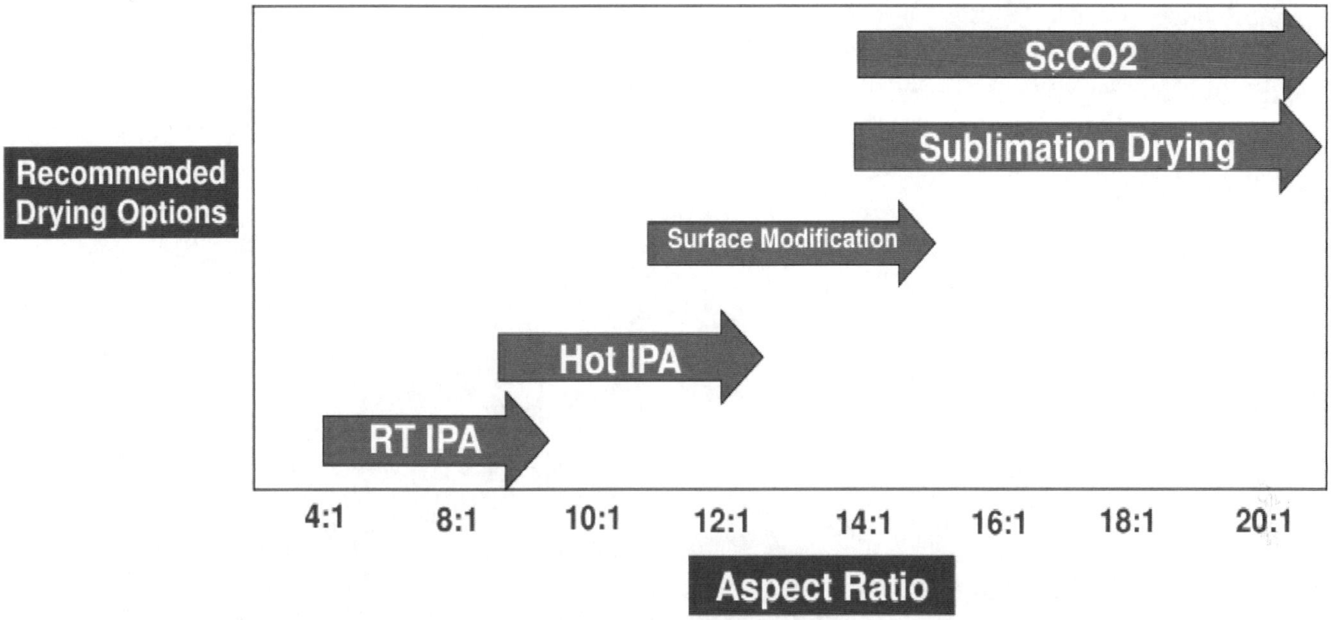

Conclusion

- Two predominant defects from post gate RIE clean identified to be condensation defects and pattern collapse
- HBr residues suspected causing condensation defects with queue time post RIE and post wets
- FOUP change and N_2 purging of FOUPs mitigated the defect formation
- Line flop-over defects arising from surface tension induced drying were addressed by SM drying and Sublimation drying
- Sublimation drying is very promising as it drastically reduced the LF defects

TEL
TOKYO ELECTRON

Superiority of Bevel Nozzle Dispense Spin over Backside Center Dispense Spin for Wafer Bevel Cleaning: Insights from Computational Fluid Dynamics

April 25, 2024

Rencheng Dong[1], Akira Fujita[2], Jun Nogami[3], Derek Bassett[1], Ihsan Simms[1], Antonio Rotondaro[1], Joel Barnett[1]
[1] Tokyo Electron America, Inc.,
[2] Tokyo Electron Taiwan Limited
[3] Tokyo Electron Kyushu Limited

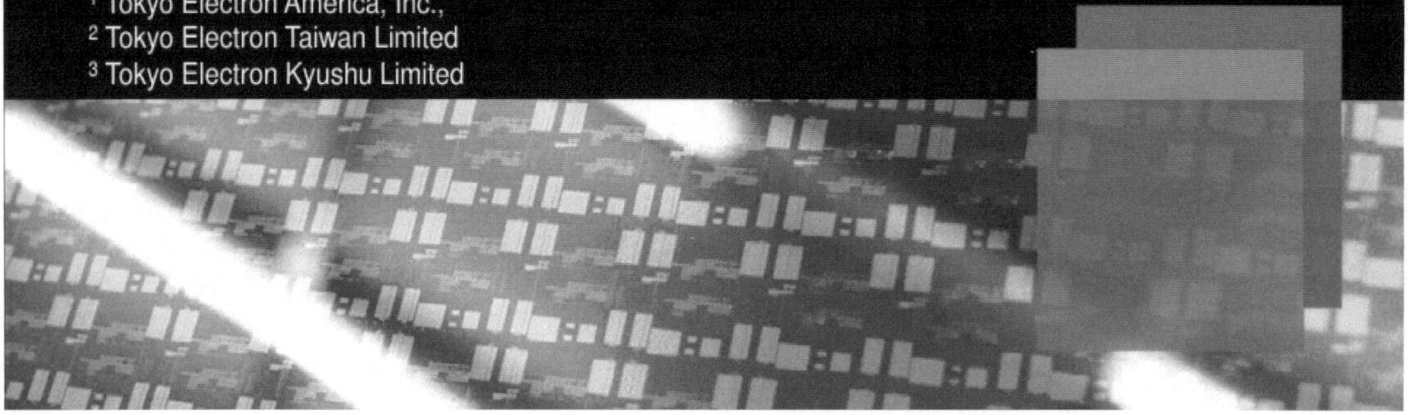

Outline

- Introduction
  - Edge yield loss
  - Wafer bevel cleaning
- Backside Center Dispense Spin (BCS) Technique
  - Liquid wrap-around flow
  - Non-uniform cut width profile
  - Etch difficulties on hydrophobic wafer
- Bevel Nozzle Dispense Spin (BVS) Technique
  - Stable circular liquid flow
  - Uniform cut width profile
- Conclusions

Introduction: Edge Yield Loss & Wafer Bevel Cleaning

Why do We Need Wafer Bevel Cleaning?

- Defects originating from wafer edge can lead to substantial yield loss.
  - Bevel film flaking/peeling may occur due to weak adhesion or thermal stress during annealing process.
  - Metal particles may form at wafer bevel region during dry etching process.
  - Detachable edge films/particles may move to the active area of wafer frontside and cause yield loss.

Failure modes of edge yield loss

Source: TEL

Wafer Bevel Cleaning

- The goal of wafer bevel cleaning is to remove undesired bevel films and particles to eliminate sources of defectivity.

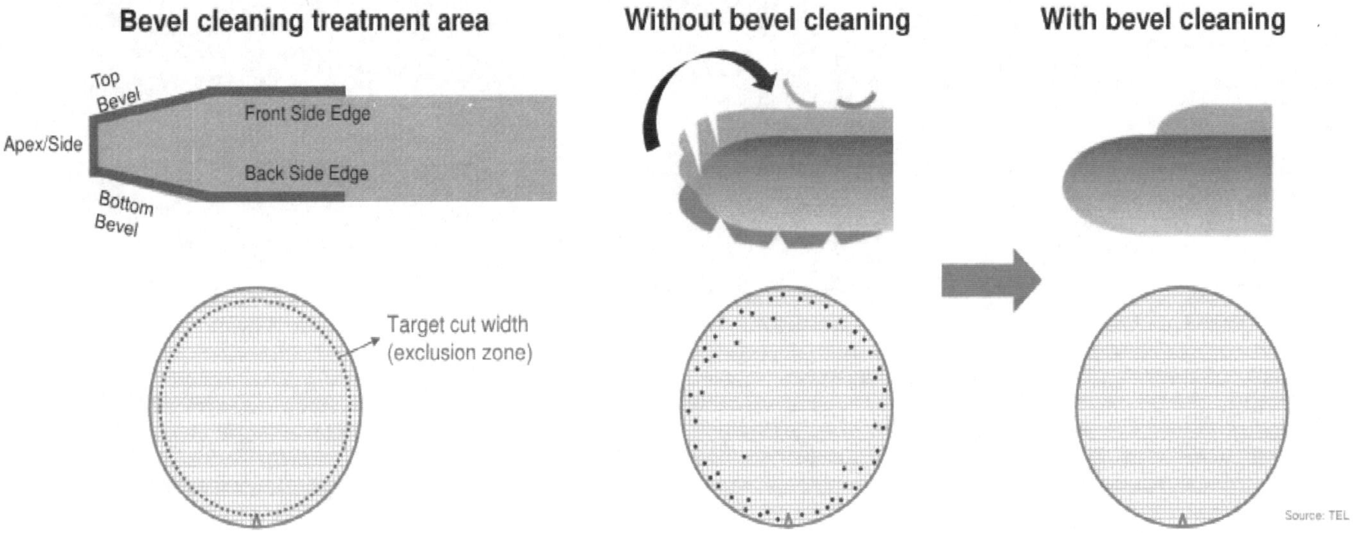

Overview of Bevel Cleaning Techniques

- Plasma Dry Etch
  - The dry etch process is to only expose wafer bevel to plasma to only etch bevel films.
  - Challenges
    - Arcing risk for metal films
    - Low etching selectivity
- Wet Etch
  - The wet etch process is to dispense chemical liquid on bevel region to etch and clean bevel films and particles.
  - Two main techniques:
    - Backside Center dispense Spin (BCS)
    - Bevel nozzle dispense Spin (BVS)

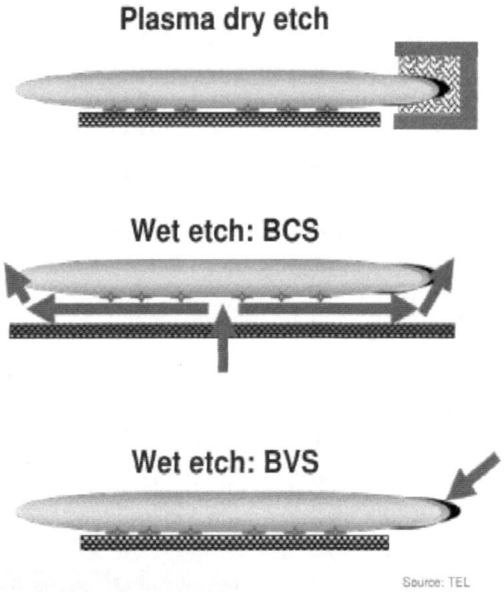

Goal: Investigate flow dynamics effects on cut width accuracy in BCS and BVS

Backside Center Dispense Spin (BCS) Technique

Backside Center Dispense Spin (BCS) Technique

- Conventionally, backside center dispense spin (BCS) is used to etch the backside films and bevel films simultaneously.
  - Liquid is dispensed at the center of the wafer backside and the liquid flows from the backside center to the backside edge due to wafer rotation.

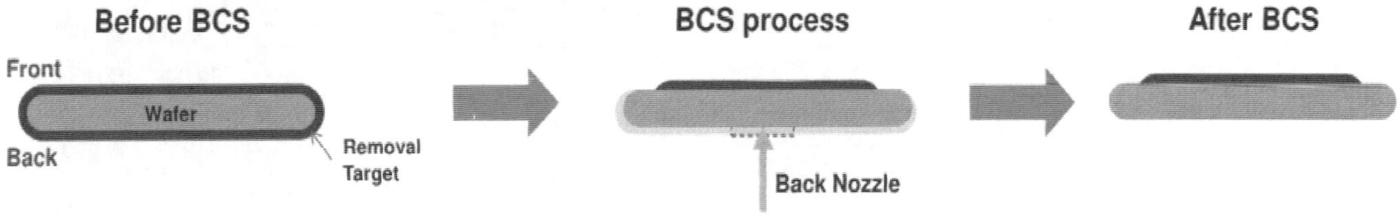

Source: TEL

Computational Fluid Dynamics (CFD) Model for BCS Process

- An axisymmetric BCS model has been developed to simulate liquid flow around the wafer edge.
  - Air-liquid two-phase flow model based on the volume of fluid (VOF) method.
  - Model considered gravity, wafer rotation, capillary force and wafer wettability.

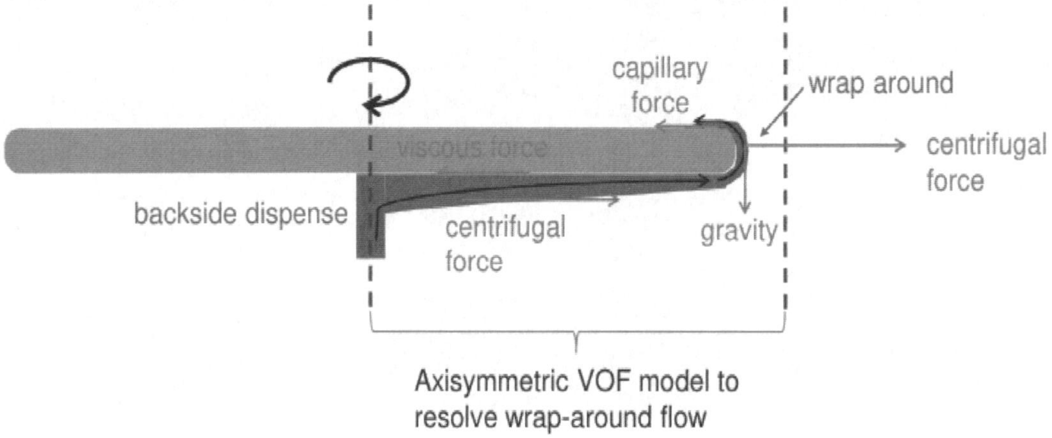

Source: TEL

Liquid Wrap-around Flow in BCS Process

- The etching on the top part of the edge relies on the liquid wrap-around flow.

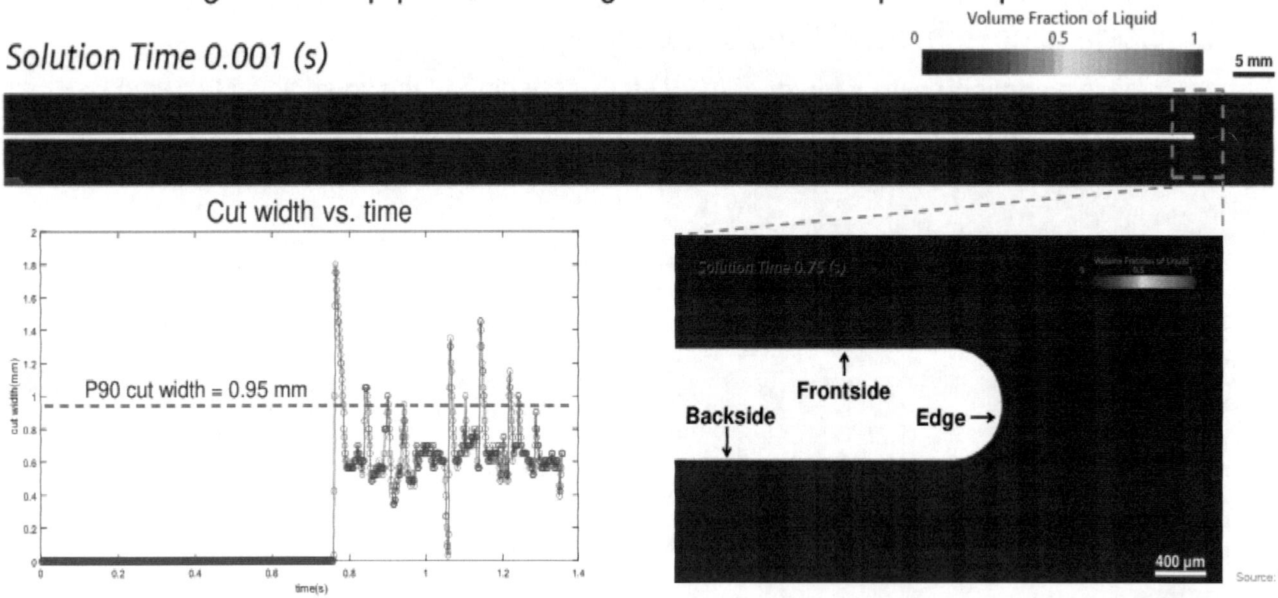

Source: TEL

Volatile Wrap-around Flow

- Liquid flow in BCS process periodically cycles.
  - The liquid first climbs from wafer backside to wafer frontside due to inertia effect and capillary force.
  - The liquid keeps accumulating around the wafer edge.
  - The extra liquid will be stripped from the liquid film when the liquid volume at wafer bevel is greater than a certain threshold.
- The leading edge of the liquid fluctuates greatly during the BCS process, which results in the less uniform etching profile.

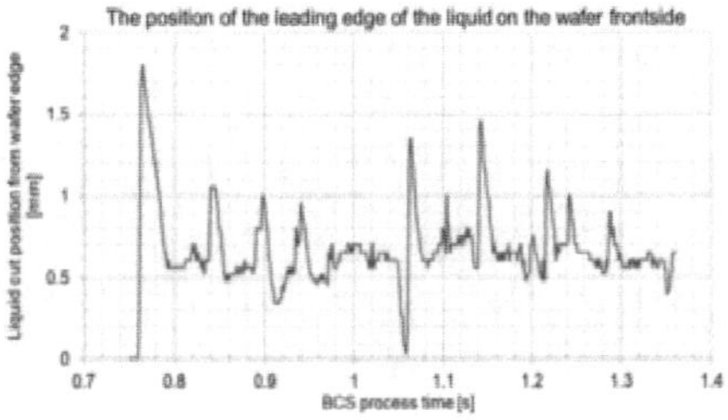

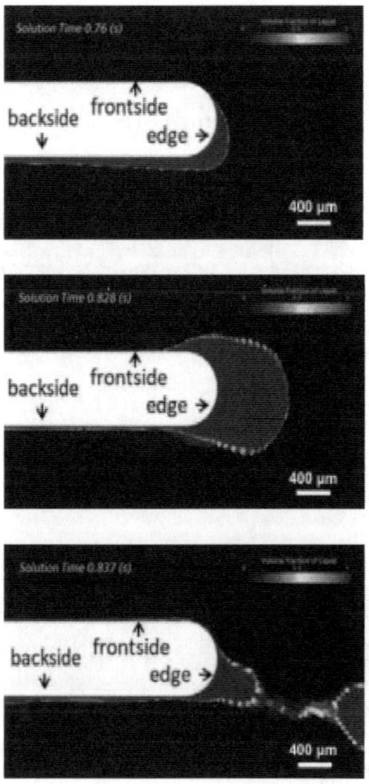

Source: TEL

Non-uniform Cut Width Profile in BCS

- The cut width after BCS process can be quite non-uniform over the wafer perimeter.
  - It can lead to yield loss on the wafer edge.

Typical cut width curve after a BCS process

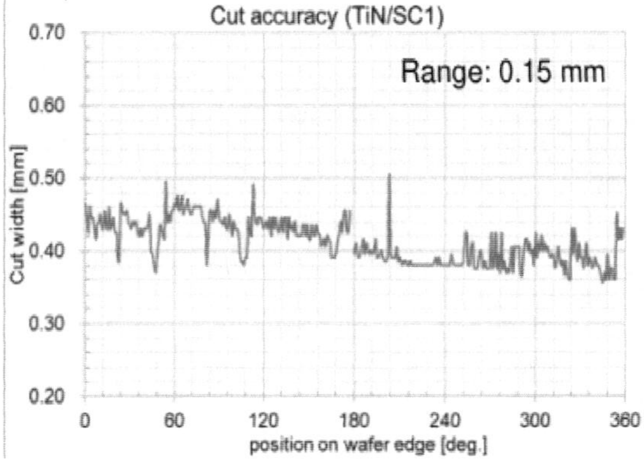

Non-uniform cut width profile illustration over the wafer perimeter

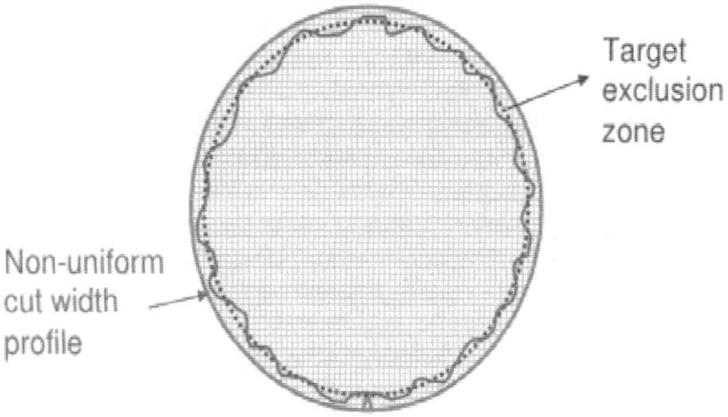

Source: TEL

BCS Etch Difficulties on Hydrophobic Wafer

- Chemical liquid barely etches the bevel frontside.
  - Liquid film breaks up on the hydrophobic wafer surface (contact angle = 70 deg.).

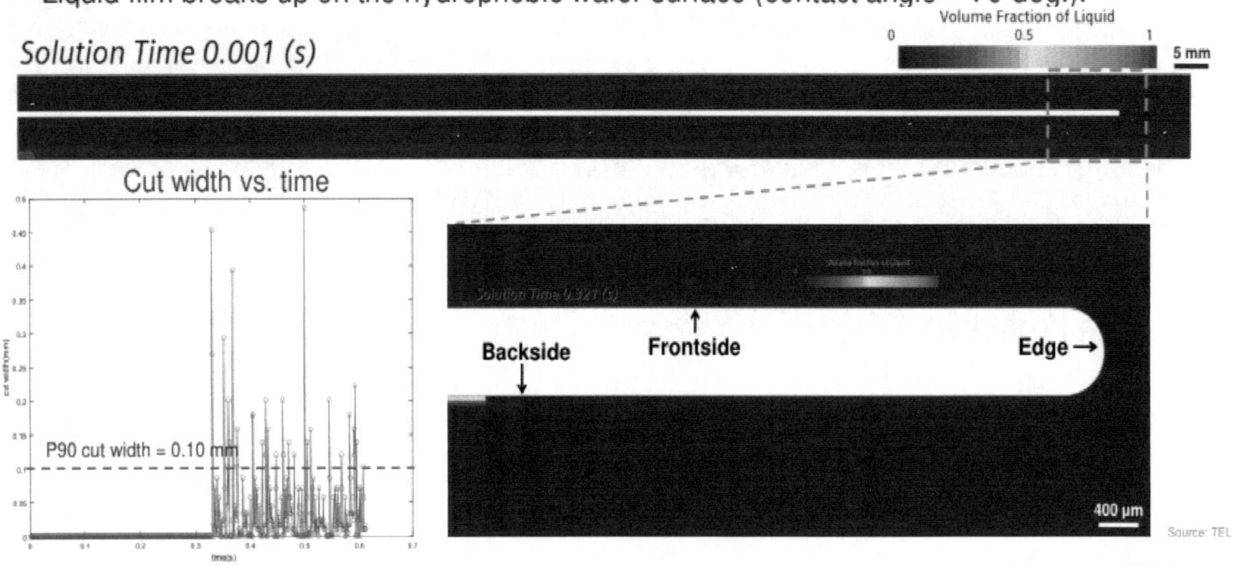

Bevel Nozzle Dispense Spin (BVS) Technique

Bevel Nozzle Dispense Spin (BVS) Technique

- In the BVS process, the chemical liquid is dispensed from a bevel nozzle to the wafer frontside near the edge.
  - The goal of BVS process is to etch and clean only the bevel films and particles.

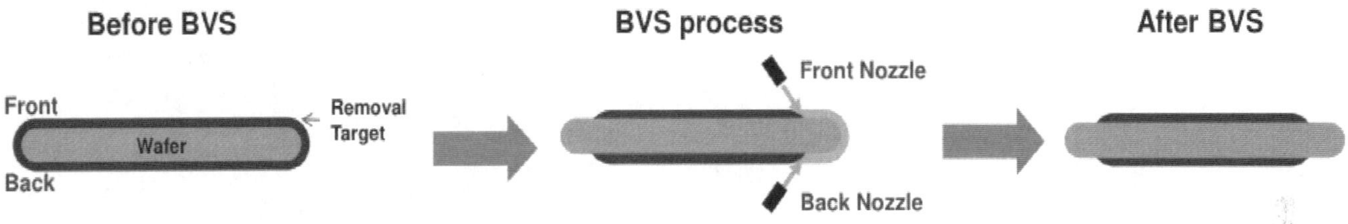

Source: TEL

Computational Fluid Dynamics (CFD) Model for BVS Process

- A 3D BVS model has been developed to simulate the liquid flow on the wafer surface.
  - Air-liquid two-phase flow model based on the volume of fluid (VOF) method.
  - Model considered wafer rotation, surface tension and wafer wettability.

Illustration of BVS model

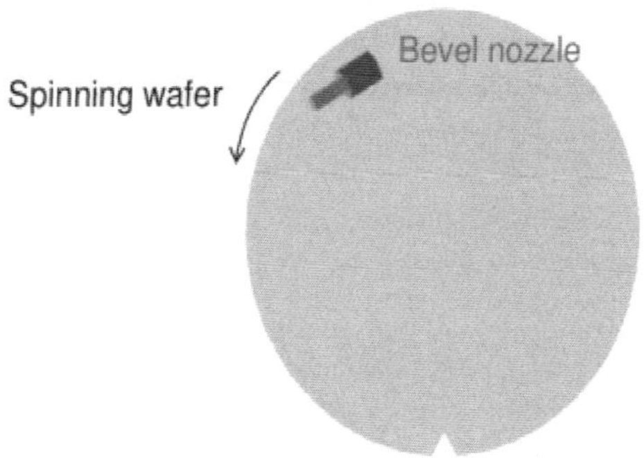

Stable Circular Liquid Flow in BVS Process

- Liquid spreads and develops a stable circular flow along the wafer edge.
  - The inner cut position of the liquid film dictates the cut width of the BVS process.

Liquid film profile in BVS (vertical cross section)

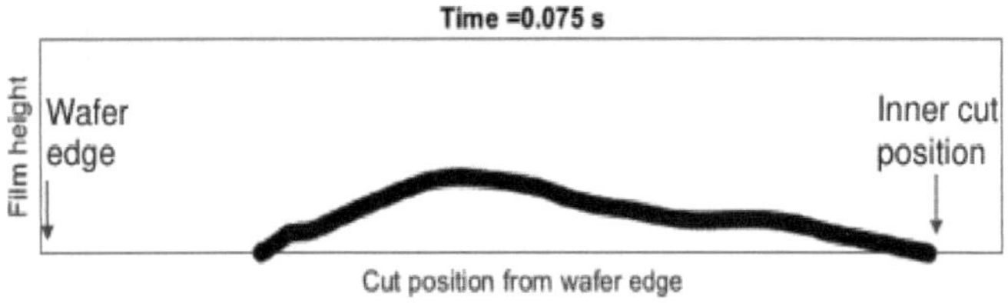

BVS simulation

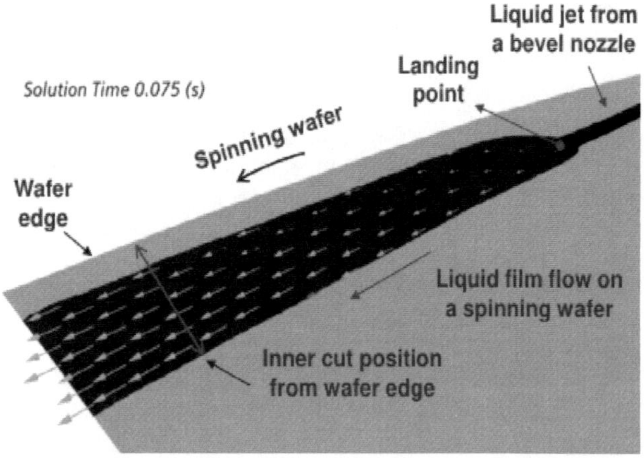

Source: TEL

Uniform Cut Width Profile in BVS

- BVS technique offers a more uniform etching profile than the BCS technique because the circular liquid flow in BVS is more stable than the liquid wrap-around flow of the BCS process.

Typical cut width curve after a BVS process

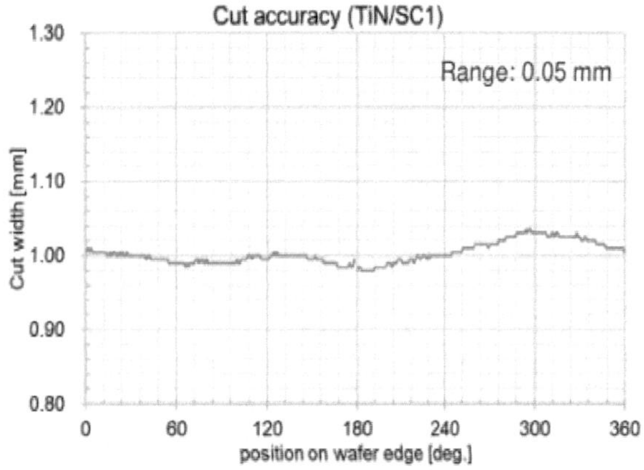

Source: TEL

Uniform cut width profile illustration over the wafer perimeter

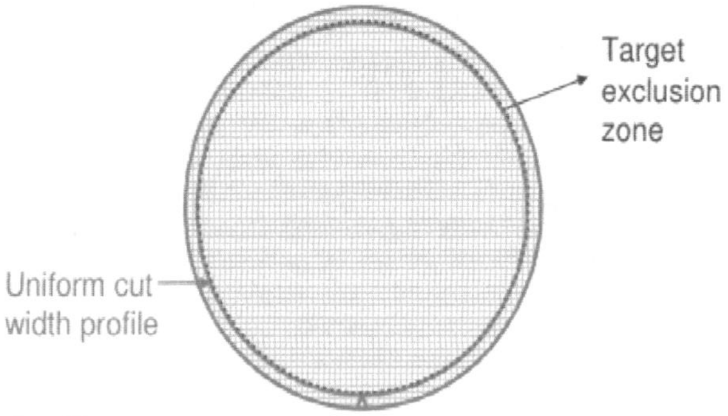

Source: TEL

Conclusions

- Backside Center Dispense Spin (BCS)
  - Bevel etching relies on the liquid wrap-around flow at the wafer edge.
  - The cut width profile after BCS is less uniform because the wrap-around flow is volatile and periodic.
  - Bevel etching on the hydrophobic wafer can be difficult because the wrap-around flow may break up.
- Bevel Nozzle Dispense Spin (BVS)
  - The liquid develops a stable circular flow along the wafer edge.
  - The cut width profile after BVS is more uniform due to the stability of liquid flow in BVS.

Typical cut width curve after a BCS process

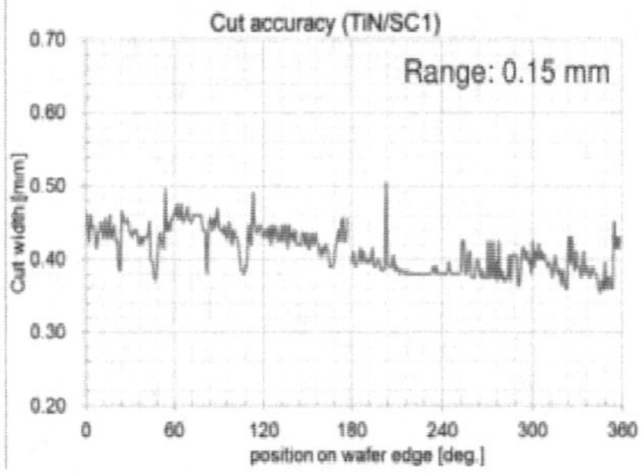

Typical cut width curve after a BVS process

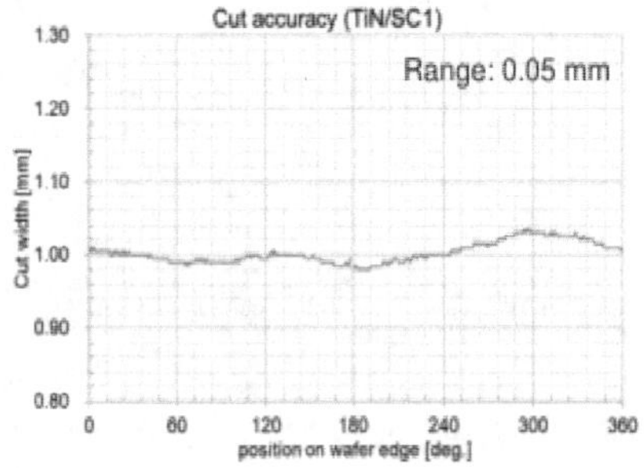

Source: TEL

Notice

You may not copy or disclose to any third party without prior written consent with TEL.

Tokyo Electron

TOKYO ELECTRON

BVS Process on Hydrophobic Wafer Surface

- Uniform cut width profile can be also achieved on hydrophobic wafer surface by using BVS technique.

Typical cut width curve after a BVS process on hydrophobic wafer surface

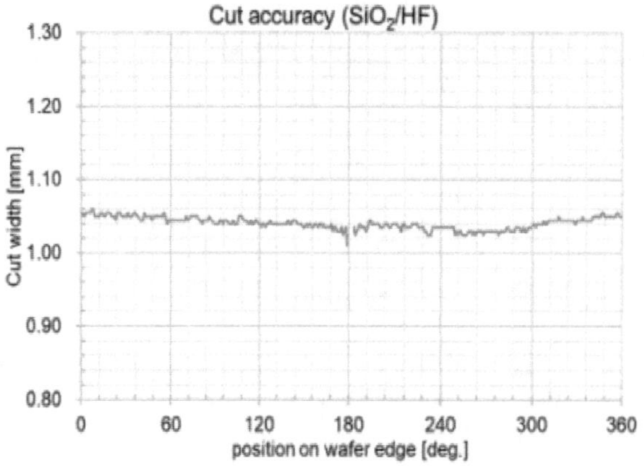

Source: TEL

Togethers SAMSUNG

Investigation of the water scatter in single cleaning process by deriving a splash parameter

'24.04.25 Samsung Electronics | **Seunghyeon Baek**

Samsung Electronics Memory Cleaning Technology Team Confidential

Togethers SAMSUNG

Agenda.

01 Introduction - Memory Cleaning Process

02 Literature Survey – Splash Parameter

03 Results – Reduction of Defect

04 Summary

Samsung Electronics Memory Cleaning Technology Team Seunghyeon Baek Confidential

01 Introduction

Togethers SAMSUNG

Change of Cleaning Equipment Type : Batch Type → Single type

「 Batch Type 」

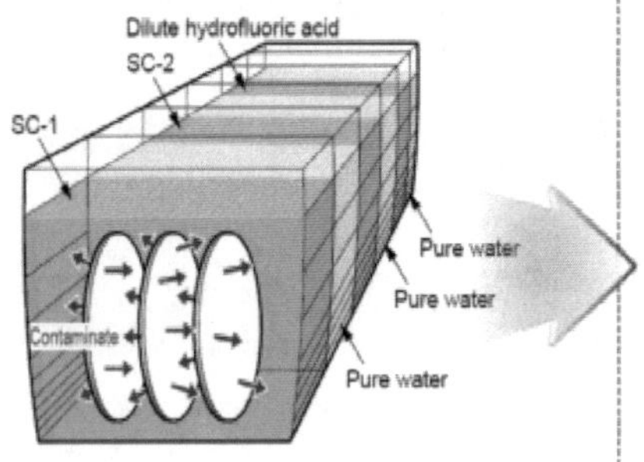

☐ Advantages

1) High Throughput
2) Low chemical consumption

☐ Disadvantages

1) Footprint ↑
2) Risk of particle (by bath flow)

「 Single Type 」

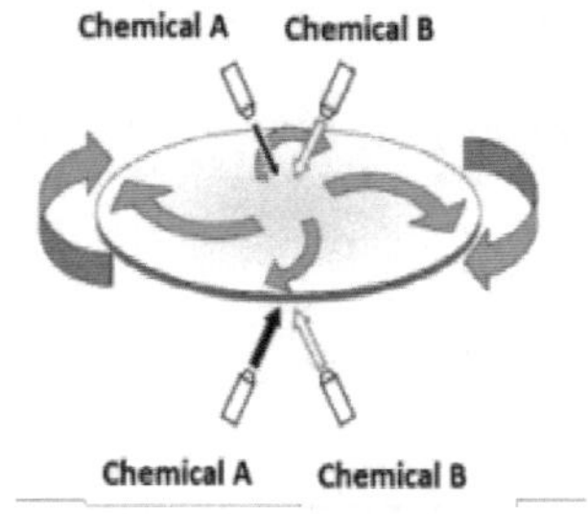

☐ Advantages

1) Resistant to particle contamination
2) Uniformity ↑ (compared to batch type)

☐ Disadvantages

1) Low throughput
2) Risk of drying issue ↑

● Global Wafer Cleaning Equipment Market

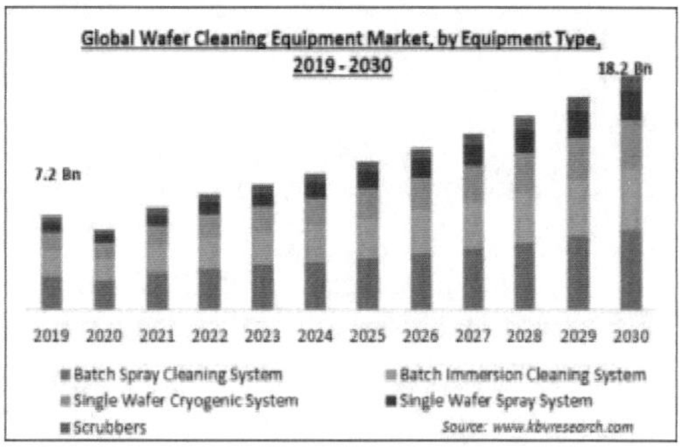

- The proportion of single type is gradually increasing

☞ Reducing problems with single type equipment is important

Samsung Electronics Memory Cleaning Technology Team 3 Seunghyeon Baek Confidential

01 Introduction

Togethers SAMSUNG

☐ Single type equipment property

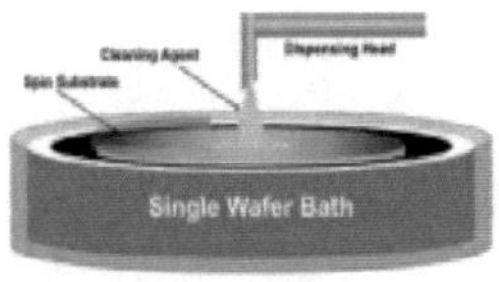

① Wafer contacts with chemical liquid with high energy (high flow rate & distance between wafer and nozzle)

② High rotating speed of wafer in single type equipment (turbulent flow is created by centrifugal force)

➔ Liquid scattering from rotating wafer and chemical ↑

☐ Defects caused by scattering

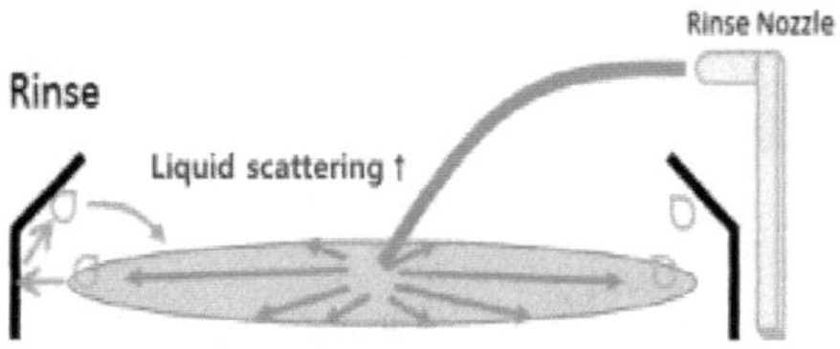

Rebounding of water droplet by Rinse Nozzle

【 Type of Defects 】

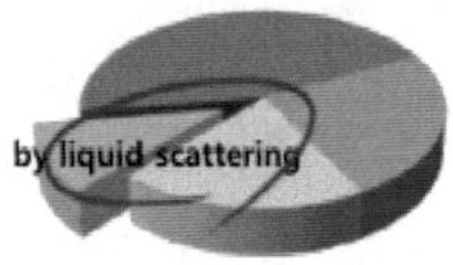

If scattering occurs, defects may occur due to ① rebounding with fume or particles and ② drying issue

➔ Particle contamination due to scattering has begun to affect semiconductor yields.

☞ It is needed to prevent the liquid scattering by analyze cause of scattering

Samsung Electronics Memory Cleaning Technology Team 4 Seunghyeon Baek Confidential

02 Literature Survey

Togethers SAMSUNG

Liquid Splash : A phenomenon in which a **liquid splits into numerous irregular droplets** after colliding with a surface

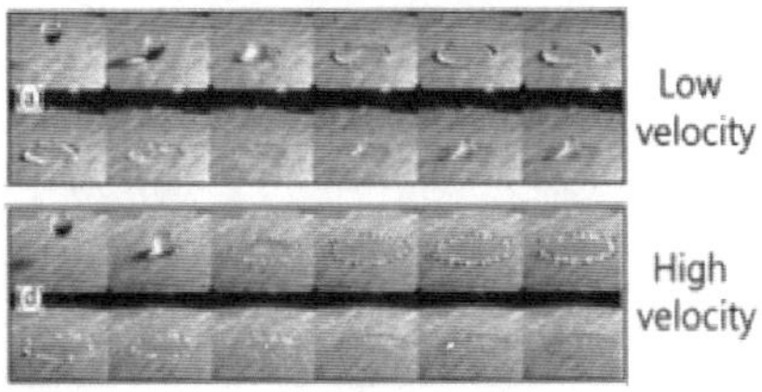

Langmuir, 2010, 26, 12, 9516-9523

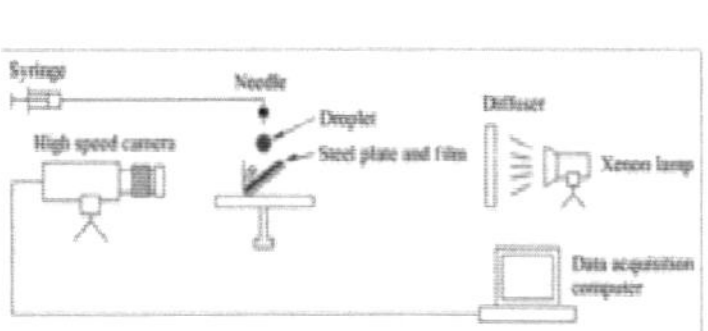

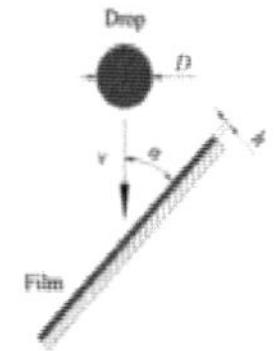

Acta Mechanica, 2013, 224, 12, 2993-3004

【 Factors causing splash 】

From liquid	Surface tension
	Viscosity
	Kinetic energy (flow rate)
From surface	Roughness
	Surface energy (Wettability)

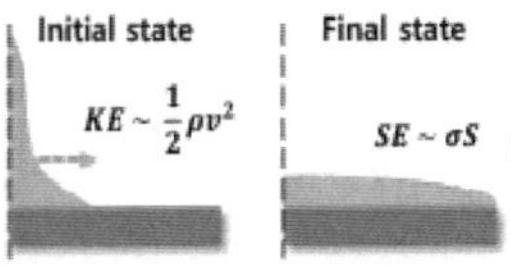

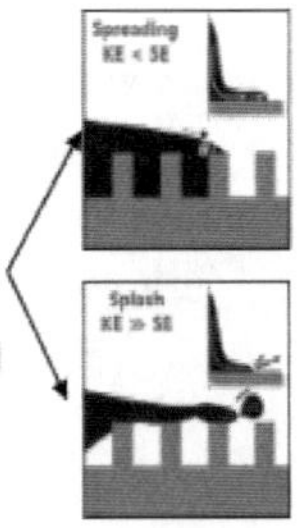

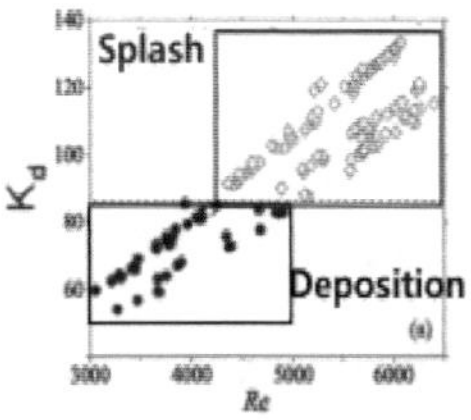

Splash $\propto Re(=\frac{\rho v l}{\mu})$

Velocity of collision ↑

→ water scattering ↑

Langmuir, 2010, 26, 12, 9516-9523

The impact of collision velocity on splash including other factors is needed for analysis

02 Literature Survey

Togethers SAMSUNG

Energy balance : Surface tension vs Kinetic energy

【 Surface tension 】

*Surface tension is the tendency of liquid surface to shrink into the **minimum surface area** possible*

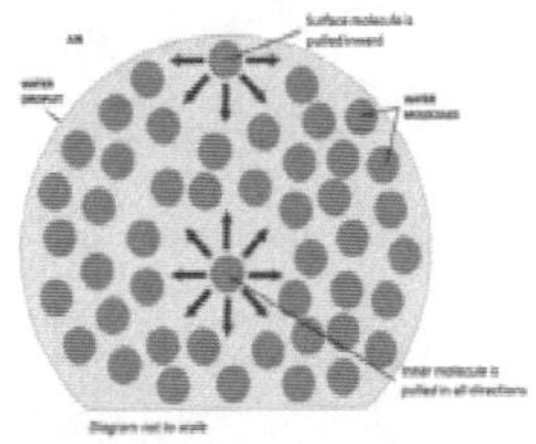

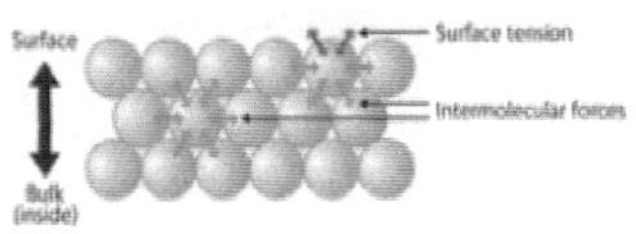

- Surface tension ↑

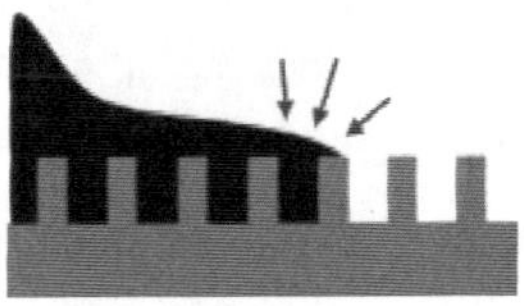

☞ **Surface tension acts as an intermolecular force and acts as a force resisting scattering.**

【 Kinetic energy 】

*Kinetic energy of an object is the energy that it possesses due to its **motion***

In fluid dynamics

▷ **Factors affecting kinetic energy**

① **liquid Flow rate**

② **Diameter of nozzle** → $v = Q/A$

③ **Height of nozzle** → $g = 9.8\,m/s^2$

- Kinetic energy ↑

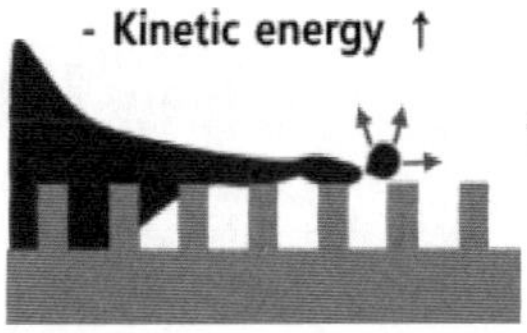

☞ **The greater the flow rate, the smaller the nozzle radius, and the higher the nozzle height, <u>the greater the KE</u>.**

02 Literature Survey

Dimensionless number : To compare the energy balance, **dimensionless number** is introduced (*We* & *Oh*)

Factors causing splash : ① Surface tension & ② Kinetic energy & ③ Viscous dissipation

【 *Weber* number 】

It is a dimensionless number representing **the ratio of surface tension and fluid inertia**

$$We = \frac{\rho v^2 D}{\sigma} \approx \frac{inertia}{surface\ tension}$$

ρ: Density v: Velocity of liquid D: Diameter σ: Surface tension

a) $We > 1$: → Effect of inertia **velocity ↑**

b) $We < 1$: → Effect of surface tension **velocity ↓**

【 *Ohnesorge* number 】

It is a dimensionless number representing **the ratio of inertia, surface tension and viscous force**

$$Oh = \frac{\mu}{\sqrt{\rho\sigma D}} \sim \frac{viscous\ force}{\sqrt{inertia \times surface\ tesnion}}$$

ρ: Density μ: Viscosity D: Diameter σ: Surface tension

$$Oh = \frac{\mu}{\sqrt{\rho\sigma D}} = 0.001885$$

In case of water droplet

= Effect of viscosity ↓ ☞ Viscous dissipation ↓

02 Literature Survey

Splash parameter : Parameter that check **whether splash occurs** by combining two dimensionless numbers (*We*, *Oh*)

$K = We \times Oh^{-0.4}$ ⟶ In case $K > K_{critical}$, splash occurs

Splash X

Velocity ↓

$K < K_{cri}$

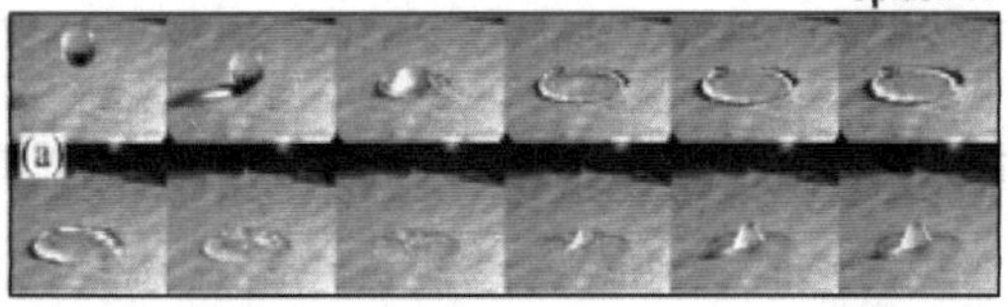

Splash O

Velocity ↑

$K < K_{cri}$

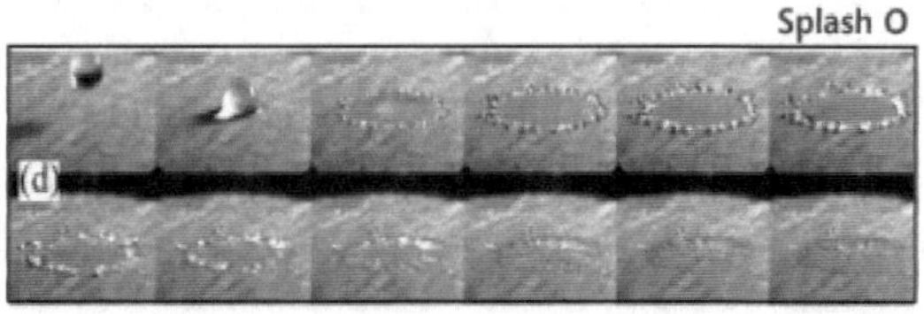

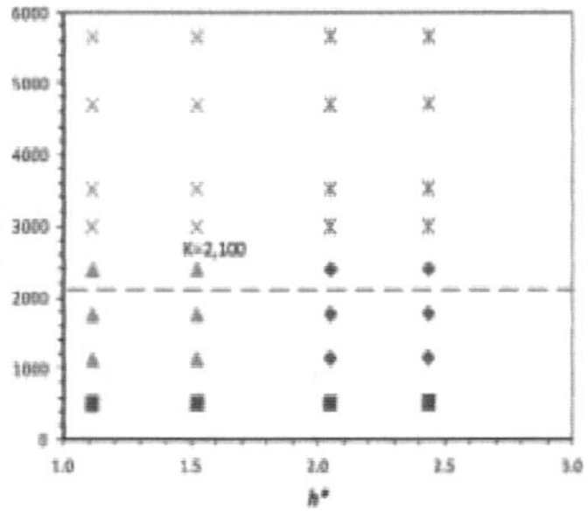

Exp Fluids (2011), 50, 1305-1316

$$K_{cri} = 2100 + 5880\delta^{1.44}$$

$$K_{cri} \sim 2100$$

($\delta \ll 1$ or in case of water droplet)

- ✓ In order to reduce scattering, the splash parameter in the current process conditions must be reduced.
- ✓ We need to analyze the factors that affect the splash parameter and reduce the splash parameter to **less than 2,100.**

03 Result

Togethers SAMSUNG

Splash parameter : Investigation of current process condition

□ Investigation of Splash parameter

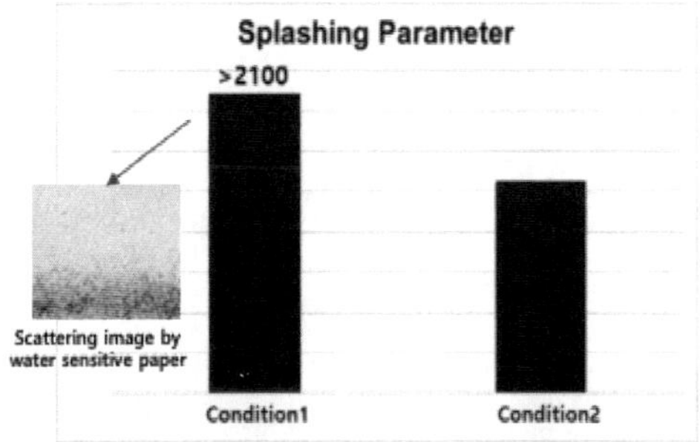

- ▶ The splash parameter value of condition1 is over 2,100, so it is vulnerable to scattering.
- ▶ The actual degree of scattering was measured using water sensitive paper, it was observed that more scattering occurred in condition1.
- ▶ Comparing two process conditions, there was a difference in the collision velocity of the fluid, which is an easily controllable factor.

□ Effect of impact velocity

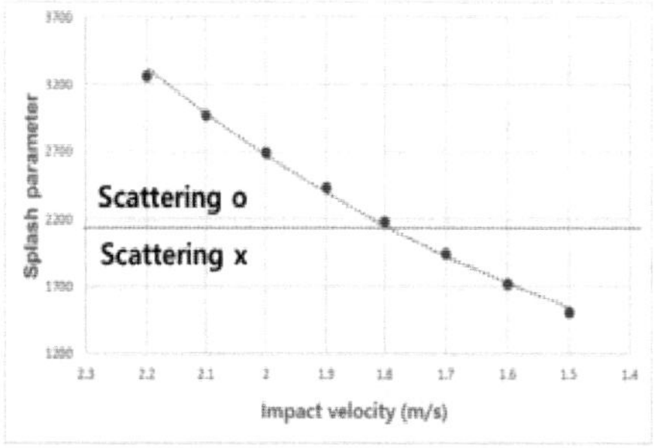

- ▶ Based on *Oh* and *We*, splash parameter was theoretically analyzed whether scattering occurred or not
- ▶ From the theoretical analysis, the splash does not occur below a collision velocity of 1.78 m/s.
- ▶ To suppress scattering, there is a need to reduce the collision velocity.

03 Result

Togethers SAMSUNG

Solution – analysis of rotating speed : In order to reduce water scattering, rotating speed should be reduced

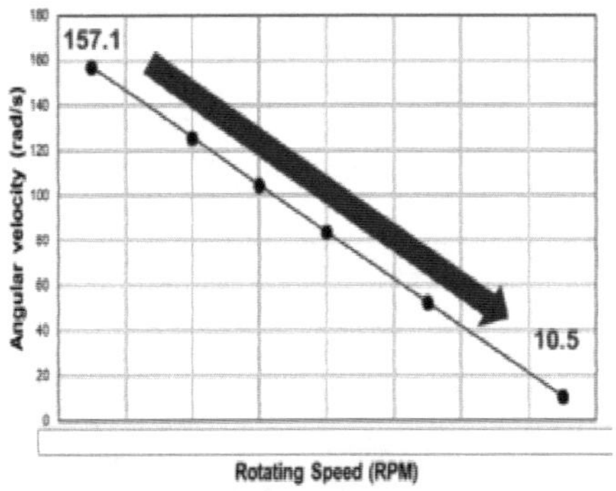

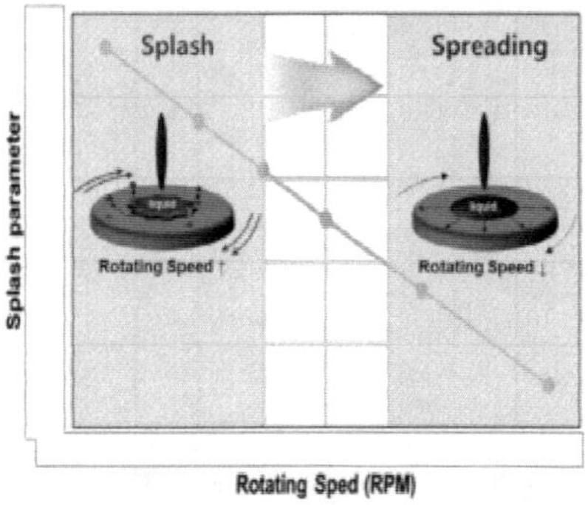

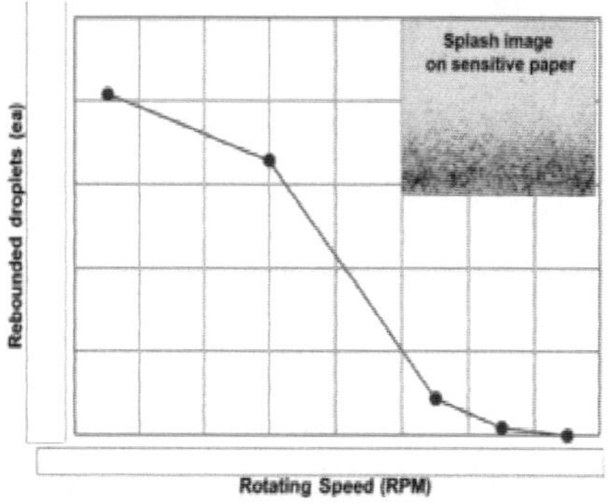

- ▶ As the rotating speed is gradually reduced, the angular velocity is also reduced from 157.1 to 10.5 rad/s. It leads the reduced impact velocity from 2.2 to 2.0 m/s.
- ▶ While the rotating speed decreases, K decreased. Based on the theoretical analysis, it is expected that the rebounded droplets gradually decreased.
- ▶ As a result of measuring the number of color-changed points by water, more than thousands of droplets were detected under existing condition, but there was no droplet detected when the rotating speed was reduced.

04 Summary

► The splash of liquid flow from distant nozzle was investigated to reduce a radial defect.

► The splash parameter was introduced to determine how well splash occurs. It is consist of two dimensionless number, *We* and *Oh*.

► The rotating speed was regulated for control a splash. As decreasing of rotating speed, the splash parameter also decreased gradually.

► The number of rebounded droplet diminished along with decreasing rotating speed.

► It is expected to prevent a scattering issue that may occur in semiconductor cleaning and reduce economic losses.

'24.04.24 (Wed)

SAMSUNG

A Study on the Causes of Carbonyl Sulfide Residue by Etching and Cleaning Process Interaction

Youngeun Kim
Samsung Electronics
young47.kim@samsung.com
+82.10.9580.5537

Youngeun Kim, Ki Dong Yang, Jiyeon Lee, Tae Kyung Lee, Joonho H. Lee, Duwon Lee, Yu Sung Lee, Dongwan Seo, Eunyoung Han, Young Jeong Kim

SAMSUNG

Agenda.

01 Introduction : Reaction byproducts during the dry patterning

02 The Particle generation on the pattern

03 Experiments : The source of the particle

04 Reducing particles

05 Summary

SAMSUNG

Reaction byproducts during the dry patterning

Reaction byproducts from the dry patterning process can cause manufacturing failure

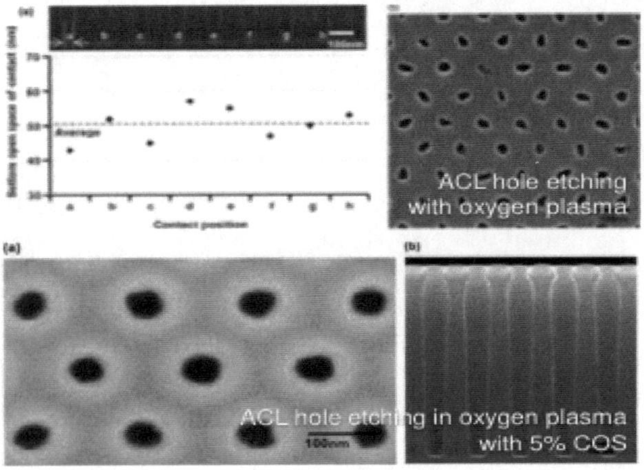

▲"Study on the etching characteristics of amorphous carbon layer in oxygen plasma with carbonyl sulfide", Jong Kyu Kim, Samsung Electronics, *J. Vac. Sci. Technol.*, 2013

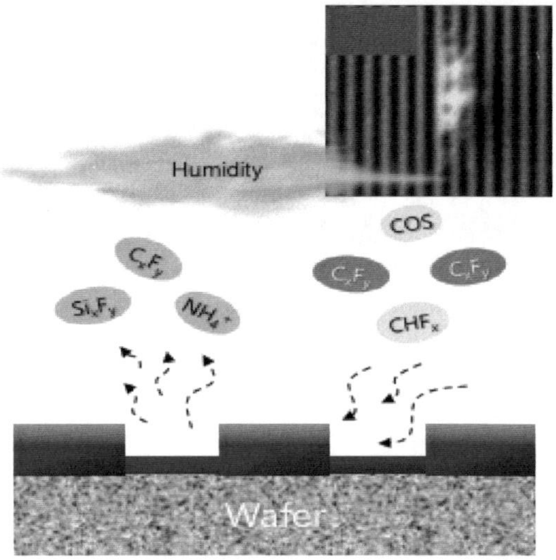

SAMSUNG

The Particle generation on the pattern

* DPT(Double Patterning Technology)

Particles induced by reaction byproducts during DPT process

- Particles generated during DPT process, inducing failure after final etching process

- The components of the particles : Carbon, Sulfur

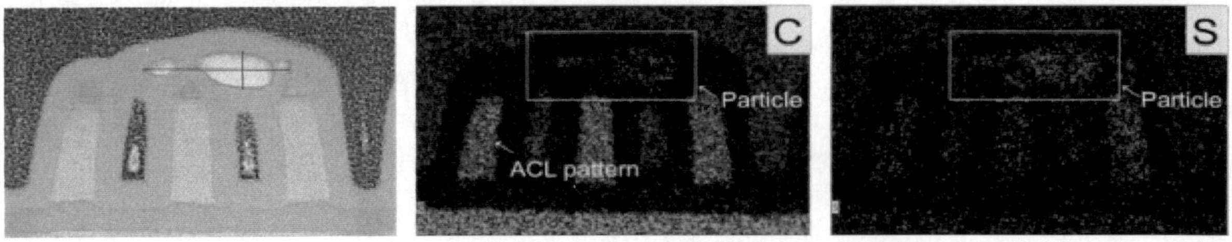

SAMSUNG

The Particle generation on the pattern

The characteristic of the detection of the particles

- The particle map shows a radiating shape, spreading out from the center of the wafer

[Particle wafer map]

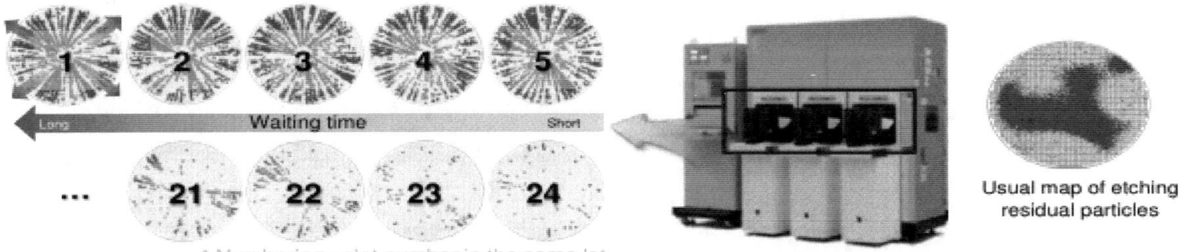

* Numbering : slot number in the same lot

- The particles appeared only after the wet cleaning, **not observed immediately after the dry patterning**

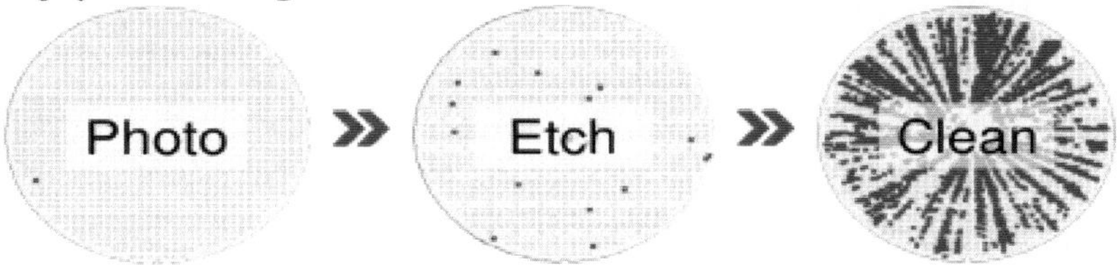

SAMSUNG

Experiment1 : Different types of wet cleaning chemicals

The order of the cleaning chemicals determine the generation of the particles

- The chemicals were solely treated at each test conditions

Test No.	1	2	3	4	5
Test condition	Original condition	DSP + drying	HF + drying	DI, IPA + drying	Test4 + Test1
Particle map	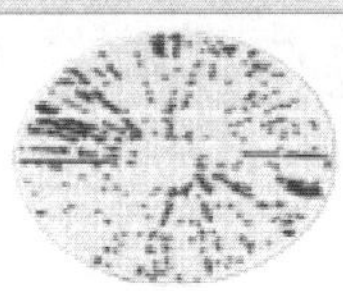	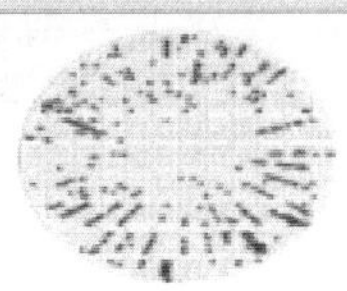		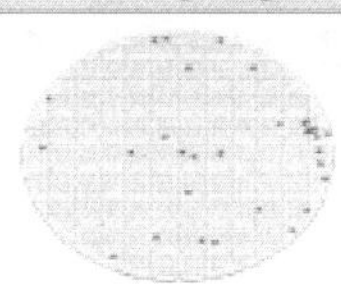	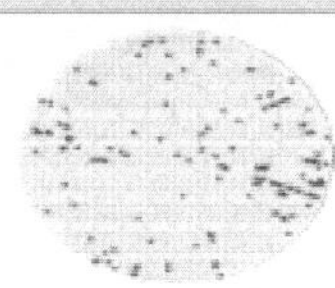

[Experimental result]

1. C and S byproducts can be condensed and grown in the DPS or HF injecting sequence.
2. After pre-treating the wafer surface with DI and IPA, no particles were detected when HF and DSP were injected.

SAMSUNG

Experiment2 : The cause of radial map of particles

Observation of fluid flowing behavior using a high-speed camera

- The identical radial map appears when the IPA drying starts

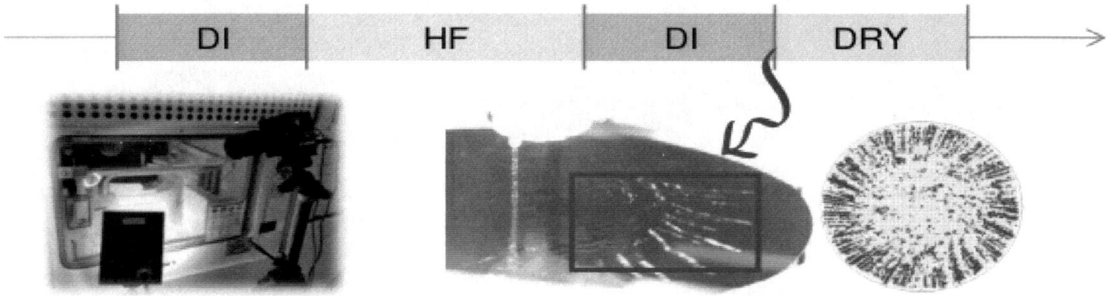

- Fingering phenomenon & Marangoni flow during IPA drying
  - Irregular IPA – DIW drying area
    : Carbonyl sulfide byproducts partially condensate in the IPA film
    → The clusters are absorbed on wafer surface as particles, leaving radial traces.

6

SAMSUNG

Experiment2 : The cause of radial map of particles

Finding condition that minimize fingering on the wafer surface

- DI spreading uniformly on the wafer surface when IPA is sprayed
  1) Shortening of idle time, 2) Increasing in RPM

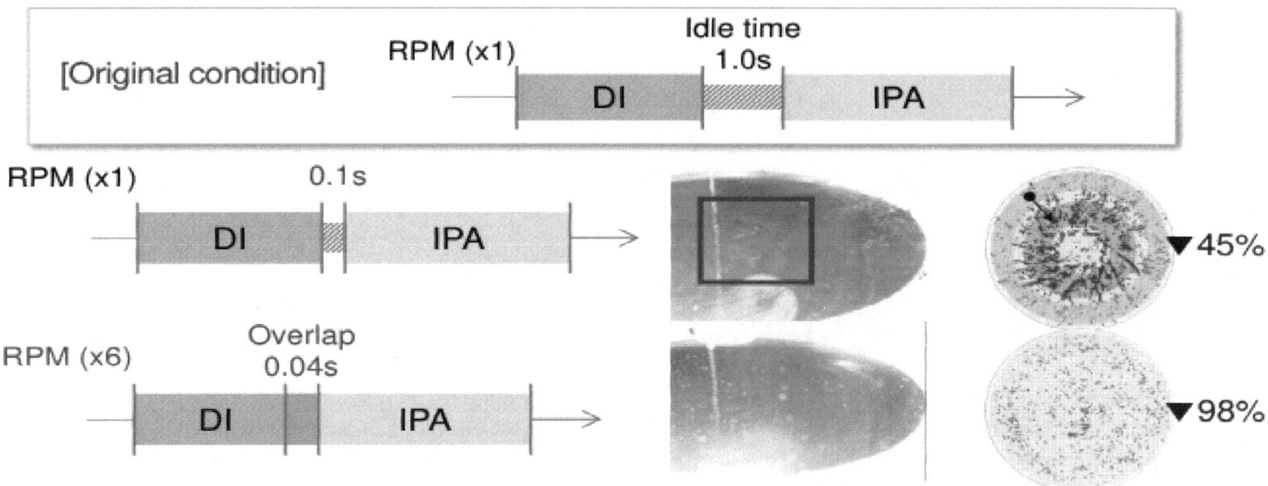

7

SAMSUNG

Modeling of the generation of the particles

Etching byproducts are aggregated during cleaning process

(1) Etching process residue of C, S components remain on the wafer surface.

(2) Process byproducts aggregated as C_xS_y compound film during the wet chemical cleaning process.

(3) Condensed C_xS_y dissolves in IPA and absorbed on the uneven drying spot of the wafer surface.

(1) After etching process **(2) Wet chemical treatment** **(3) After IPA drying**

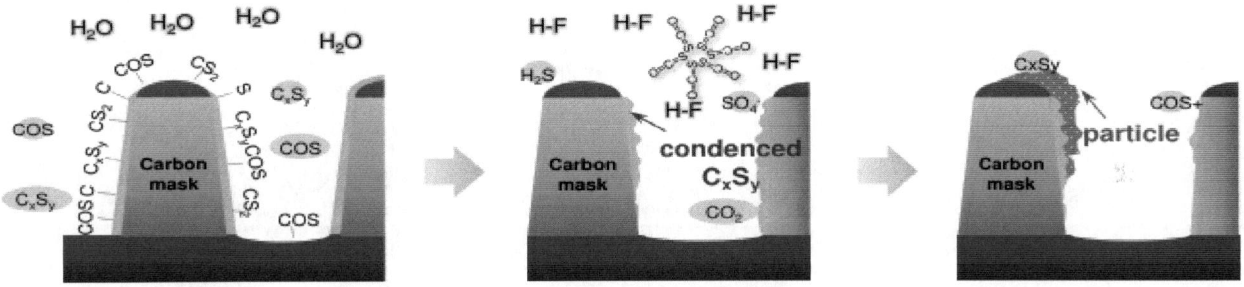

8

SAMSUNG

Reducing the particles

Removal processing byproducts in the etching chamber

- O_2 plasma treatment and exhaust strengthening help reducing residues on the wafer surface and the increasing in particle counts with the staying time after etching was lowered

* High performance ion chromatography

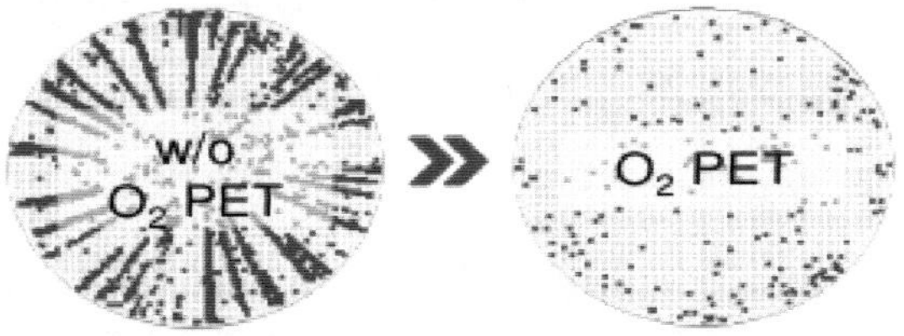

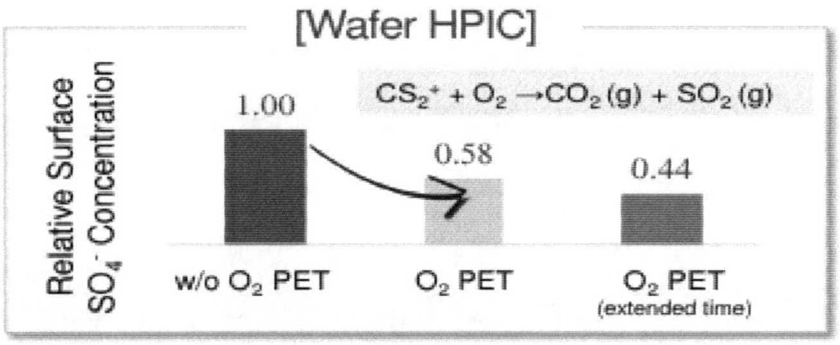

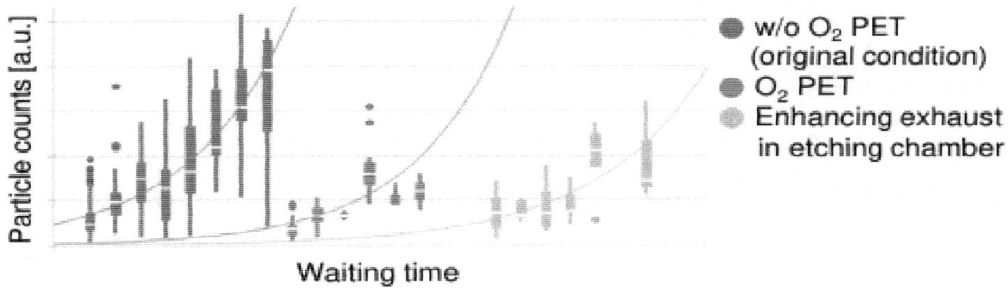

9

SAMSUNG

Reducing the particles

»» Lowering surface byproducts before being clustered by wet chemical

- IPA rinsing effectively reduces S, C residue on the ACL pattern, ultimately resulting in no radial particles being generated

* Time of flight secondary ion mass spectrometry

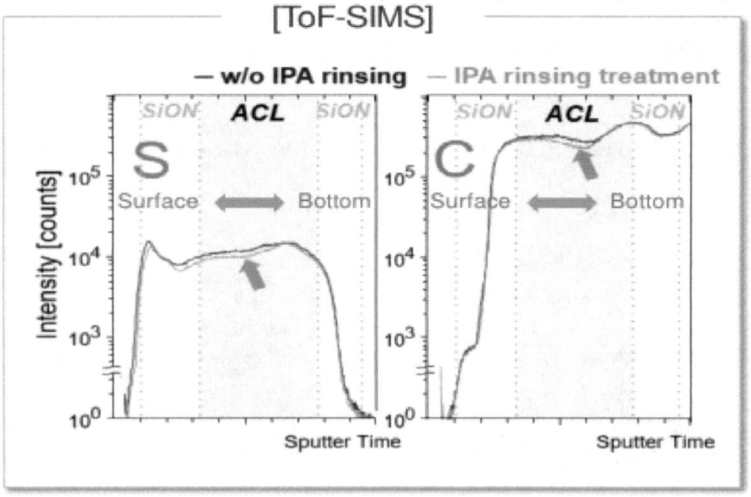

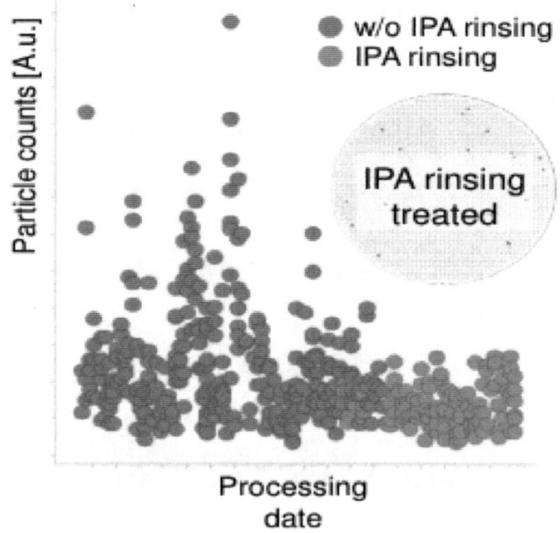

10

SAMSUNG

Summary

- The particle that occurs when a cleaning process follows an etching process using COS gas during the DPT process is studied.
- The source of the particle was etching byproducts, but particles were generated only after cleaning, and they had the characteristics of both etch and clean-induced particles.
- Wet chemicals such as DSP or HF cause byproducts to condensed, which are then absorbed onto the wafer in the form of particles during the IPA drying.
- This is caused by the fingering phenomenon due to the uneven drying of IPA on the wafer surface.
- Particle generation can be reduced by removing the byproducts in the etching chamber or by optimizing the order of exposure to the cleaning chemicals.

11

SAMSUNG

THANK YOU.

April 24th, 11.30 Etching of tungsten via a combination of thermal oxide formation and wet chemical oxide dissolution

Antoine Pacco[1*], Teppei Nakano[2], Jana Loyo-Prado[1]
Kenneth Lai[1], Hikaru Kawarazaki[2], Efrain Altamirano Sanchez[1]
[1]imec, Belgium, [2]SCREEN, Germany, [3]SCREEN, Japan
***antoine.pacco@imec.be**

OVERVIEW

- Controlled metal recess: needs & applications
- Challenges for controlled metal recess

- Wet etching of Tungsten in PAN
- Oxidation of Tungsten in O_3 (STEP1)
- Oxide dissolution (STEP2)
- [THERMAL – WET]$_x$ (STEP1+2)

- Application: CFET

CONTROLLED METAL RECESS: NEEDS & APPLICATIONS

LOGIC & MEMORY APPLICATIONS

METAL WORD LINE ISOLATION FOR **3D NAND**

REQUIREMENT:

- ISOTROPIC ETCH with excellent top-to-bottom uniformity in HAR trench

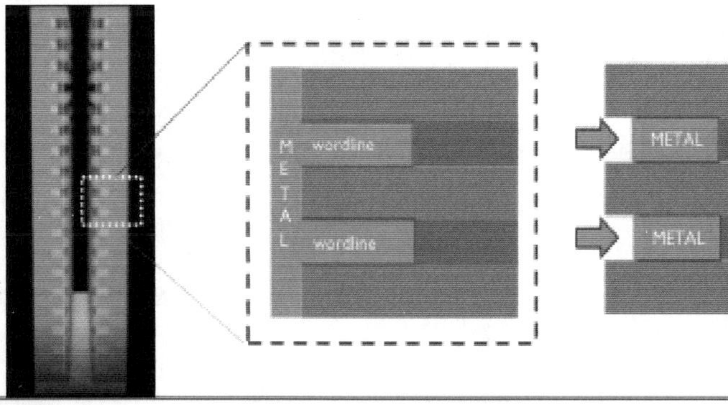

METAL RECESS FOR **BURIED POWER RAILS** (BPR)

- ULTRA DEEP RECESS (>100 nm) WIW uniformity & flat profile

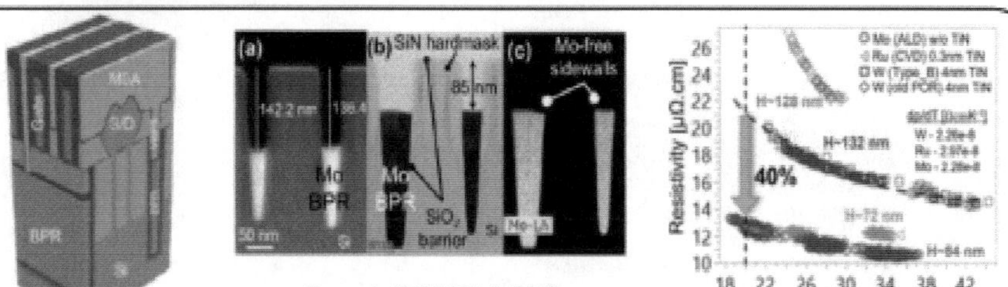

Gupta, A. IITC 2021 & 2022

CONTROLLED METAL RECESS: NEEDS & APPLICATIONS

LOGIC & MEMORY APPLICATIONS

BOTTOM CONTACT (BC) ISOLATION FOR **CFET**

REQUIREMENT:

- ISOTROPIC ETCH with excellent local uniformity* and accuracy in HAR trench, selective to SiN

- *VEPE: **V**ertical **E**tch **P**lacement **E**rror

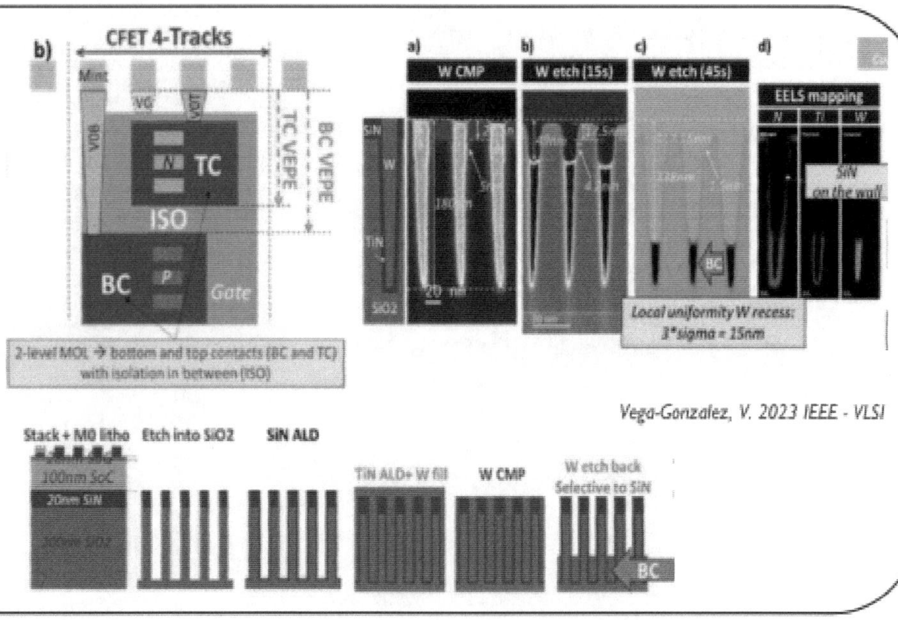

Vega-Gonzalez, V. 2023 IEEE - VLSI

imec SCREEN

CHALLENGES FOR CONTROLLED METAL RECESS

WET RECESS

SURFACE ROUGHNESS CONTROL

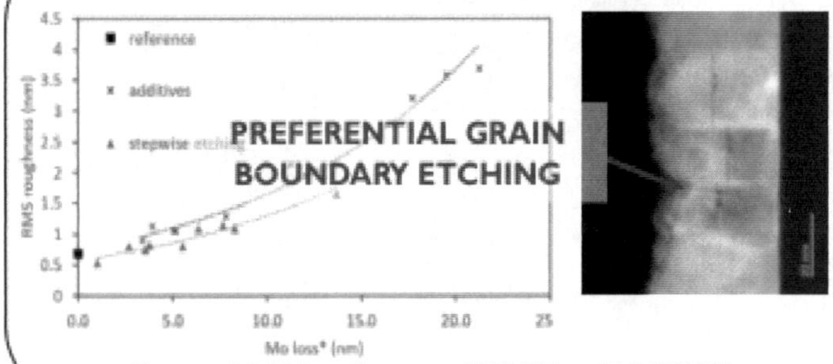

Pacco et al. Solid State Phenomena 2021 (Volume 314) 295-301

TOP-TO-BOTTOM NON-UNIFORMITY & RESIDUES

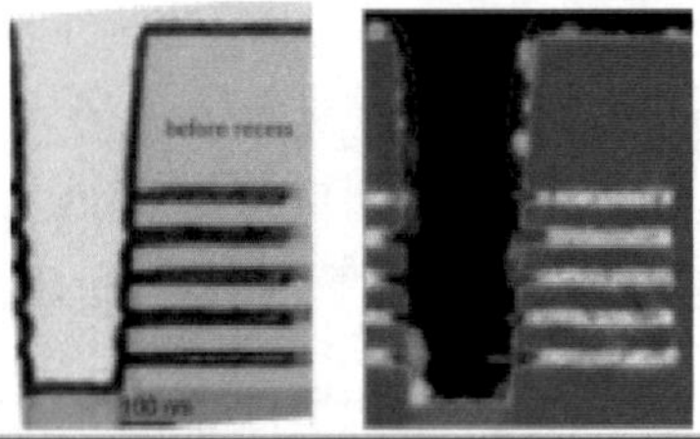

ROUGH SURFACE AND UNEQUAL RECESS DEPTH

DRY PLASMA RECESS

HAR ETCHING OVERALL CHALLENGE

TOP-TO-BOTTOM UNIFORMITY

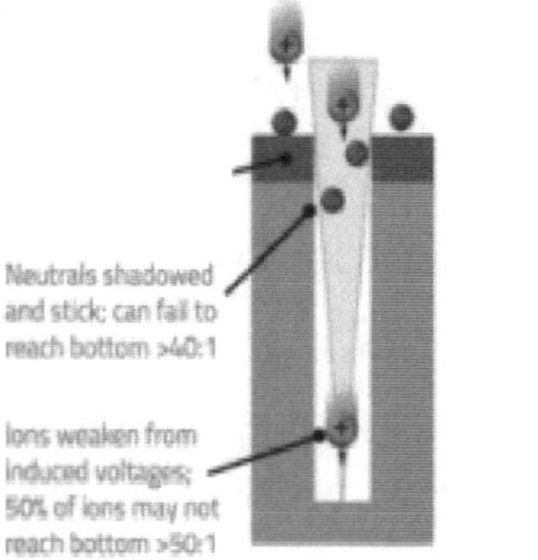

(Source: Lam Research)

HAR ETCHING OVERALL CHALLENGE

WIW UNIFORMITY

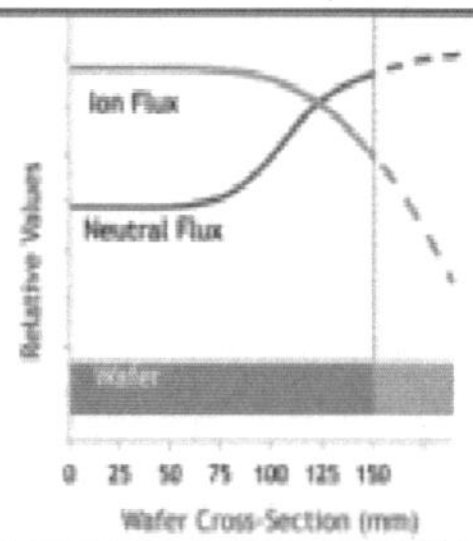

(Source: Lam Research)

imec SCREEN

Etching of Tungsten

PAN

- INDUSTRY SOLUTION for 3D NAND = PAN (**P**hosphoric-**A**cetic-**N**itric acid)

Suggested mechanism*

- W oxidizes in the **nitric acid** forming a by-product $W(NO_3)_x$
- $W(NO_3)_x$ dissolves in the **phosphoric acid.**
- **acetic acid** acts as a wetting agent, facilitates the etch process by removing the H_2 by-product.

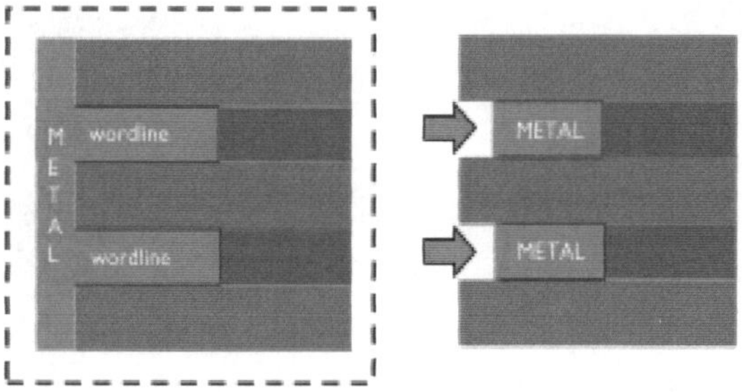

Alternative mechanism:

$W+2HNO_3+6H^+(Acid)+2H_2O \rightarrow [W(H_2O)_6]^{6+}+2NO$

* ref: https://www.ecitechnology.com/tag/tungsten-etch/

imec SCREEN

FIGURE OF MERIT W ETCH BACK

	Etch rate	Select. to SiN	VEPE*
(1) DRY RIE	+++	-	-
(2) WET PAN	+	+++	-
(3) DRY RAD	++	-	+

*VEPE: **V**ertical **E**tch **P**lacement **E**rror

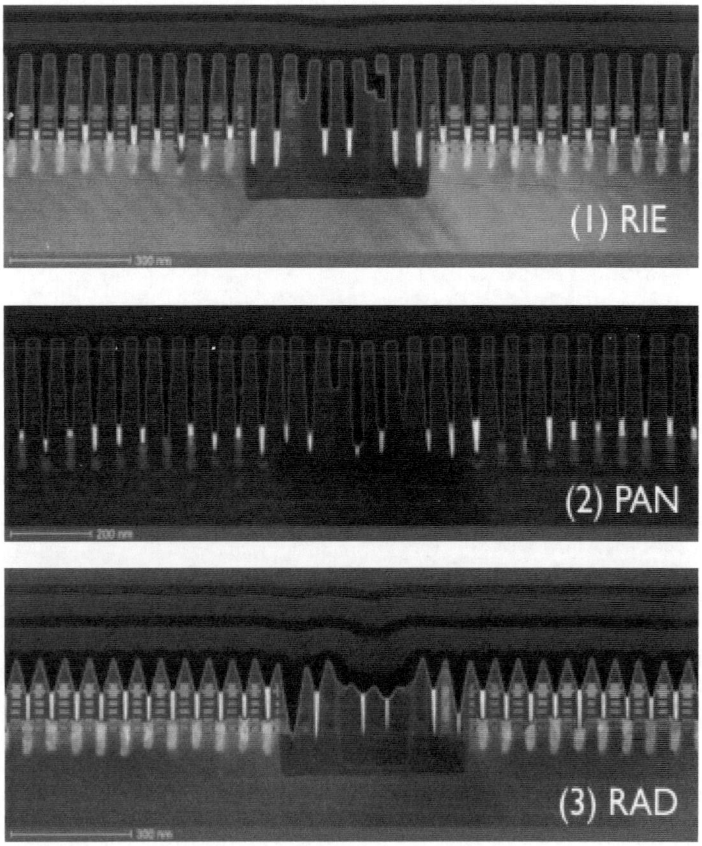

imec SCREEN

[THERMAL – WET]$_x$ ALE
SCHEMATIC

STEP 1:
THERMAL OXIDATION
(Self-limiting oxide growth)

STEP 2:
OXIDE REMOVAL
(Selectively towards metal)

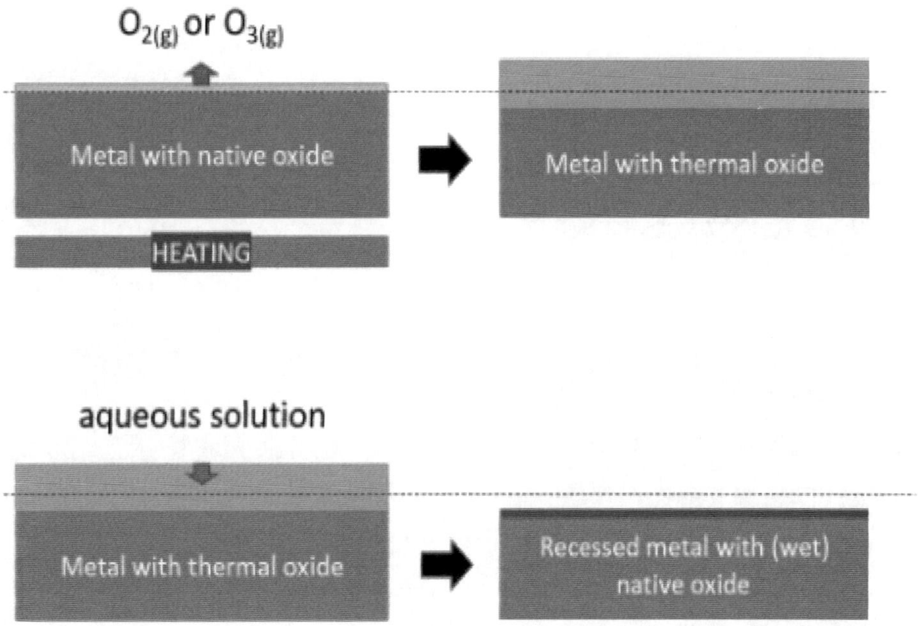

imec SCREEN

Thermal oxidation of W in ozone

(quasi) self-limiting oxide growth

oxide thickness ~ f(T)

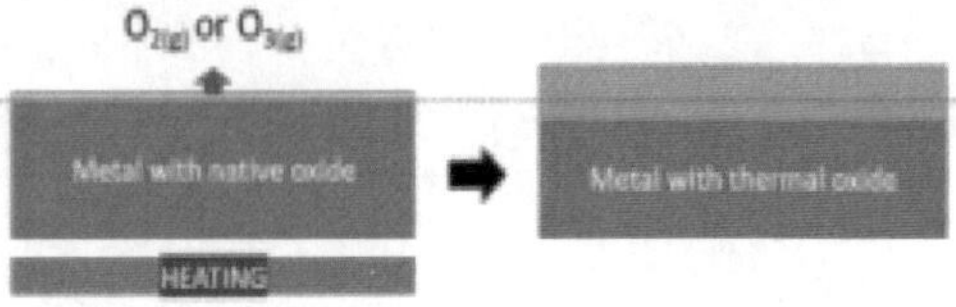

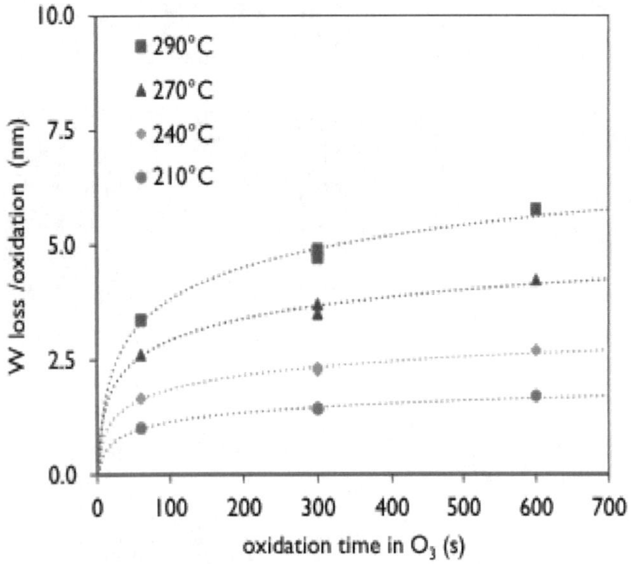

- The observed parabolic growth of the oxide is associated with the formation of thin, (quasi) self-limiting oxide film.

Thermal oxidation of W in O_3

Elemental analysis of oxide grown at 290°C in O_3 by TEM/EDS → WO_3

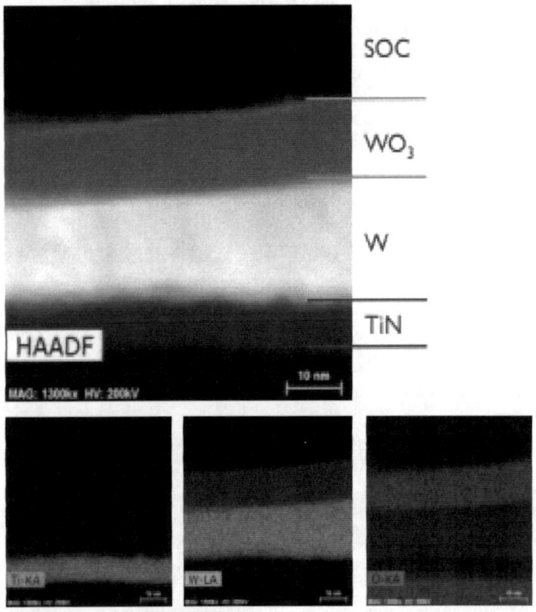

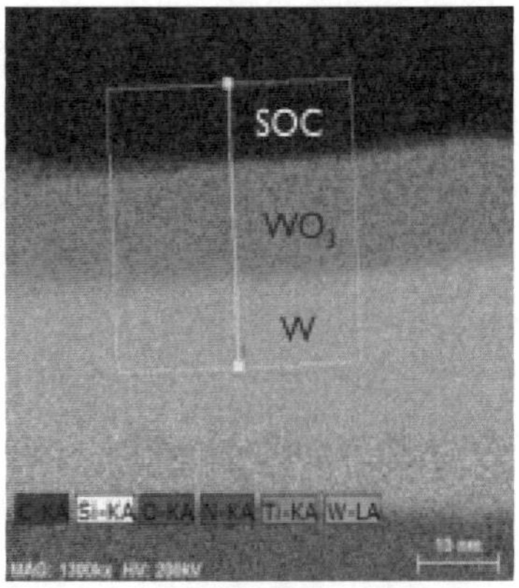

-EDS quantification after averaging and normalization:
W (M series)
$W_{0.23}O_{0.77}$ → close to WO_3

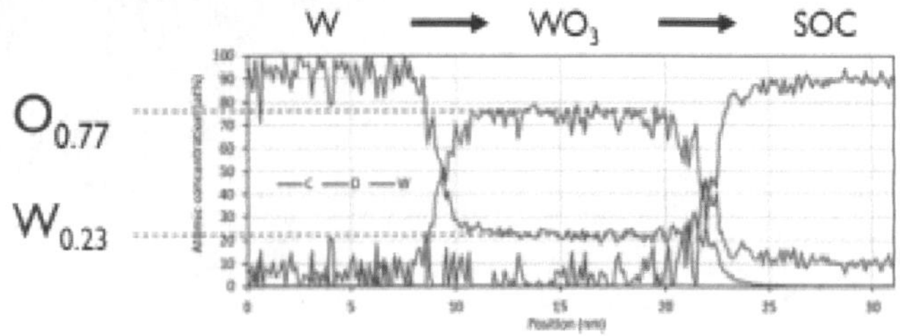

WO_3 dissolution

Effect of solution type (pH)

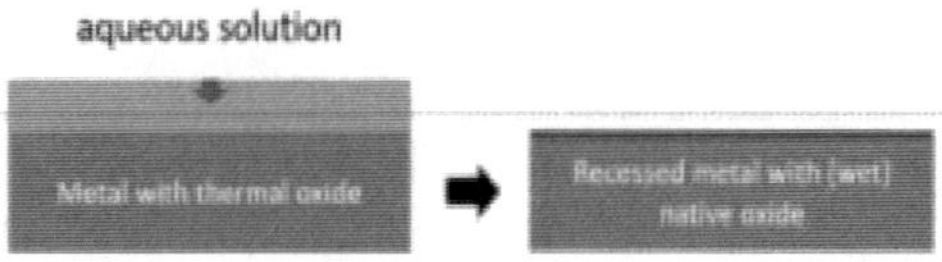

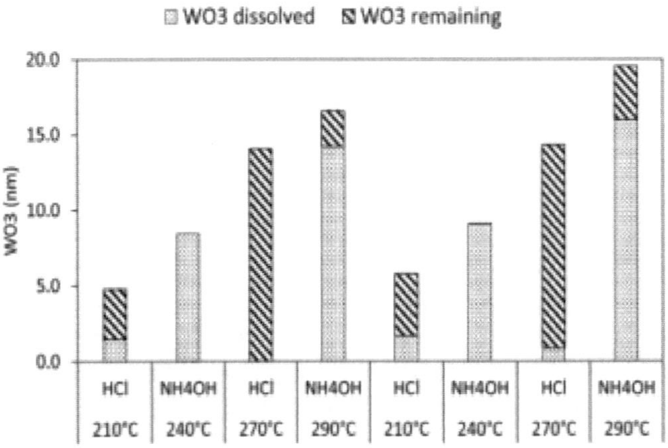

- WO_3 not/barely soluble in acidic solutions like HCl (or HF – not shown here)
- WO_3 soluble in alkaline NH_4OH solutions

→ ammonia good candidate for step 2

imec SCREEN 11

WO_3 dissolution (rate)

Monitoring of WO_3* dissolution

(WO_3 grown at 290°C in O3)*

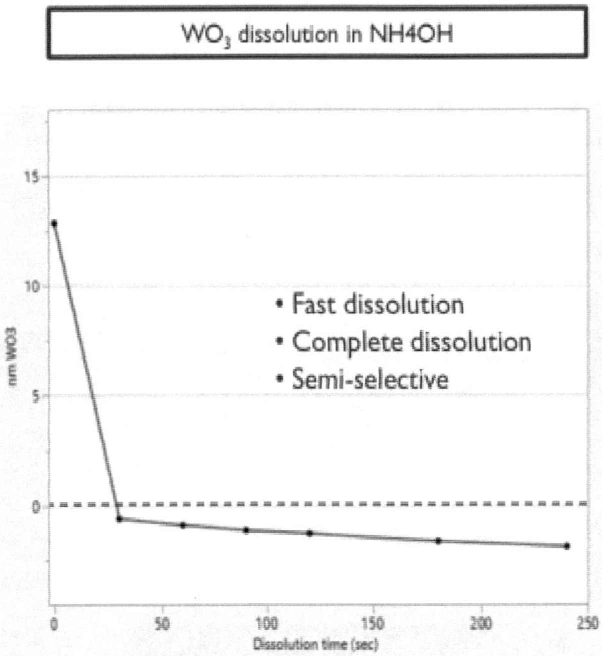

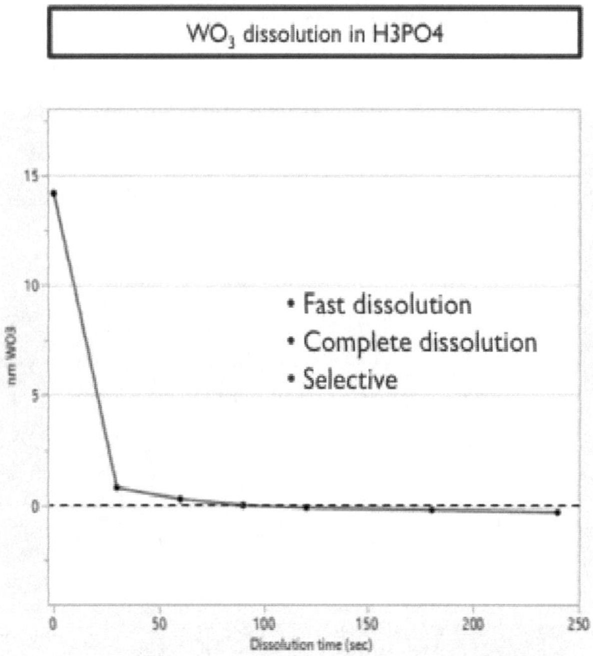

imec SCREEN 12

ALE-type Tungsten recess

Stepwise oxidation and oxide dissolution

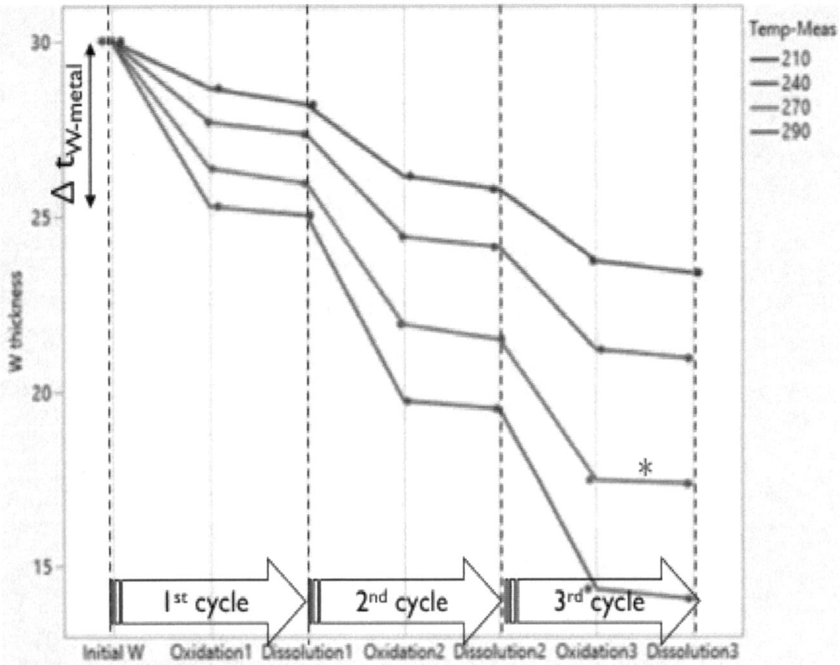

* Selective oxide dissolution/ removal towards the Mo metal

→ Allows controlled recess etching of the Mo metal

Δ $t_{W\text{-metal}}$ ~ 2 to 5 nm / cycle
(dependent on O3 bake temperature)

imec SCREEN

ALE-type Tungsten recess

EA / cycle ~ f(T)

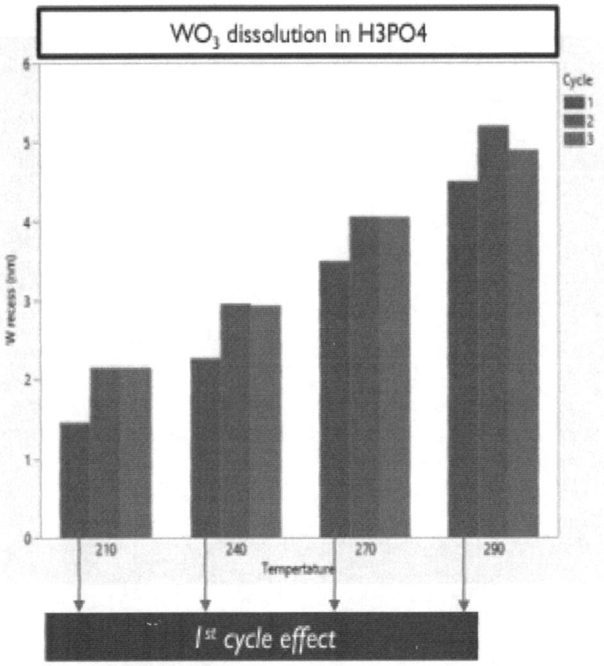

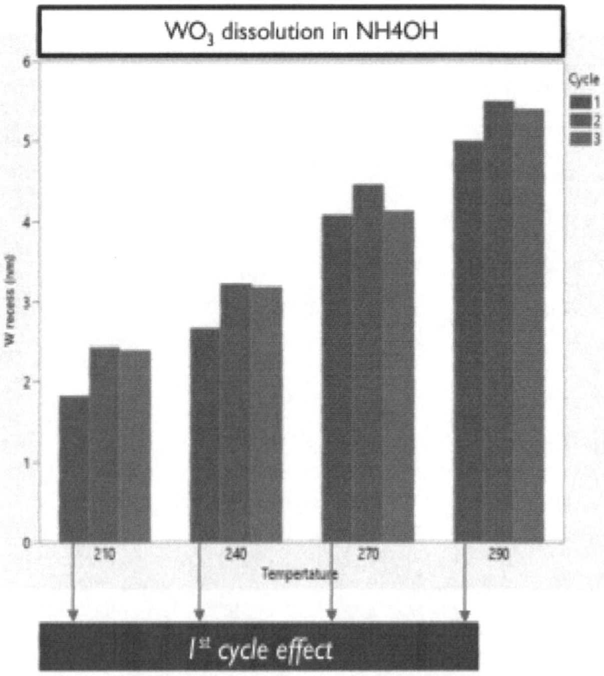

imec SCREEN

	# cycles	NH4OH	H3PO4
O3 240°C	x = 10	A	C
O3 290°C	x = 5	B	D

W recess in narrow trenches

CYCLIC [thermal oxidation in O_3 / wet WO_3 dissolution]$_x$

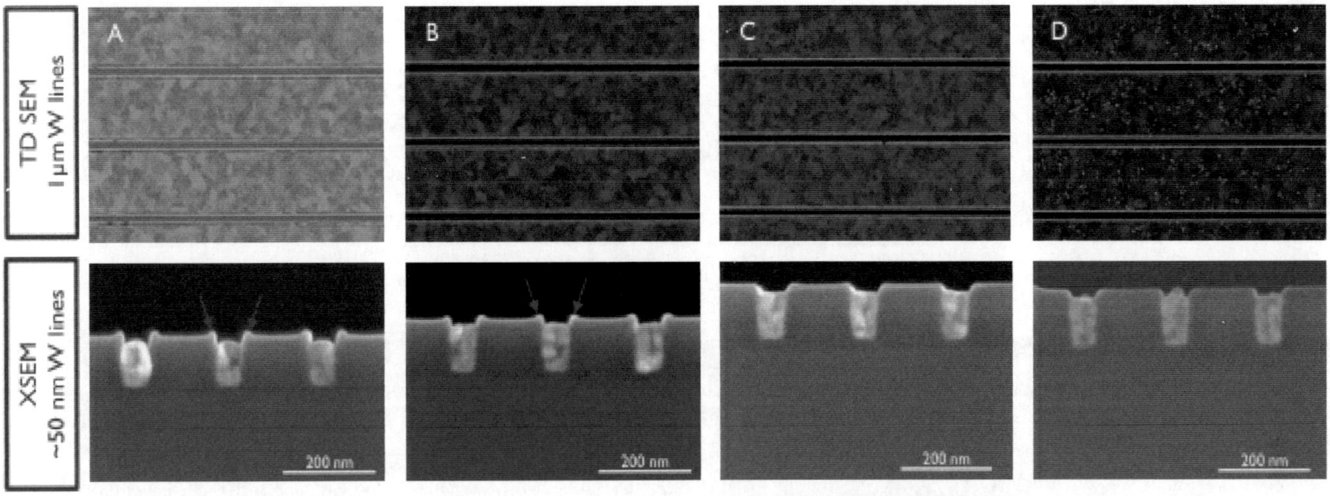

\+ best option

− remaining TiNOx liner "ears" when using NH4OH | remaining cryst. WOx

imec SCREEN

15

COMPARISON CYCLIC (O_3/NH_4OH) vs. wet (PAN)

~50 nm W trenches

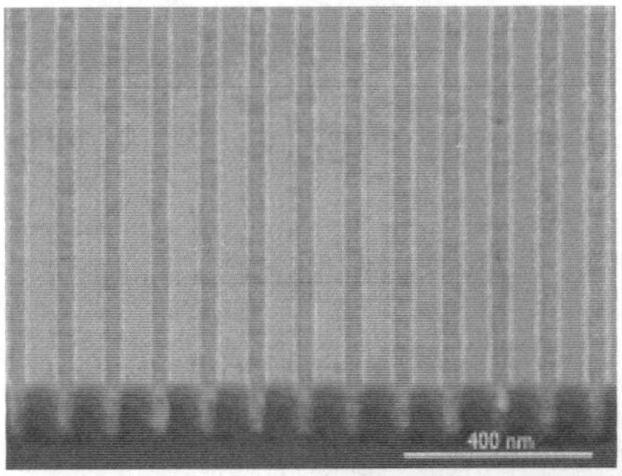

W surface is smoother

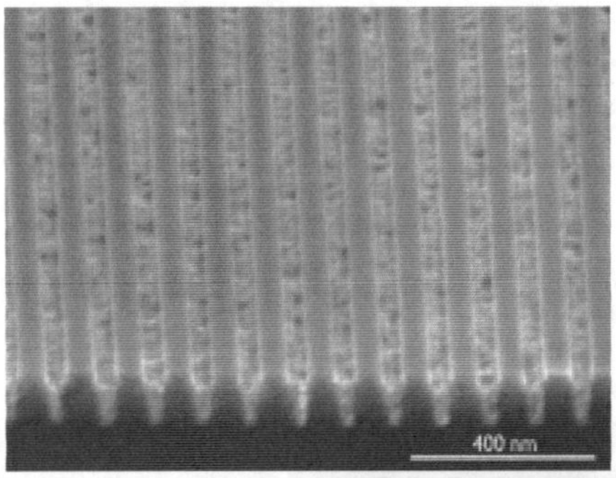

W surface is rougher ~ GB

(oxidized) TiN liner remains

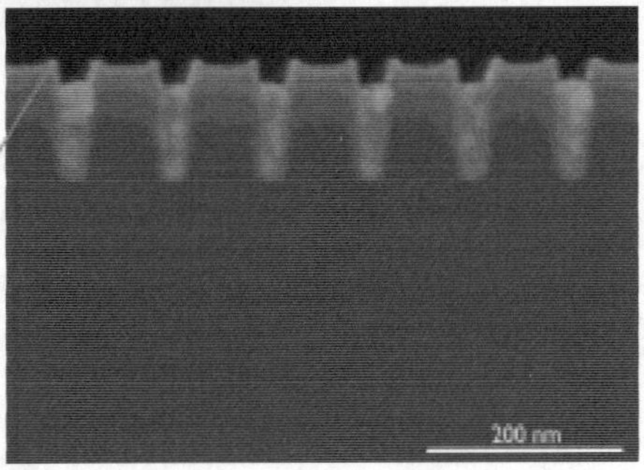

TiN liner removed

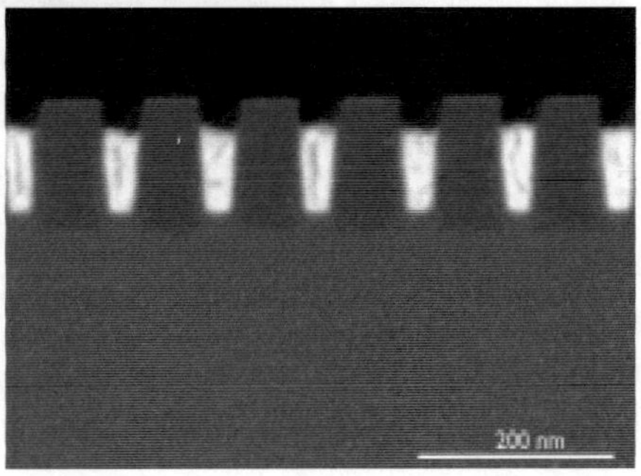

imec SCREEN

16

Vertical recess for BC isolation for CFET

W trenches with CD ~20 nm

X-cut (through the lines)

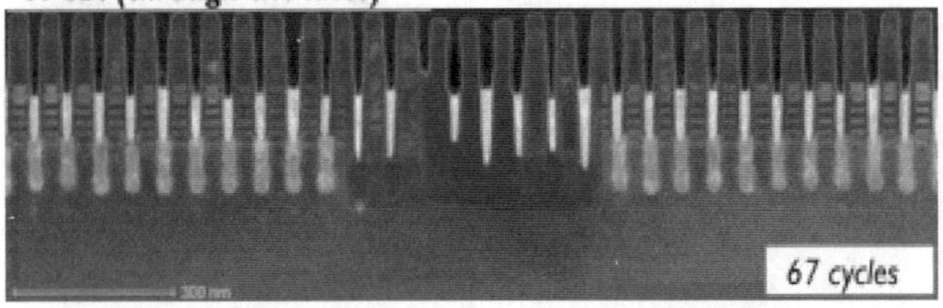

X-cut (through the lines)

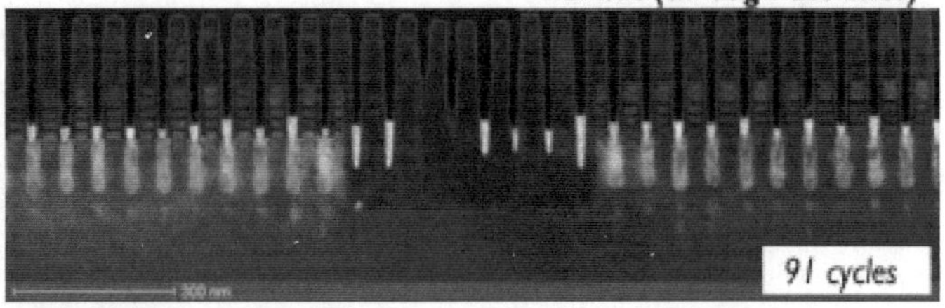

Average (nm) =	130.6
Sigma =	5.0

Average (nm) =	183.7
Sigma =	6.1

Y-cut (along the lines)

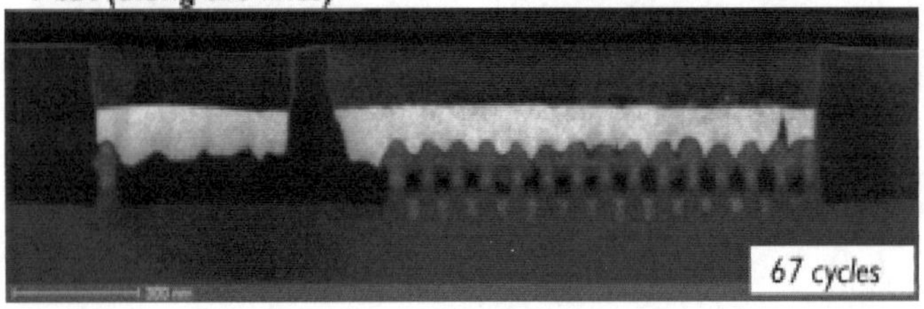

Y-cut (along the lines)

Extremely deep & uniform W recess (~ 180 nm) demonstrated in ~ 20 nm narrow trenches

imec SCREEN

17

FIGURE OF MERIT W ETCH BACK

	Etch rate	Select. to SiN	VEPE
(1) DRY RIE	+++	-	-
(2) WET PAN	+	+++	-
(3) DRY RAD	++		+
(4) CYCLIC	--	+++	+

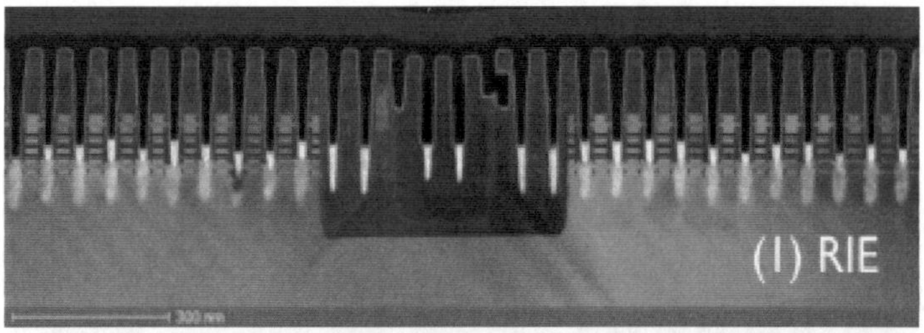

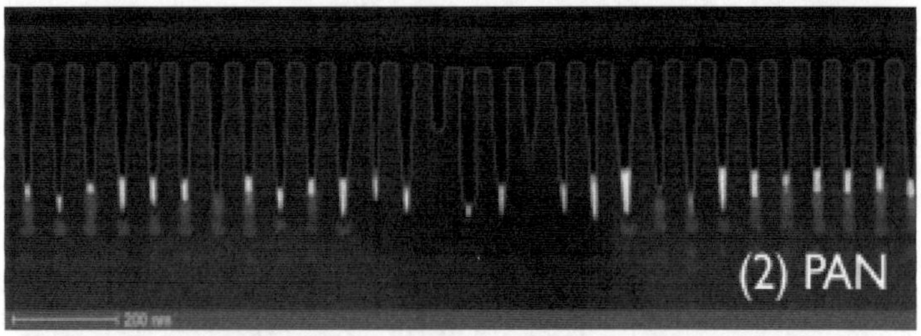

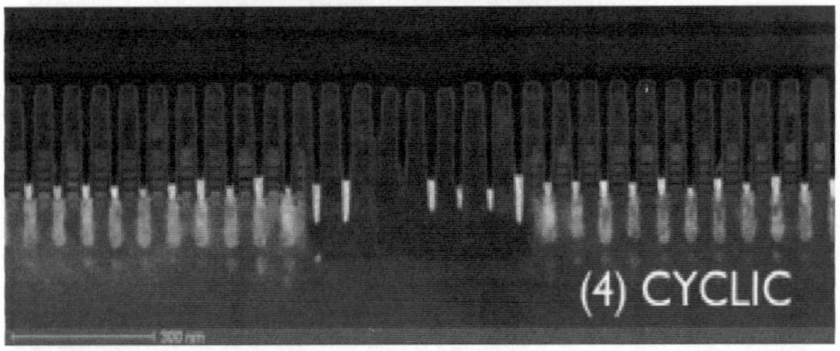

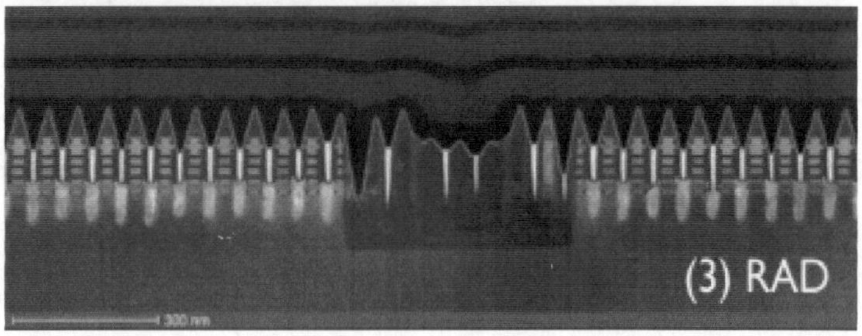

imec SCREEN

18

SUMMARY

- The thermal oxidation of W metal in an oxidizing atmosphere resulted in the controlled growth of a thin, quasi self-limiting oxide film with $t \sim f(T_{ox})$.
- The formation of this oxide film was used for the development of the first step of an ALE-type of recess etch.
- In a next step, WO_3 was dissolved in NH_4OH or H_3PO_4 solutions, selectively towards the underlying W metal, resulting in an ALE-type of recess sequence.
- Vertical & lateral recess of ALD W with this process was demonstrated down to ~20 nm narrow lines.

Acknowledgements

- Surface and Interface Processing Group / imec: Jana Loyo-Prado, Kenneth Lai, Pallavi Puttarame Gowda, Efrain Altamirano Sanchez
- SCREEN: Teppei Nakano, Hikaru Kawarazaki

imec

embracing a better life

SCREEN

W/TiN recess for 3D DRAM

Metal Gate recess

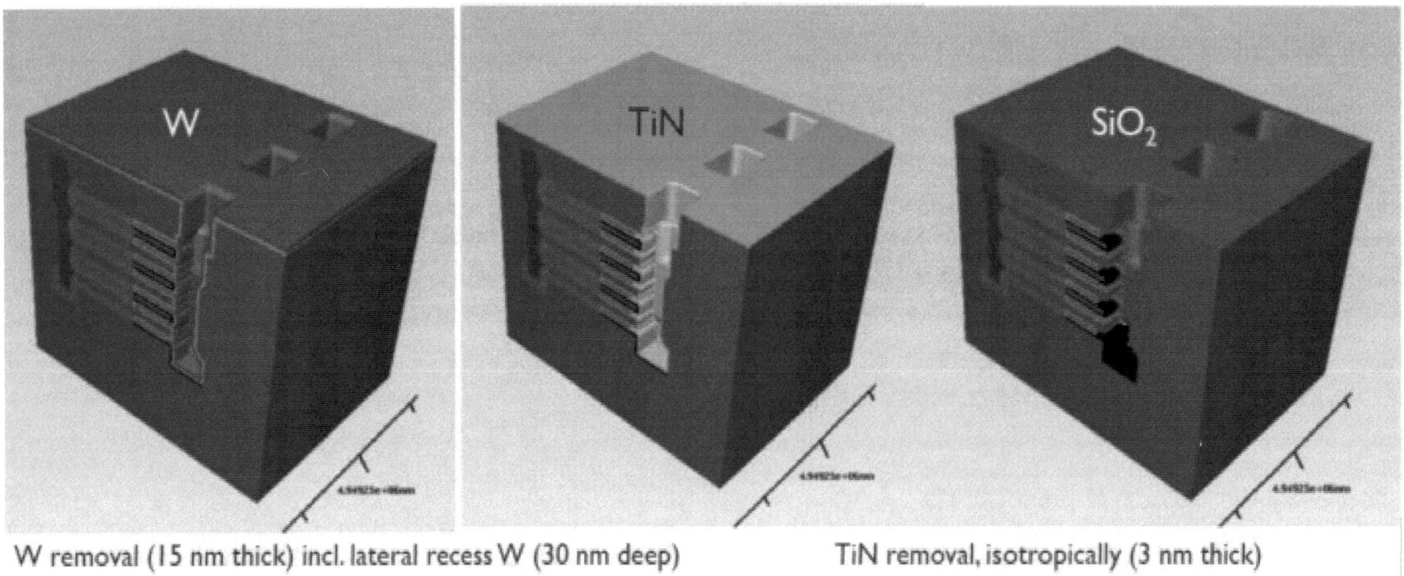

W removal (15 nm thick) incl. lateral recess W (30 nm deep)
Selectively to TiN or combined with TiN

TiN removal, isotropically (3 nm thick)
Selectively towards W & oxide

REMAING WOx RESIDUES
TEM

- nanocrystals are locally distinguished in the WOx layer.
- nanocrystals too small to perform SAED

imec SCREEN

23

Enhanced Surface Activation Through Advanced Hydrogen Peroxide Delivery

Adrian Alvarez
Product Line Manager
aalvarez@rasirc.com
RASIRC

Background

- H_2O_2 vapor use has been limited due to challenges with stable delivery
- Stable delivery achievable with BRUTE Peroxide
- H_2O_2 easily dissociates into two –OH groups
- Alternative to O_2 and water for high density, low temperature plasma activation

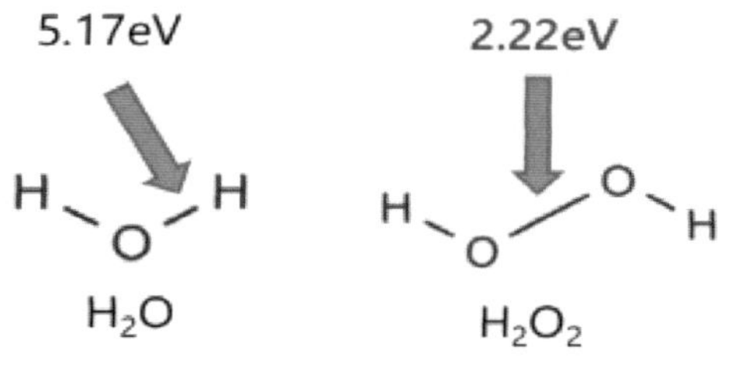

Figure 1: Dissociation energies for water and H_2O_2.

Figure 2: BRUTE Peroxide vessel.

Delivery Method with BRUTE Peroxide

RASIRC

- Anhydrous H_2O_2 is sorbed onto solid polymer matrix and loaded into an ampoule
- Carrier gas flow 0 – 10slm flows through the vessel and is saturated by H_2O_2 vapor
- Source concentration does not change with use and vapor pressure remains stable until depletion

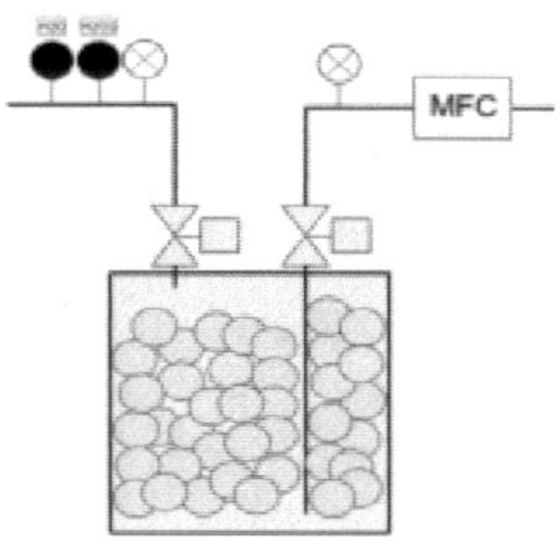

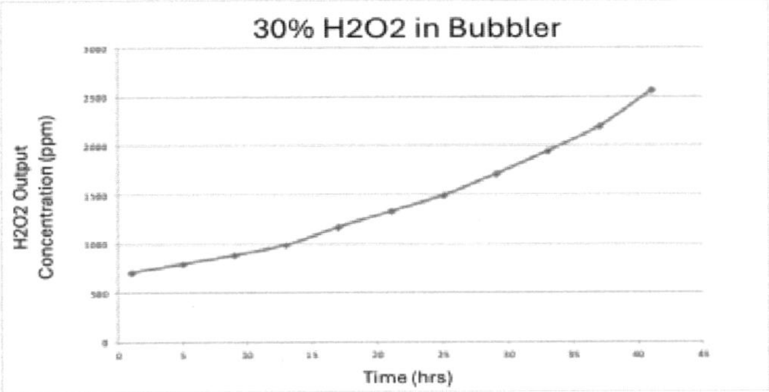

Figure 3: Rault's Law: H2O2 vapor pressure changes with source concentration.

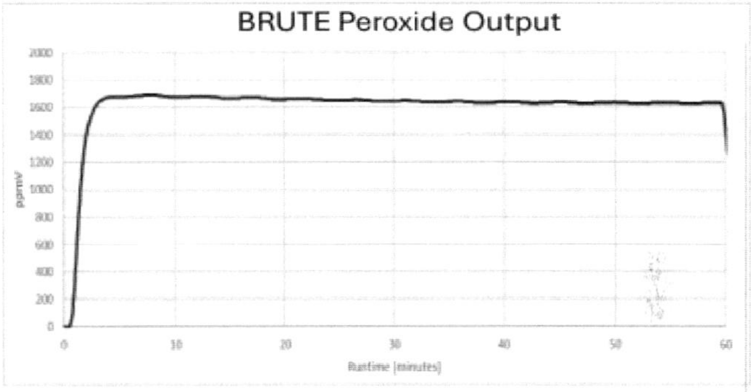

Figure 4: Rault's Law: H2O2 vapor pressure changes with source concentration.

Reviewing Functionalization Mechanisms

RASIRC

- Oxygen radicals can dehydroxylate to for form siloxane (Fig. 5)
- Water can leave non-reactive –H terminated sites
- H_2O_2 functionalizes via ligand exchange (vs combustion with –O radicals or O_3)

contamination. Assuming O atoms are the primary species responsible for oxidation, then it is proposed that they dehydroxylate the surface by the following reaction:[38–40]

$$\mathrm{Si(OH)} + \mathrm{Si(OH)} \xrightarrow{2\mathrm{O}} \mathrm{Si{-}O{-}Si} + \mathrm{O_2} + \mathrm{H_2O}$$

Figure 5: *The Surface Chemistry of Amorphous Silica, Zhuravlev*

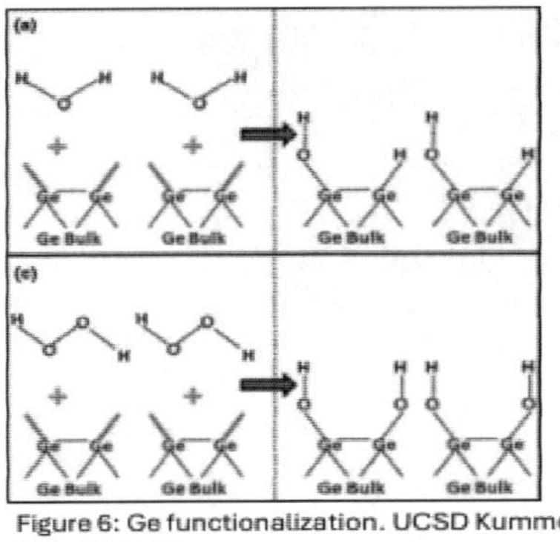

Figure 6: Ge functionalization. UCSD Kummel Lab, ALD2018

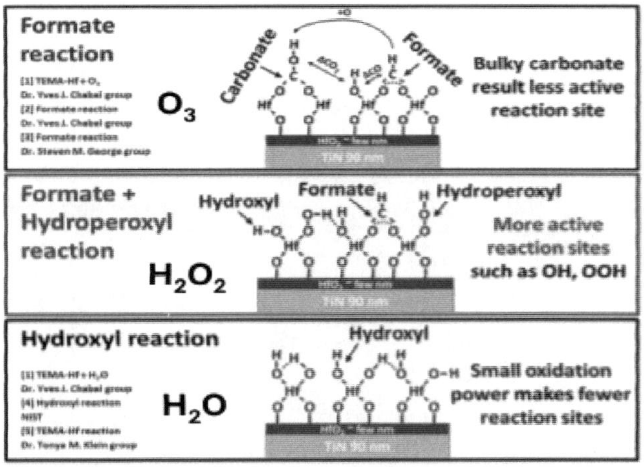

Figure 7: Reaction mechanisms. O3 vs H2O2 vs H2O. UT Dallas Prof. Jiyoung Kim Lab, ALD2023 H2O2 ligand exchange.

Experimental Setup

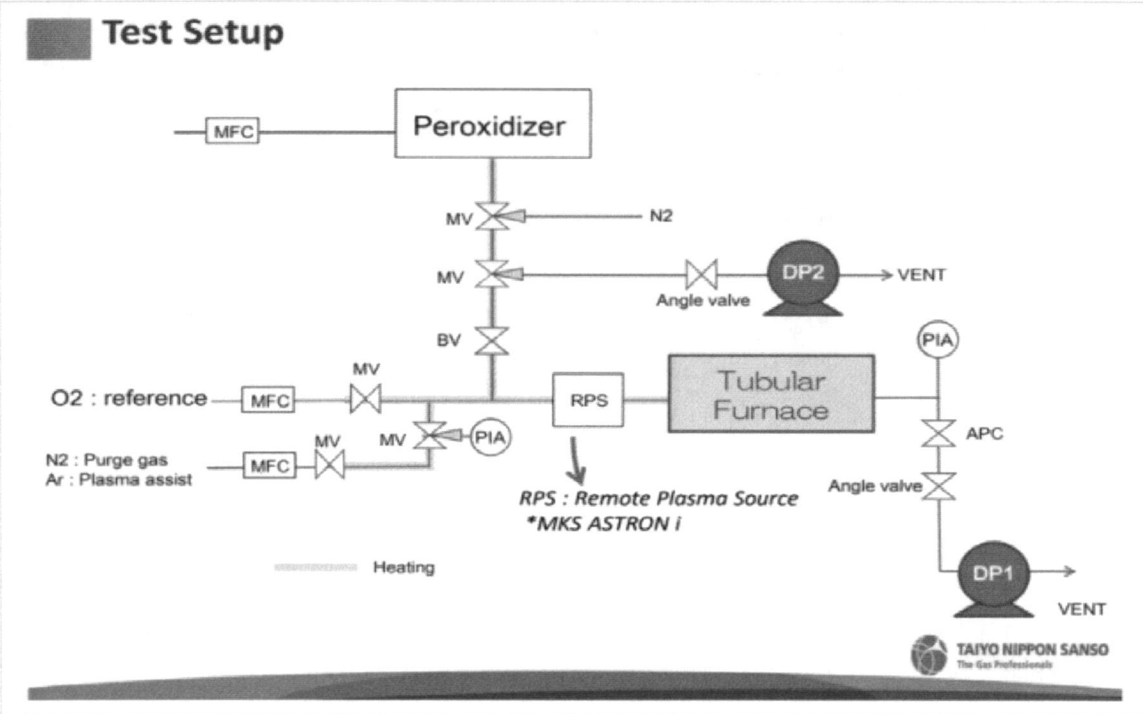

Figure 8: Plasma system at TNSC Tsukuba Lab.

OES Characterization

- OES performed with different gas mixtures
- H_2O_2 plasma –OH density is 10x higher than water plasma
- -OH and other functional radical density dependent on carrier gas type

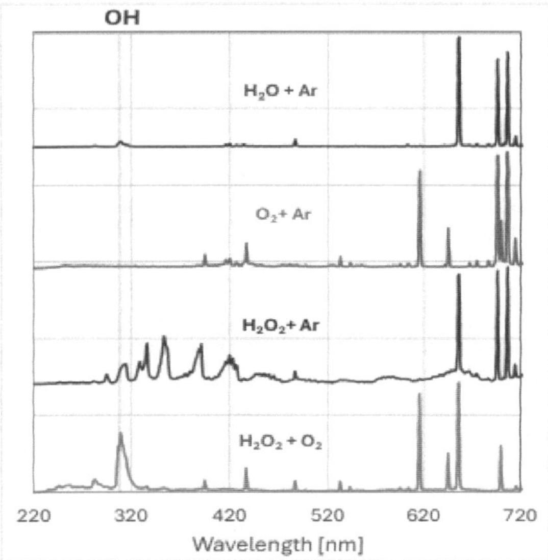

Figure 9: OES data. H_2O_2 plasma contains 10x higher hydroxyl density than water plasma.

Figure 10: Ar + H_2O_2 plasma.

Experimental: Si Functionalization with H_2O_2 Plasma

- HF last Si
- All samples run at 200°C
- H_2O_2 plasma provides higher hydrophilicity than water and O_2

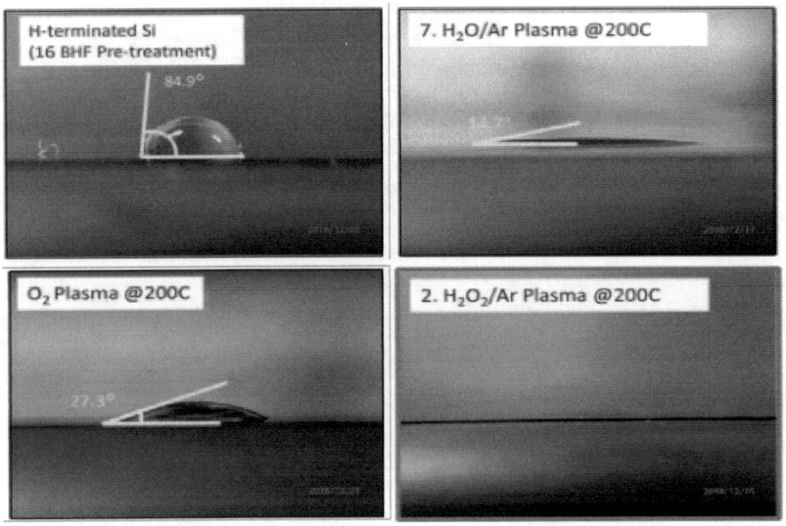

Figure 11: Si contact angles after treatment.

Si Treatment	Contact Angle
HF Last	84.9°
O2 Plasma	27.3°
H2O/Ar Plasma	14.7°
H2O2/Ar Plasma	< 4°

Figure 12: Si contact angles after treatment.

Experimental: Effects of O_2 Plasma

- Hydrophilicity <u>decreases</u> as O_2 doping <u>increases</u>
- Same results are not seen with variation of Ar flowrate (e.g. not dilution driven)
- -O radicals may dehydroxylate the surface to for siloxane

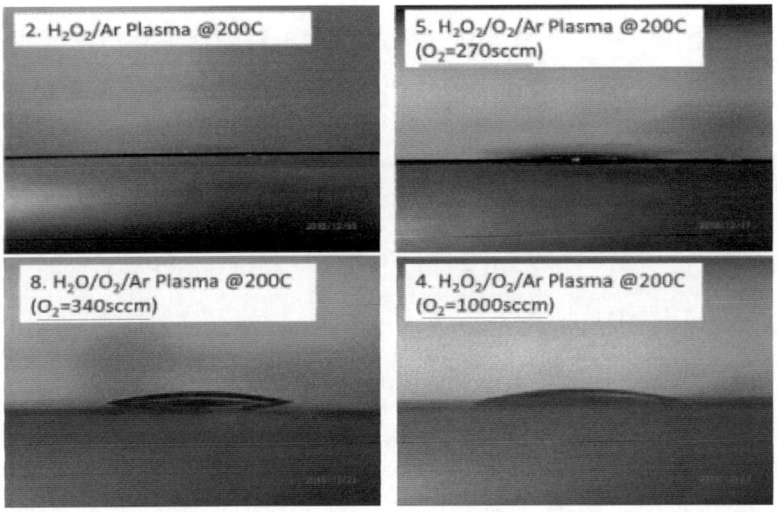

Figure 13: Si contact angles after plasma treatment with O2 doping.

Gas Mixture	O_2 Flowrate (sccm)	Contact Angle
H2O2/Ar	0	< 4°
H2O2/Ar/O2	270	8.1°
H2O2/Ar/O2	340	16.2°
H2O2/Ar/O2	1000	21.6°

Figure 14: Si contact angles after treatment. O2 plasma reduces surface activation.

Experimental: H_2O_2 Plasma Ash

RASIRC

- AHM samples provided by major OEM
- H_2O_2 plamsa removes AHM at all tested plasma source powers
- Surface is hydrophilic after ashing

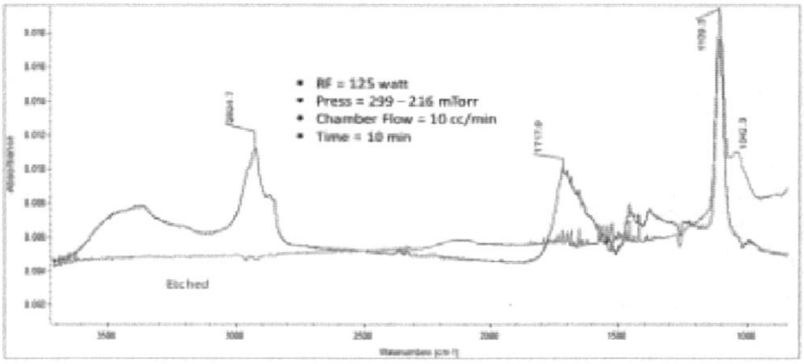

Figure 15: FTIR before and after H2O2 ashing.

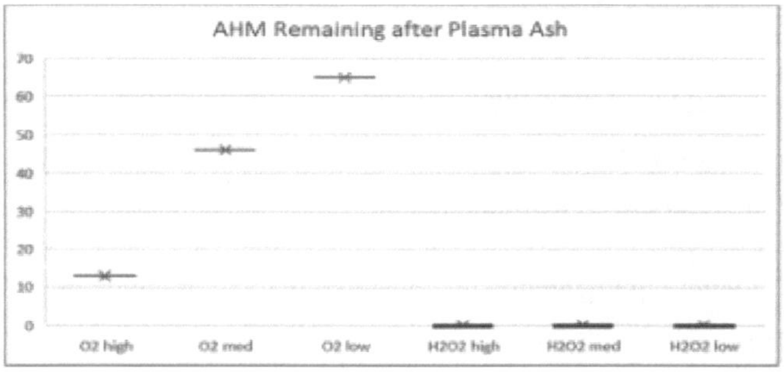

Figure 16: Normalized counts of AHM remaining after O2 and H2O2 plasma ash.

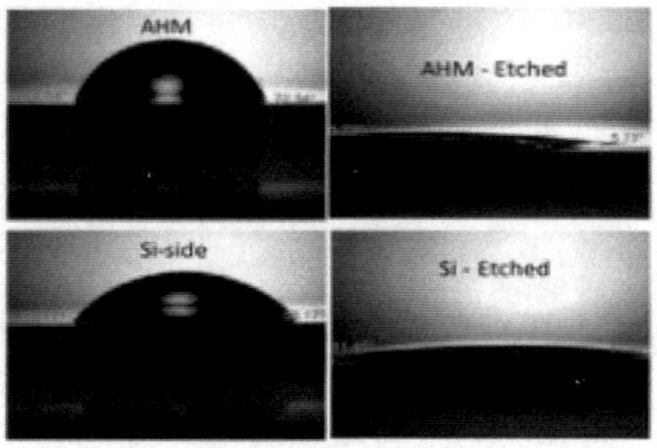

Figure 17: Si substrate activation during AHM plasma ash with H2O2.

Conclusion

RASIRC

- Stable delivery of H_2O_2 vapor is possible with BRUTE Peroxide
- Peroxide plasma provides 10x higher –OH density than water plasma
- O_2 plasma can limit surface activation by removing hydroxyl groups to form siloxane
- H_2O_2 shows strong potential as an alternative to traditional oxidants for leading edge surface activation applciations

TECHCET

Electronics Materials Information

The Intersection of Semiconductors and Governments:

Increased Funding, Support…and Regulations

Kevin J. McLaughlin, PhD
Senior Business Development Director
TECHCET CA LLC

Surface Preparation and Cleaning Conference
April 25, 2024 | Chandler, AZ

Agenda

- Industry overview and growth drivers
- Funding, fabs, and supply chain
- Challenges (the boundary conditions)
- Summary

TECHCET

Semiconductors in the Spotlight

Suddenly everyone knows us!

The Big Picture

- This may look like trickle-down economics, but it's exactly the opposite.
- Materials make semiconductors make electronics which in turn power a major part of the global economy.

Materials comprise <15% of semiconductor value and only a tiny part of global GDP, but they enable the entire value chain!

Global Economy and the Electronics Supply Chain (2023)

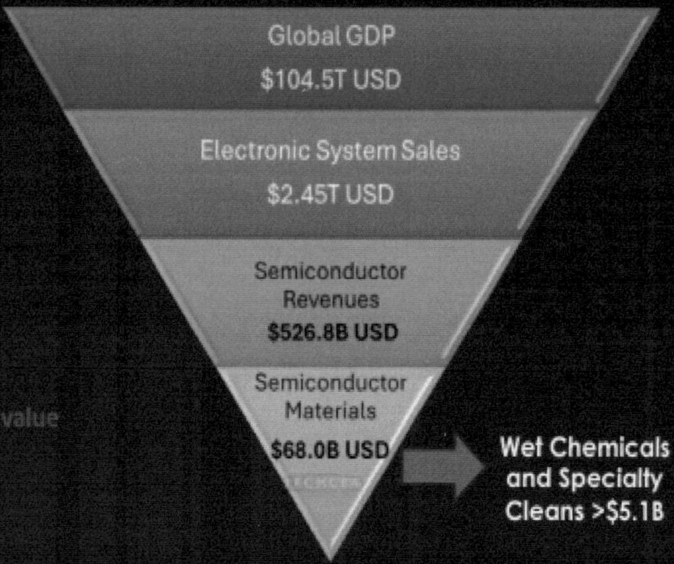

HERE'S WHERE WE'RE GOING

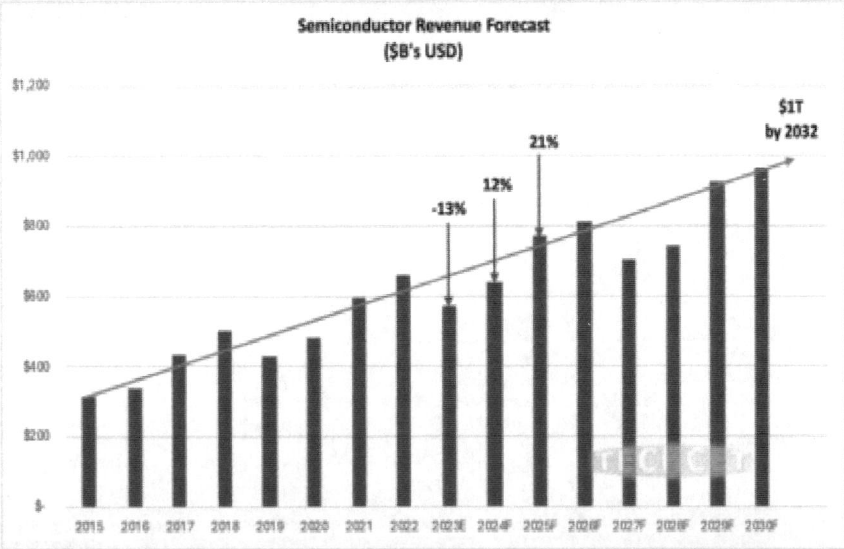

www.techcet.com
TECHCET

HERE'S WHY

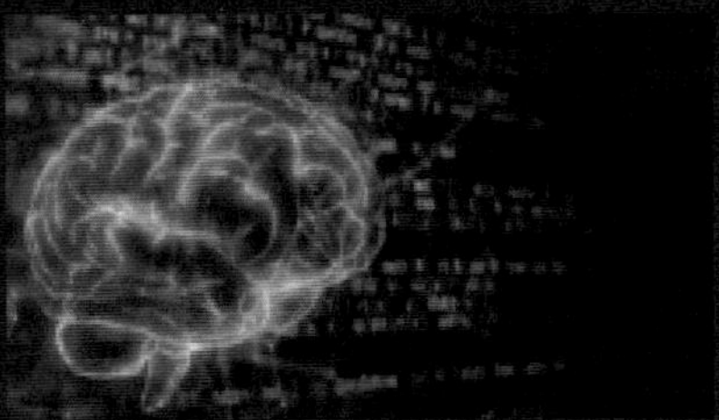

AI / ML

HPC / HBM

Automotive

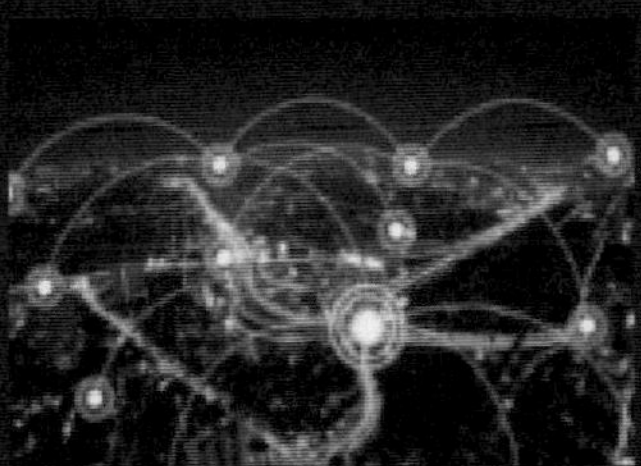

5G / IoT

Disruption-proofing
Supply Chain

www.techcet.com
TECHCET

HERE'S WHAT'S HAPPENING

From foreignpolicy.com: "…the world is future-proofing (and China-proofing)…" semiconductor supply chains

- Show me the money…
  - ✓ Government funding abounds (at least for big, shiny fabs)
  - ✓ About that supply chain…
- Fabs, fabs, everywhere new fabs!
  - ✓ 109 new fabs coming online 2022-2026
- Where are all the workers going to come from?
  - ✓ As many as 1M additional needed by 2030

TECHCET
SPCC Conference 042524 7

NEW GOVERNMENT FUNDING

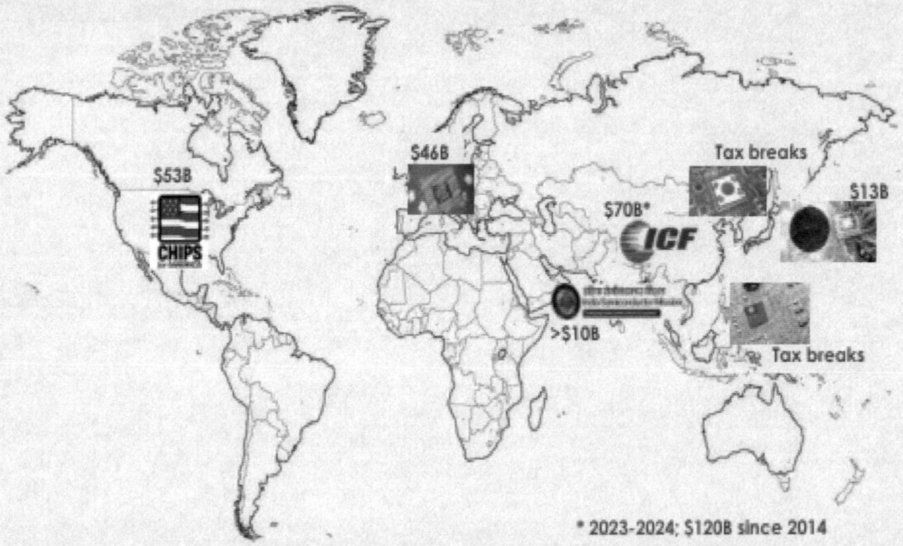

* 2023-2024; $120B since 2014

TECHCET
SPCC Conference 042524 8

NEW FABS 2022-2026

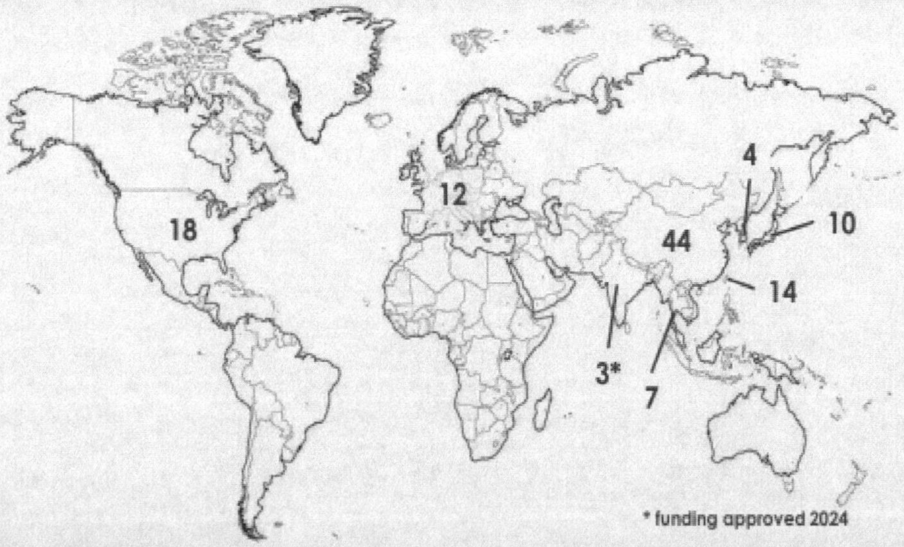

* funding approved 2024

CHIPS ACT(S) STARTS TO TAKE SHAPE

CHIPS Incentives Announcements

Preliminary memoranda of terms (PMTs) announced since December 2023

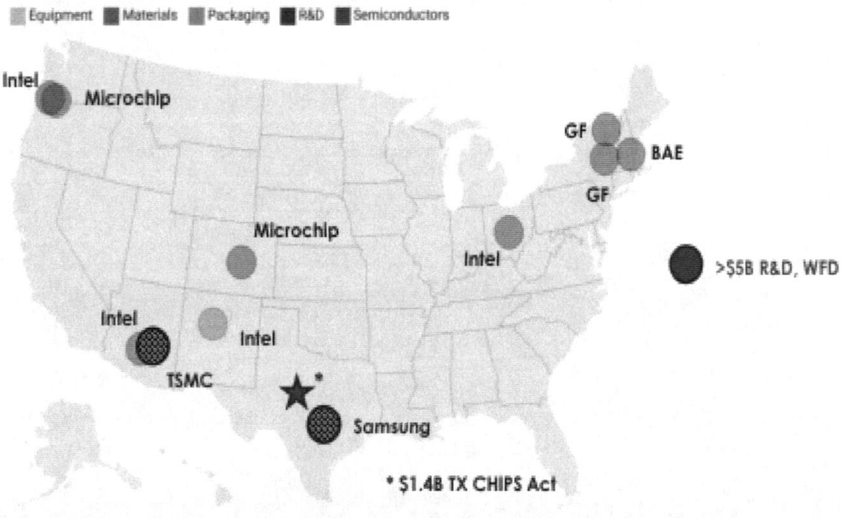

Source: CHIPS Program Office · Created with Datawrapper

MATERIALS REMAIN A QUESTION MARK

- CHIPS Act funds directed through Notices of Funding Opportunity (NOFOs)
  - ✓ NOFO 1: Commercial Fabrication Facilities (CFF)
    - Leading-edge fab applications accepted starting 3/31/2023
    - Current-generation and mature-node fab applications starting 5/1/2023
  - ✓ NOFO 1a: CFF Amendment
    - Materials and Equipment manufacturing applications (>$300M) starting 10/23/2023
  - ✓ NOFO 2: Facilities for Semiconductor Materials and Manufacturing Equipment (SMME)
    - Concept plans (<$300M) accepted for review 12/1/2023-2/1/2024
    - Concepts that are invited to submit full applications <u>will be notified, timing TBD</u>
- CHIPS Act also authorized 25% investment tax credit (ITC)
  - ✓ Materials manufacturers not included in this provision

> Section 48D provides a 25% investment tax credit (ITC) for an investment in an advanced manufacturing facility, defined in the statute as a facility for which the primary purpose is the manufacturing of semiconductors or semiconductor manufacturing equipment.

A DYNAMIC LANDSCAPE

TARIFFS AND TRADE RESTRICTIONS

25% tariff – both directions

Japan-Korea trade dispute (ended mid-2023)

Trade restrictions ON China (electronics, manufacturing equipment, and equipment support)

Trade restrictions BY China (minerals)

PFAS REGULATIONS

- Per- and Poly-fluoroalkyl Substances
  - ✓ If it has a C-F bond, it's under scrutiny
- 2021 EPA PFAS Strategic Roadmap
  - ✓ Confronting human health and environmental risks of PFAS
- SIA – Semiconductor PFAS Consortium
  - ✓ Non-PFAS alternatives, if they can be found, will require years to decades to develop and qualify
    - Wet chems: 3-15 years
    - PAGs: more than 25 years
    - Fluorocarbon-based plasmas: no known substitutes, could require different semiconductor device technology
- PFAS in Semiconductor Manufacturing
  - ✓ Process chemicals
    - Lithography (PAG, ARC, CAR)
    - Formulated wet chems (PERR, HAR pattern collapse, surface conditioning)
    - Plasma processing (etch, clean, deposition)
  - ✓ Ancillary chemicals
    - Fluoropolymers
    - Fluoroelastomers
    - Heat transfer fluids
    - Lubricants

Sustainability

US SEC Climate Disclosures

Washington D.C., March 6, 2024 —

SEC Adopts Rules to Enhance and Standardize Climate-Related Disclosures for Investors

EU Corporate Sustainability Reporting Directive

Merck Sustainability Strategy: 3-Goal Plan

SEMI Sustainability Initiative

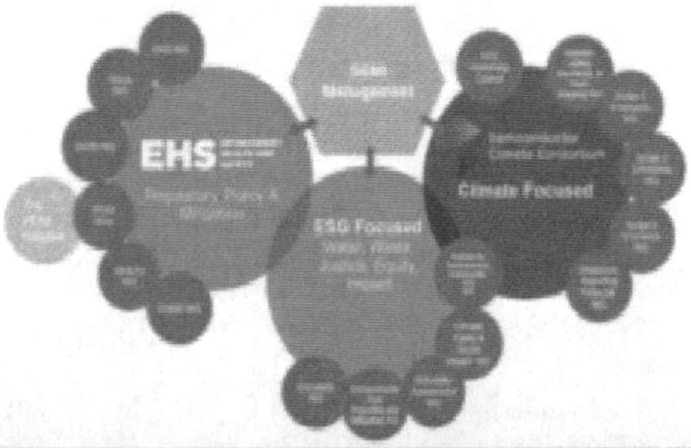

Samsung Semiconductor Sustainability Plan: Minimize Environmental Impact

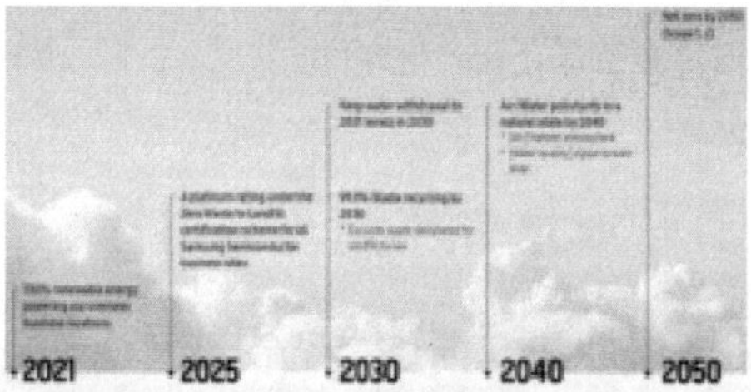

Diversity, Equity and Inclusion

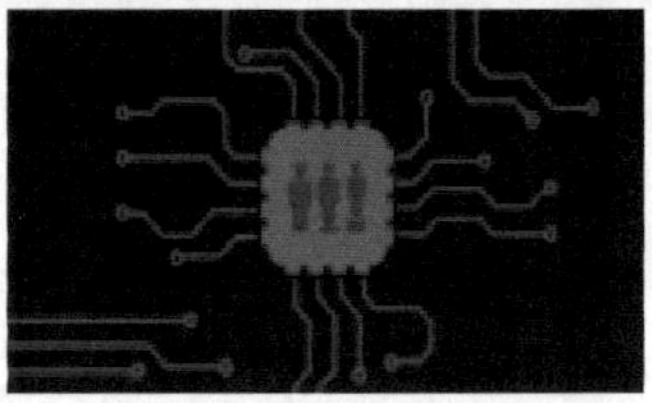

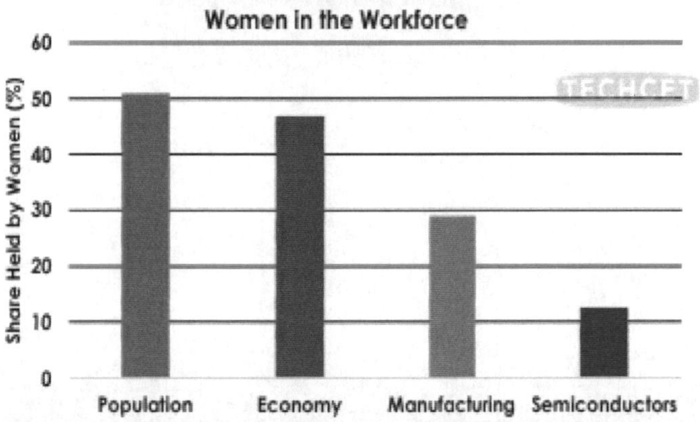

DEI In Semi: we need a million people!

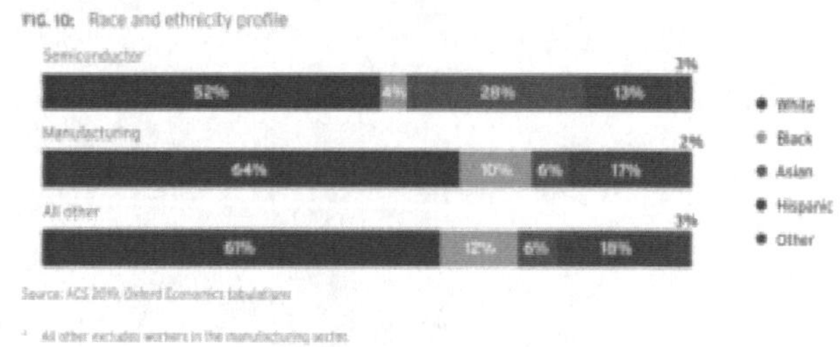

US CHIPS Act Support

- CHIPS Act legislation mentions "diversity" 64 times
  - ✓ Equity – 8 times
  - ✓ Inclusion – 24 times
- Designates Chief Diversity Officer at NSF
- $200M NSF Workforce Fund
- Funding applications required to address workforce requirements

Plain Old Economics

- TSMC
  - ✓ Fab 21 Phase 1 scheduled 2024, now 2025
  - ✓ Phase 2 originally 2025, now 2027-2028
- Intel
  - ✓ Ohio One facility groundbreaking September 2022
  - ✓ Originally to be complete in 2025
  - ✓ Now planning 2026-2027 completion, operational 2027-2028
- Suppliers report being "expansion-ready"
  - ✓ Awaiting "demand signals" before further investment
- CHIPS Act supports capex, harder to address opex
  - ✓ TSMC projects chip production costs to be up to 50% higher in US than in Taiwan

Intel Ohio One as of February 2024

Source: Intel

www.techcet.com
TECHCET

Summary

- Semiconductors have moved out of the shadows and into the limelight.
  - ✓ High technology, harnessing quantum mechanics, using solid state physics, all enabled by chemistry!
  - ✓ $1T and more is on the way
  - ✓ BUT...it's a path function, not a state function
- We cannot ignore or avoid the harsh realities that form the boundary conditions:
  - ✓ Innovating to angstrom levels and in 3 dimensions
  - ✓ Geopolitics
  - ✓ Societal and environmental stewardship
  - ✓ Making the (billions of) dollars make sense
- We will get there...
  - ✓ The trillion-dollar question: what will our industry, our society, and our planet look like when we do?

www.techcet.com
TECHCET

Thank You!

Questions?

TECHCET 19

Is ultrahigh vacuum technology potential approach to prepare industrial Si surfaces?

Zahra (Elmira) Jahanshah Rad
zsjara@utu.fi
+358 413198103

UNIVERSITY OF TURKU

24.4.2024

The Surface Preparation and Cleaning

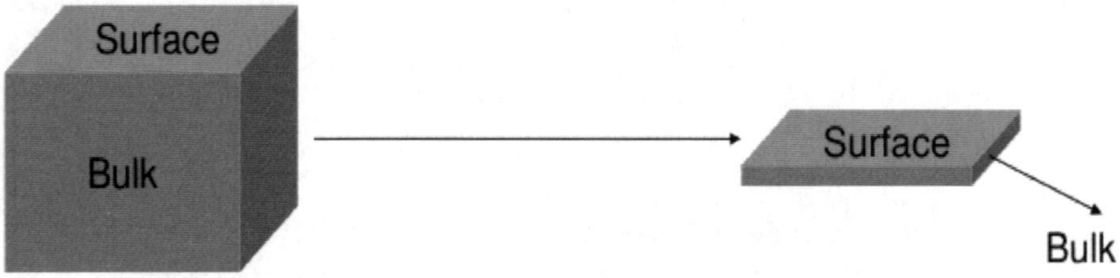

$$\frac{Bulk}{Surface} \downarrow$$

The role of surface preparation and cleaning becomes more important.

UNIVERSITY OF TURKU

The Surface Preparation and Cleaning

Wolfgang Pauli

"God made the bulk; surfaces were invented by the devil."

UNIVERSITY OF TURKU

The Surface Preparation and Cleaning of Silicon

Wet chemical methods used for several decades to clean Si surface and to produce ultrathin SiO2

There is a need to Reduce the consumption of chemicals, water, and energy

Developing dry cleaning methods

UNIVERSITY OF TURKU

The Surface Preparation and Cleaning of Silicon

Dry or wet-chemically cleaned Si surface has low degree of crystallinity

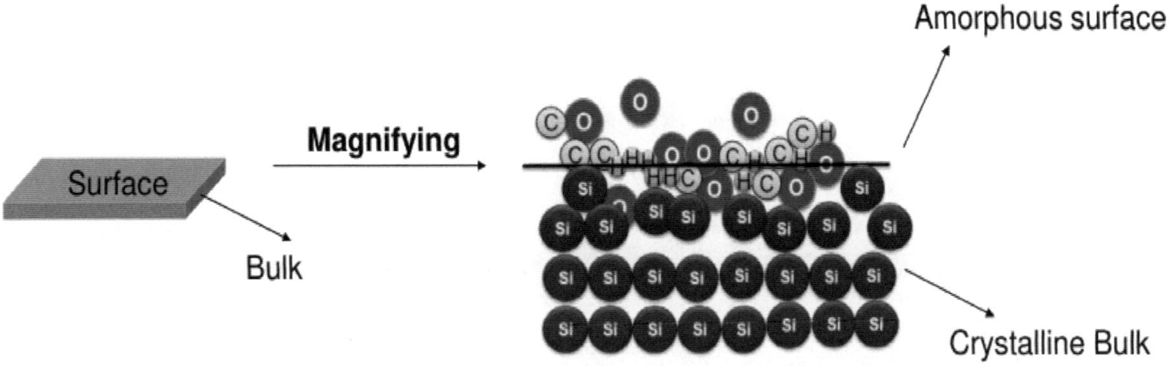

UNIVERSITY OF TURKU

The Surface Preparation and Cleaning of Silicon

A clean Si surface does not imply a crystalline surface

Amorphous Si surface includes high density of point defects affecting:

- **Electrical and optical properties (including power consumption)**
- **Manufacturing yield**
- **Working lifetime**

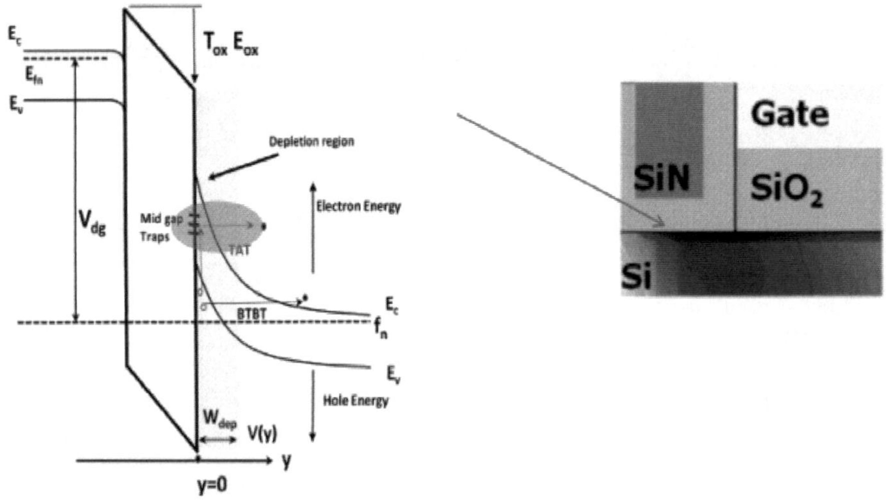

The Surface Preparation and Cleaning of Silicon

One solution

Using an ultrahigh vacuum step in the cleaning process of Si surface

Results in:

- Reduced point defects
- Improved performance of devices made of Si
- Increased yield

The Surface Preparation and Cleaning of Silicon

Most surface characterization methods are not sensitive to crystalline quality of the surface

We have used characterization methods that are sensitive to crystalline degree of the surface:

- Low energy electron diffraction (LEED)
- Scanning tunneling microscopy (STM)

UNIVERSITY OF TURKU

The Surface Preparation and Cleaning of Silicon

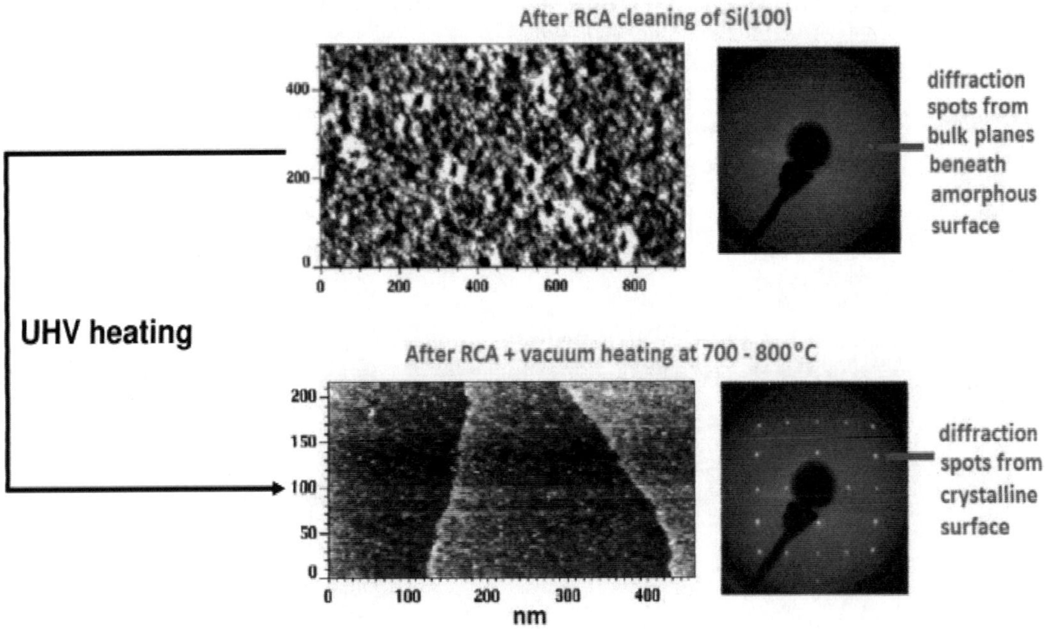

UNIVERSITY OF TURKU

The Surface Preparation and Cleaning of Silicon

UHV systems are considered complex and are not often used in silicon industry

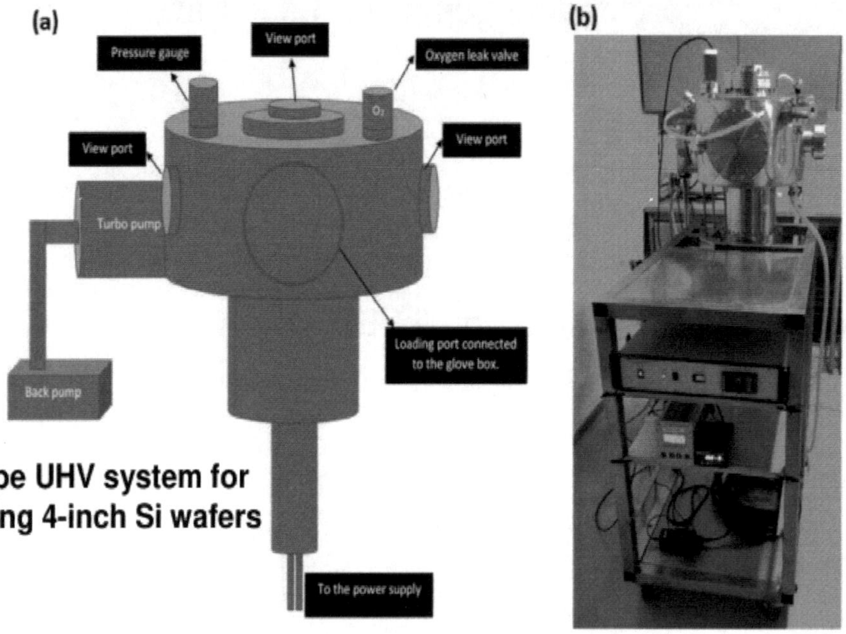

Prototype UHV system for processing 4-inch Si wafers

UNIVERSITY OF TURKU

The Surface Preparation and Cleaning of Silicon

Controlled oxidation of Si surface in UHV environment

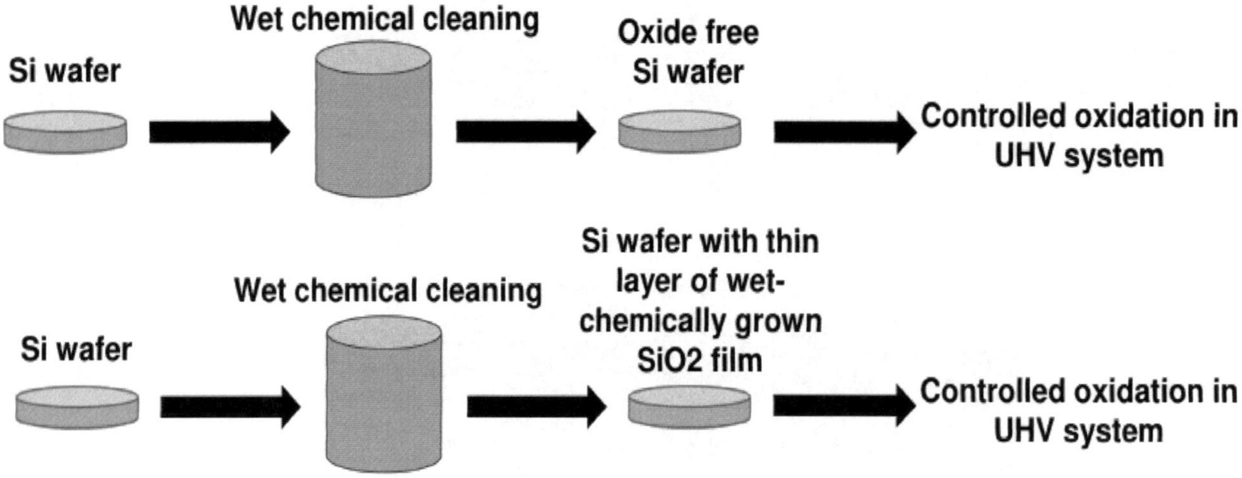

UNIVERSITY OF TURKU

The Surface Preparation and Cleaning of Silicon

Controlled oxidation of Si surface in UHV environment

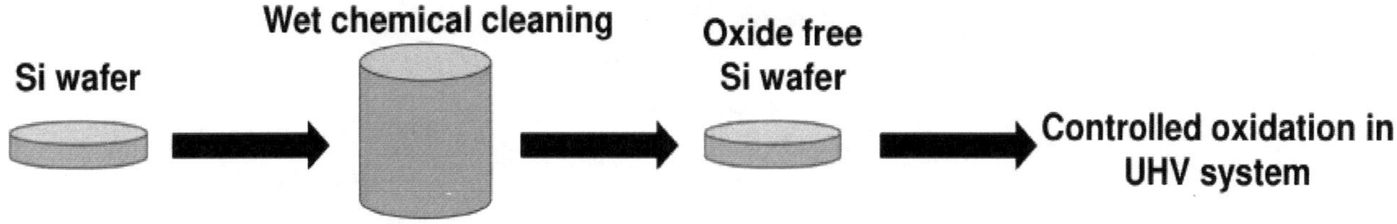

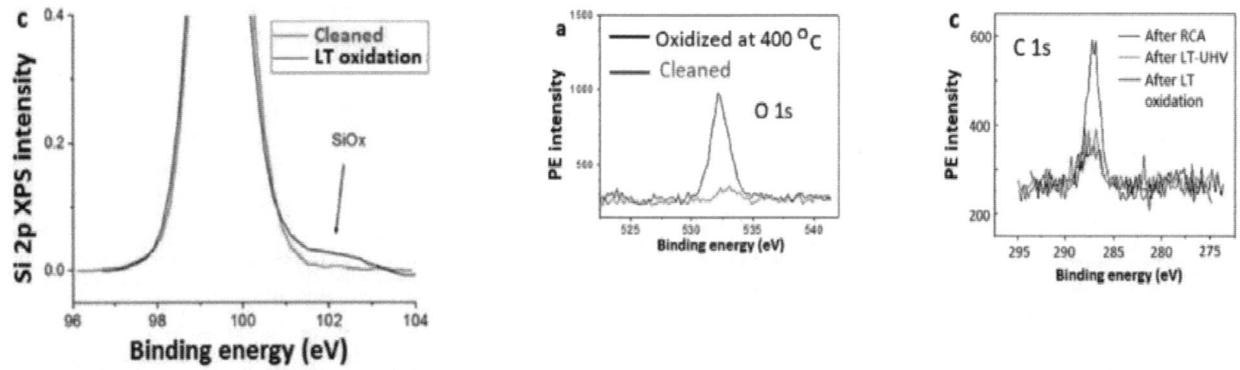

UNIVERSITY OF TURKU

https://dx.doi.org/10.1021/acsami.0c12636

The Surface Preparation and Cleaning of Silicon

Controlled oxidation of Si surface in UHV environment

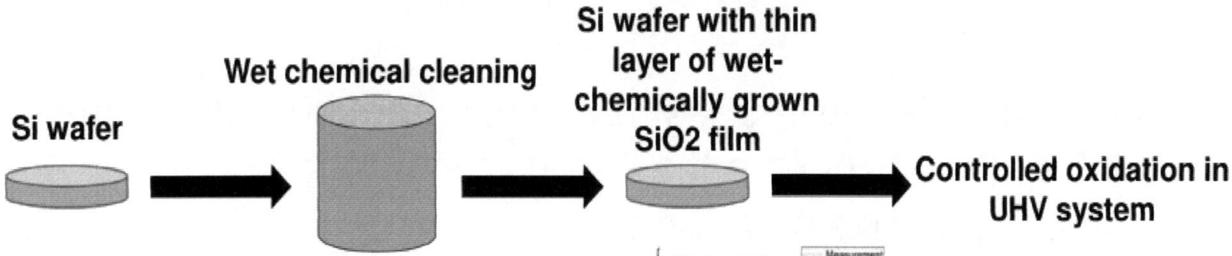

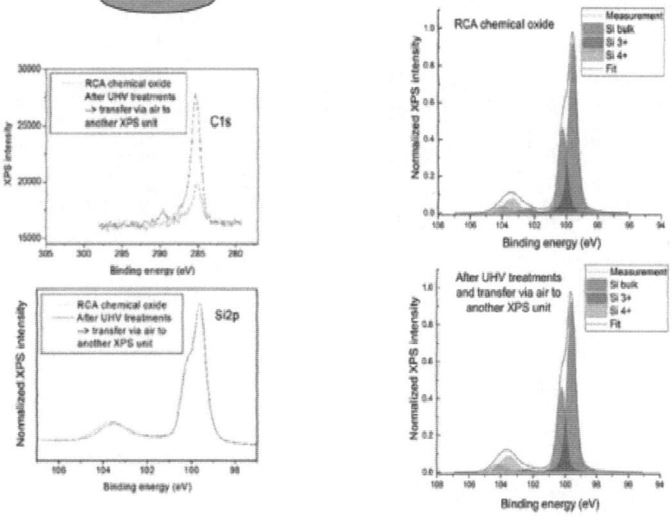

Vacuum 202 (2022) 111134

UNIVERSITY OF TURKU

The Surface Preparation and Cleaning of Silicon

Controlled oxidation of Si surface in UHV environment

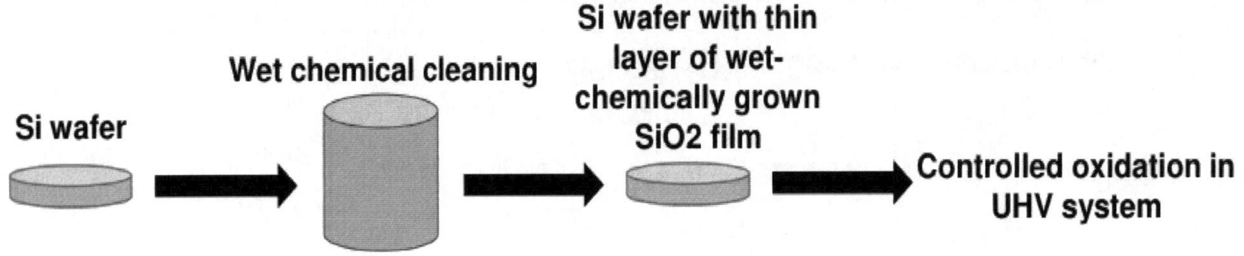

	Reference sample with chemically grown silicon oxide	Chemically grown silicon oxide + LT-UHV treatments
D_{it} ($eV^{-1}cm^{-2}$) for planar p-Si	$2.6 \cdot 10^{11}$ ($4.1 \cdot 10^{11}$)	$1.5 \cdot 10^{11}$ ($2.2 \cdot 10^{11}$)
Q_{tot} (q/cm^2) for planar p-Si	$-1.7 \cdot 10^{12}$ ($-1.6 \cdot 10^{12}$)	$-2.3 \cdot 10^{12}$ ($-2.3 \cdot 10^{12}$)

	Reference sample with chemically grown silicon oxide	Chemically grown silicon oxide + LT-UHV treatments at 350 °C	Chemically grown silicon oxide + LT-UHV treatments at 450 °C
τ for planar Si	2.2 ms	2.9 ms	1.0 ms
τ for black Si	0.9 ms (1.0 ms)	2.4 ms (2.1 ms)	0.7 ms

Vacuum 202 (2022) 111134

UNIVERSITY OF TURKU

The Surface Preparation and Cleaning of Silicon

Based on the results, we suggest that semiconductor community conders integrating UHV with the current technology of Si surface cleaning and preparation to:

- **Increase the degree of crystallinity of the surface and reduced point defects**
- **Reduce the amount of contamination on the surface**

UNIVERSITY OF TURKU

UNIVERSITY
OF TURKU
zsjara@utu.fi

Insights into Hydrophobicity or Hygroscopicity of Non-Volatile Residue

Derek Oberreit PhD, Kanomax FMT
Derek.Oberreit@kanomaxfmt.com
Michael Walker, Kanomax FMT
Jikku Thomas PhD, Kanomax FMT

The Surface Preparation and Cleaning Conference 2024
Phoenix, AZ

KANOMAX FMT
A Kanomax Company
NANOPARTICLE MEASUREMENT SOLUTIONS

Motivation

- Particles formed through aerosolization of a process chemical mimics formation on wafers after rinse and dry
- Condensation Particle Counting of particles formed by aerosolization has shown correlation with wafer defects
- Current technology utilizes Butyl alcohol as a condensing fluid
  - Reduced material dependence
  - Low miscibility with water, which reduces fouling
  - Raises flammability, AMC, and health concerns
- Water-based CPCs are an alternate option
  - Introduces some material dependence due to Hydrophilicity and Hygroscopicity of particle material, which can raise/lower detection threshold size

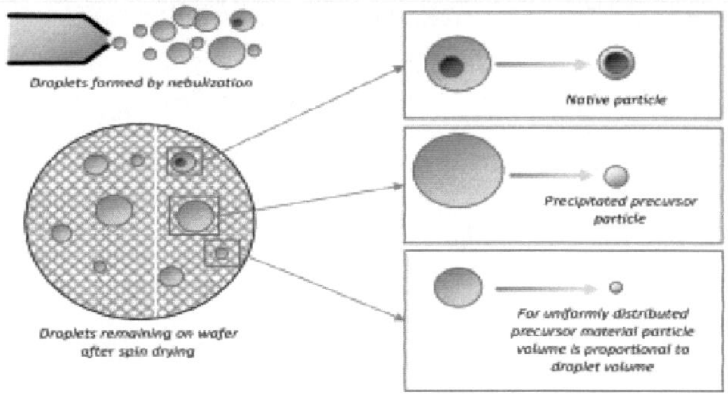

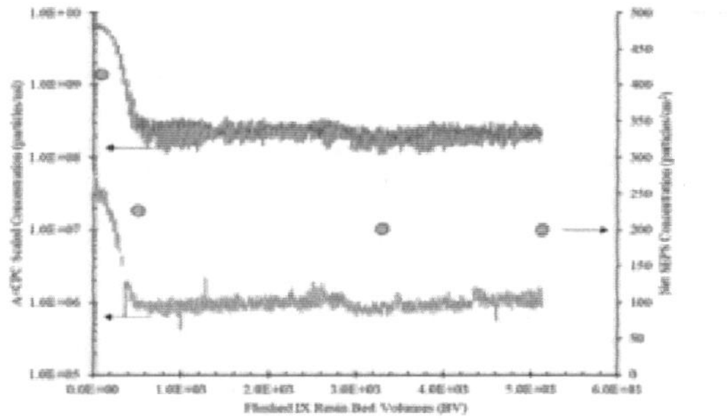

Rinse curve of an ion exchange resin showing corrected volume concentration measured A+CPC at 3nm threshold diameter (top trace) and 9 nm threshold diameter (bottom trace) and particle surface concentration measured using SSC + SEPS > 8 nm (circles), where Bed Volumes (BV) is the total amount of water flushed through the resin divided by the volume of the resin tank. (*Ultrapure Micro Conference*, Phoenix, 2022)

Page 2

Relevant Thermodynamics - Sorption

- Water Activity
  $a_w = \frac{p}{p^*}$
  where p = partial water vapor pressure in "solution" and p^* is the partial pressure for pure water
- Lower water activity reduces surface vapor pressure which lowers the size threshold for nucleation
- Deliquescence
  - Occurs when the water/particle mixture is more thermodynamically stable as a saturated solution
  - Causes a step change in growth factor in water uptake studies
  - Deliquescence Relative humidity (DRH) ≈ a_w for large particles
  - DRH increases inversely with particle diameter below ≈ 50 nm

Material (saturated aqueous solution)	a_w @ 25 C
Sodium chloride	0.753
Potassium chloride	0.843
Potassium sulfate	0.973
Lithium bromide	0.064
Sucrose	0.843

DOI 10.1002/9780470376454.app1
DOI: 10.1080/10942910903474393

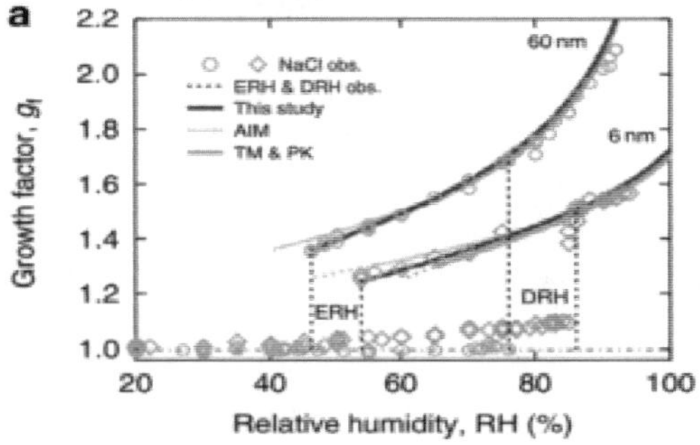

Cheng, Y., Su, H., Koop, T. et al. Size dependence of phase transitions in aerosol nanoparticles. Nat Commun 6, 5923 (2015). https://doi.org/10.1038/ncomms6923

Relevant Thermodynamics – Condensation

- Condensation of a supersaturated vapor onto particle surface largely driven by particle size and saturation ratio

$$S_{\max} = \frac{P_{\mathrm{vap}}}{P_{\mathrm{sat}}} \propto e^{\frac{1}{D_t}}$$

- The peak saturation ratio is determined by the operating temperatures of the saturation and condensation sections of the device.
- The threshold particle size for detection can be adjusted by varying the peak saturation ratio.
- Wettability/contact angle of a condensing vapor onto the particle surface can also affect the threshold diameter

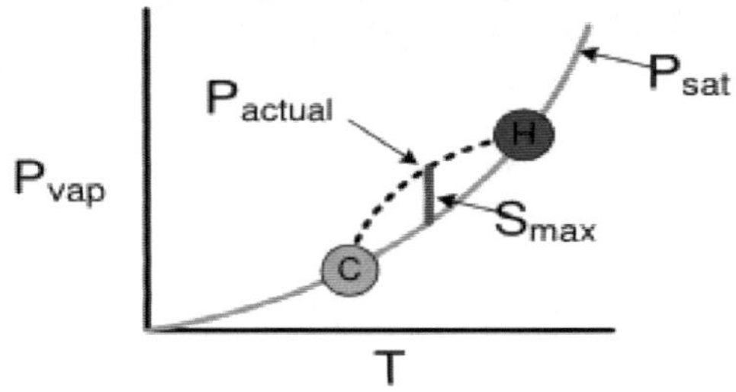

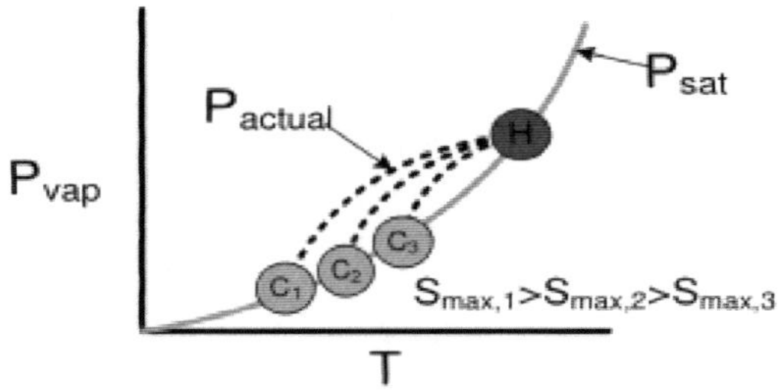

Experiment

Objective: Challenge aerosolization-based instrument with hygroscopic and hydrophilic material to compare the response of CPCs using water or butyl alcohol as the condensing fluid.

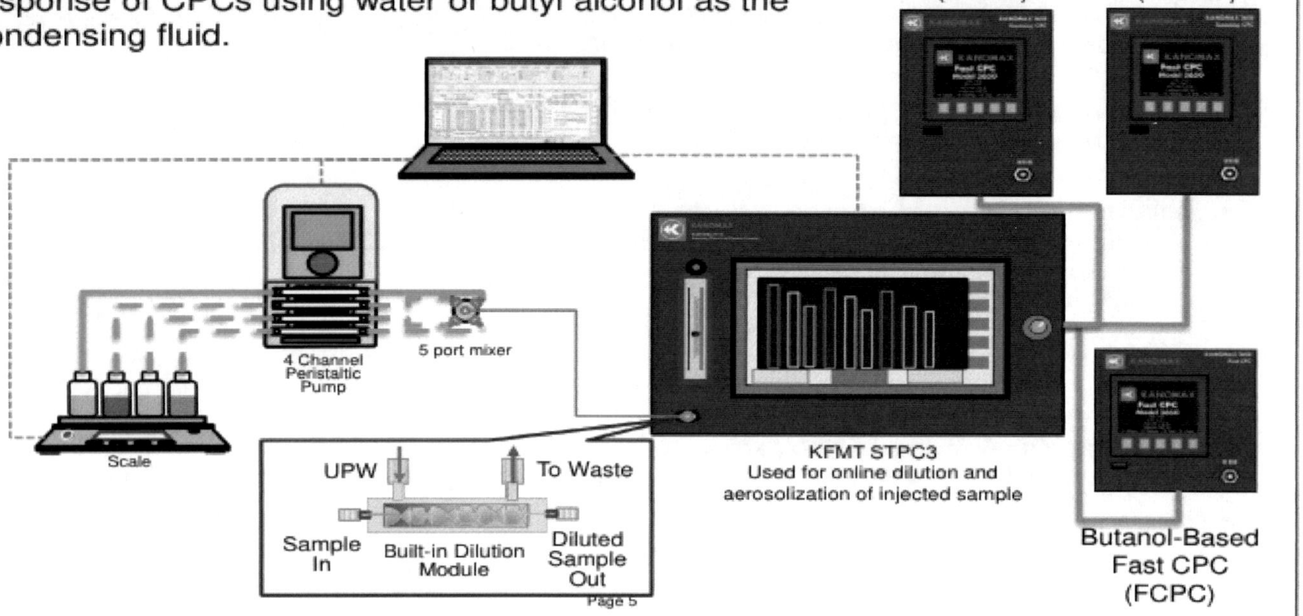

Results – Sodium Chloride

- Lowest water activity of materials tested
- Water-based CPC showed increased response on all size threshold channels
  - 9 nm @ 0.3 ppb
  - 15 nm @ 1 ppb

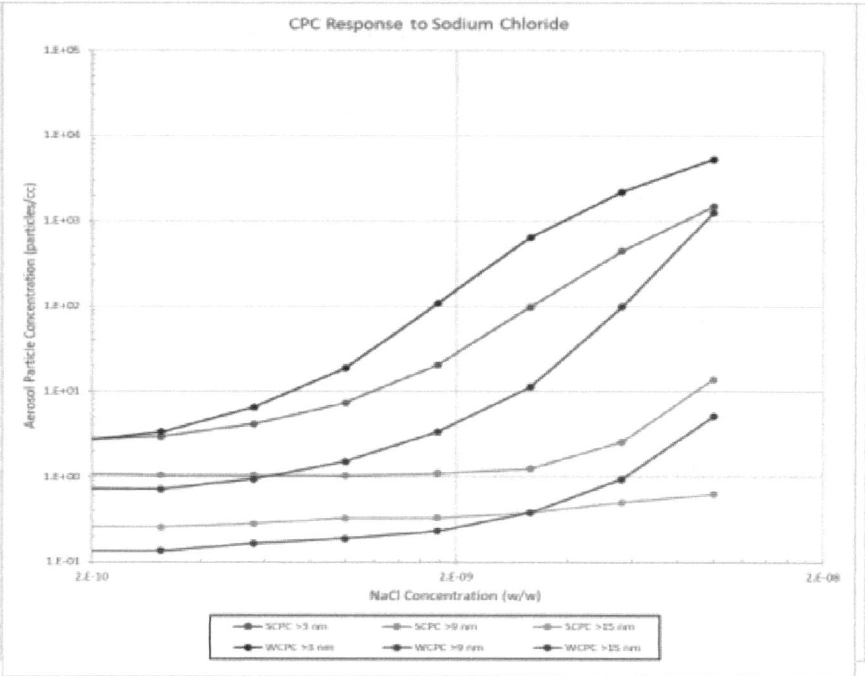

Results – Potassium Chloride

- Mid-range water activity
- Water-based CPC showed increased response on all size threshold channels
  - 9 nm @ 0.6 ppb
  - 15 nm @ 1 ppb
- Detection onset concentration slightly higher than Sodium chloride
- Peak concentration for water CPC slightly lower than NaCl for water-based CPC

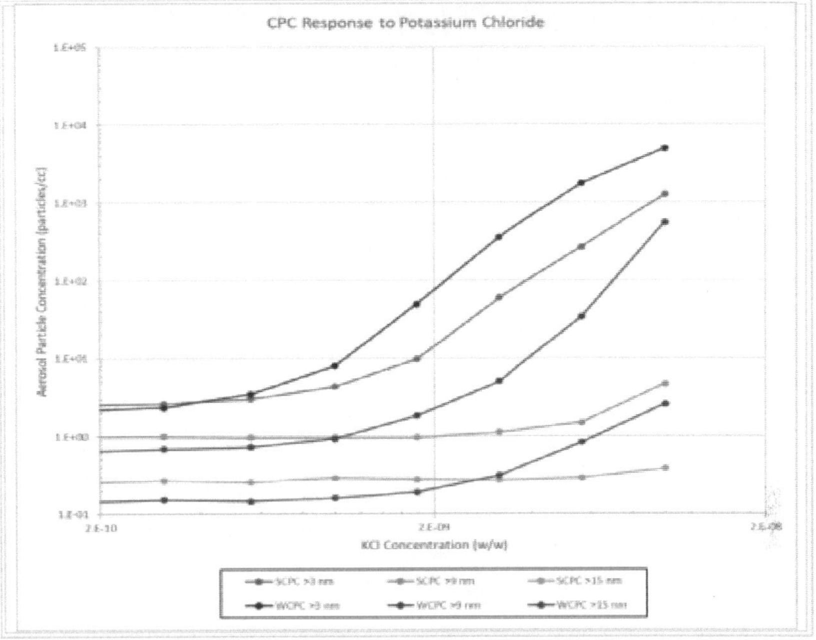

Page 7

Results – Potassium Sulfate

- Highest water activity for salts tested
- Water-based CPC showed increased response on all size threshold channels
  - 9 nm @ 0.6 ppb
  - 15 nm @ 1 ppb
- Peak concentration for 9 and 15 nm size thresholds on the order of 10X higher than KCl and NaCl
  - Shows saturation of detector at > 10^4 particles/cc
- Would expect less hygroscopic particle growth
  - Salt purity?

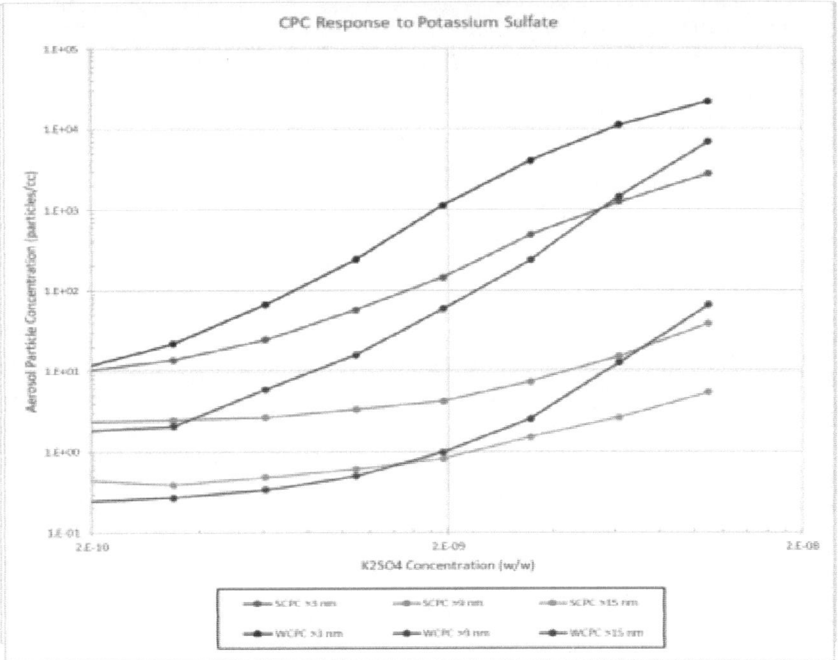

Page 8

Results - Sucrose

- Showed limited response for all working fluids. Not expected based on DRH published in literature
- Possibly due to formation of anhydrous crystalline sucrose; however, does not explain overall lower response

J. Agric. Food Chem. 1956, 4, 1, 72–77
Publication Date:January 1, 1956
https://doi.org/10.1021/jf60059a010

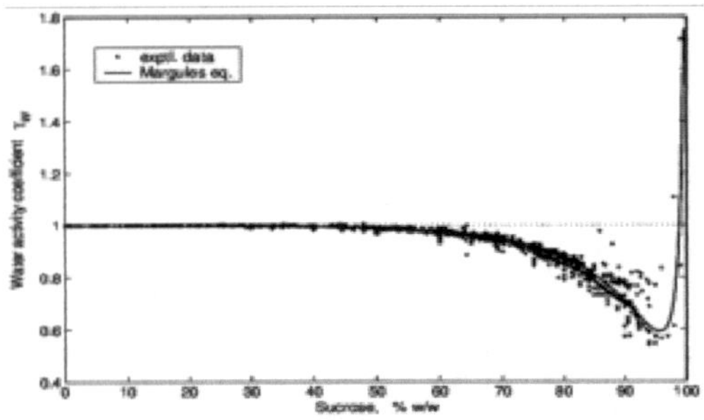

Starzak, Maciej & Mathlouthi, Mohamed. (2002). Zuckerindustrie. 127. 175-185.

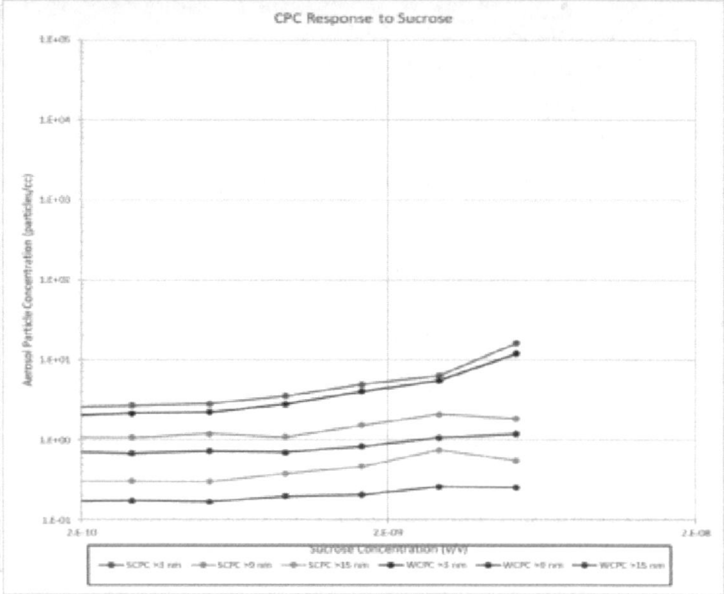

Results – PFA Extract

- Generally little material dependency observed
- All CPCs saw "spike" at the start of injections, which has been observed by others for similar experiments

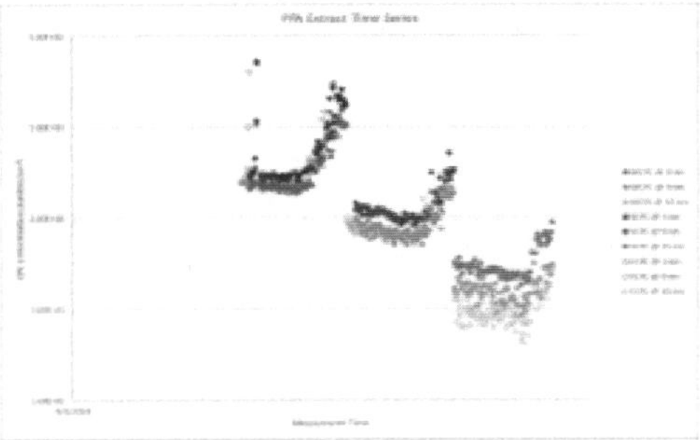

PFA extract provided by CT Assoiates, Inc..

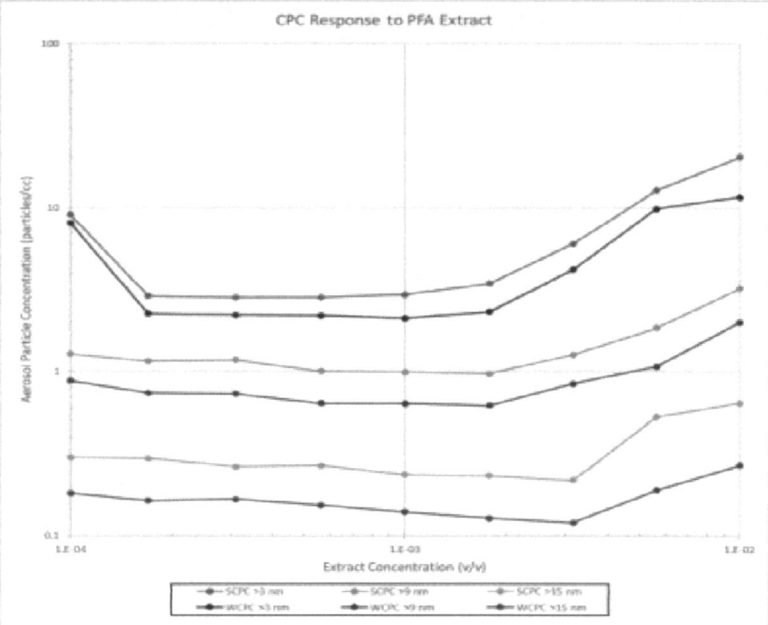

Results – IX Resin Spool Rinse

- Water CPC saw higher readings on 9 and 15 nm threshold diameters

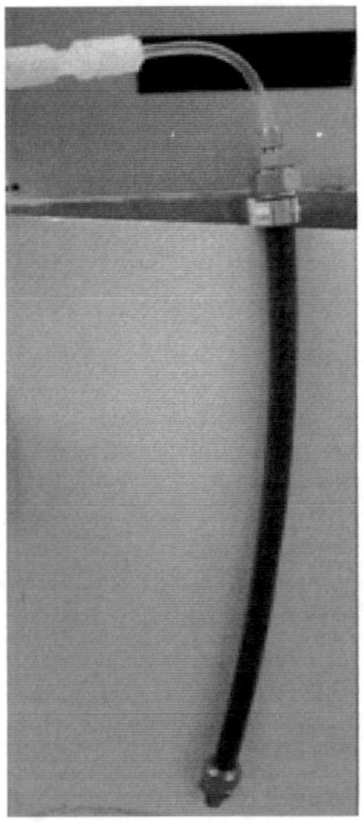

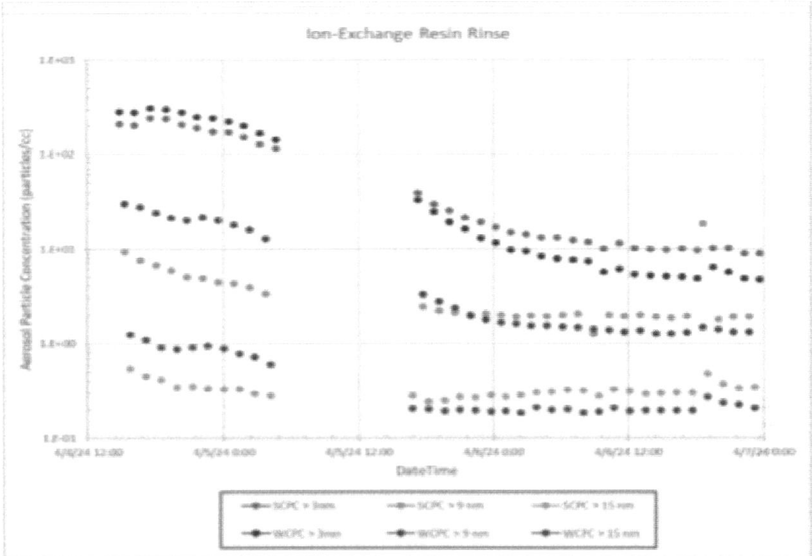

Results – HCl injection

- 3% Semiconductor Grade HCl injected at 260 µl/min into 100 ml/min UPW resulting in 80 ppm total HCl
- CPCs showed a similar response regardless of working fluid during HCl injection
- Post-injection rinse observed much more strongly on WCPC following pH change at the nebulizer

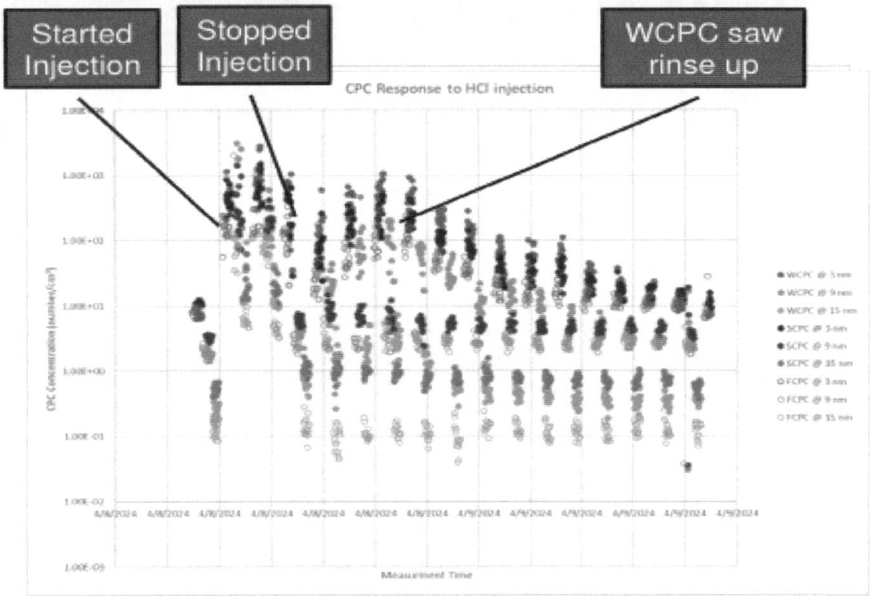

Results – UV On/Off

- UV TOC lamps turned off to investigate material dependence on organic material
- Water-based CPC detected fewer particles under this excursion condition; however, both CPCs captured the event

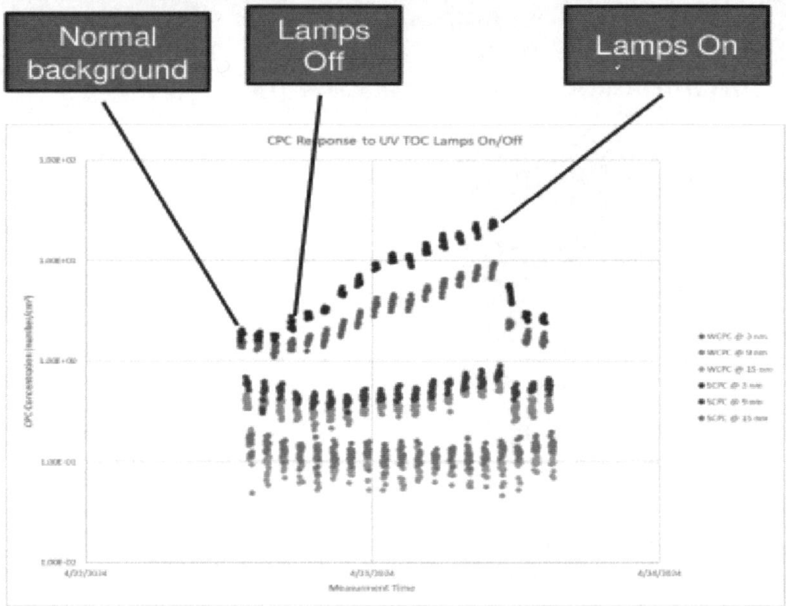

Page 13

Results – Colloidal Silica

- Confirms general counting efficiency
- Differences in low concentrations between butanol CPCs likely due to differences in background concentration
- Lower reading by water-based CPC at low silica concentrations similar to prior observations
  - Possibly due to lower amounts of precipitated residue on silica surface

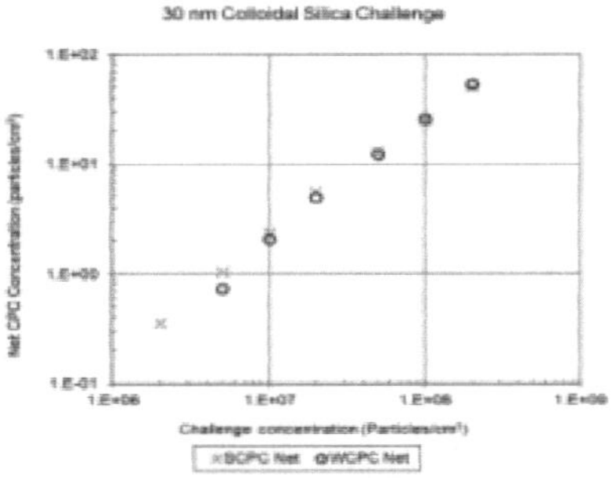

Data collected December 2023

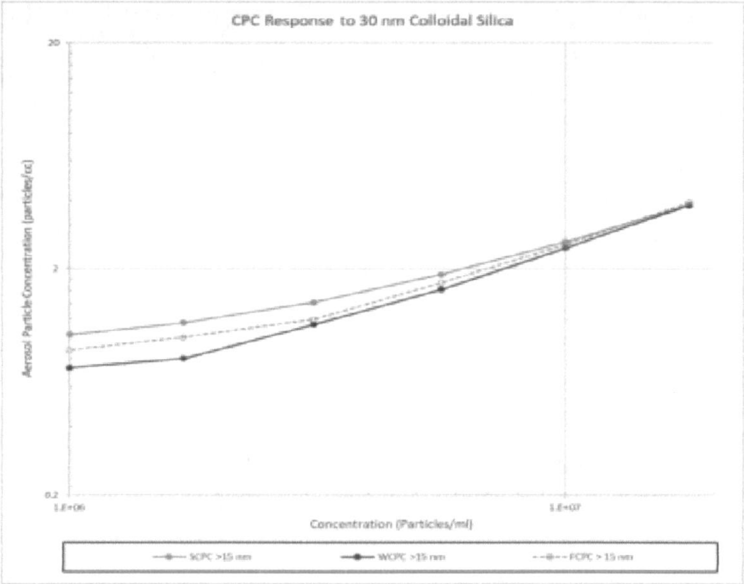

Page 14

Discussion and Takeaway

- The Condensation Particle Counter working fluid does have an effect on the response to some materials
  - –Demonstrated in this work for hygroscopic salts
- Water activity from bulk measurements may not be sufficient factors for estimating material dependence
- Material state/structure may contribute to condensation efficiency
  - –Is this what we observe with Sucrose?
- Real-world examples such as resin rinse, background monitoring, and HCl injection also show some variability between Water-based and Butanol-based Condensation Particle Counters.

Follow on work

- Repeat injections of Sucrose and Potassium Sulfate to confirm unexpected observations
  - –Increase sucrose concentration
- Inject materials with relevant water-specific properties
  - –Amphipathic molecules (organic acid)
  - –Hydrophobic molecules (oil)
- Perform resin rinse with alternate IX resin source
- Wafer defect correlation
  - –Hydrophilic/phobic wafer surfaces

Thank you!

The Surface Preparation and Cleaning Conference 2024
Phoenix, AZ

KANOMAX FMT
A Kanomax Company
NANOPARTICLE MEASUREMENT SOLUTIONS

Supplemental Materials

Vapor Condensation Overview

The Surface Preparation and Cleaning Conference 2024
Phoenix, AZ

KANOMAX FMT
A Kanomax Company
NANOPARTICLE MEASUREMENT SOLUTIONS

When does vapor condense onto particles?

First: Describe the effect of surface curvature

- $Work_{(surface)} = Surface\ Tension \times dArea_{(surface)}$
- Liquid Surface ± 1 Molecule → $|dArea_{(surface)}|$

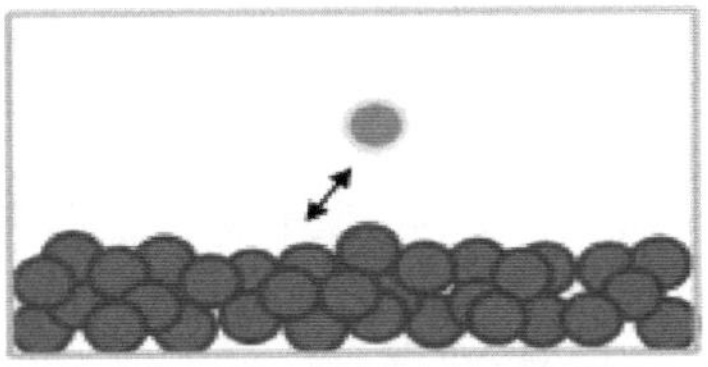

$$|dArea_{(flat)}| \rightarrow 0 \quad |dArea_{(flat)}| < |dArea_{(sphere)}|$$

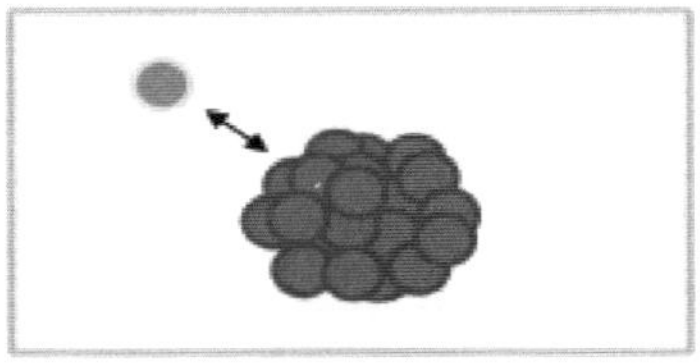

Surface Energy$_{(flat)}$ < Surface Energy$_{(sphere)}$

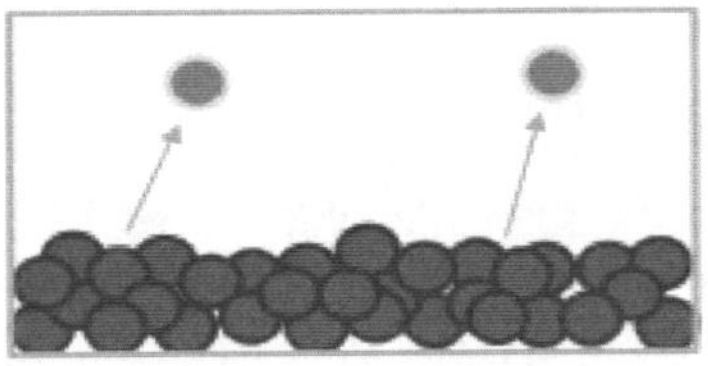

$$Flux_{(evap,flat)} < Flux_{(evap,sphere)}$$

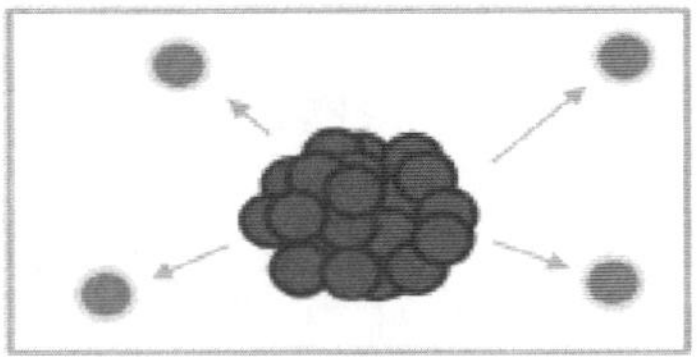

Page 19

When does vapor condense onto particles?

Second: Describe equilibrium vapor pressure

- Vapor pressure (P_{vapor}) and Vapor flux (F_{vapor}) are both functions of the vapor molecule concentration (n),
  - $P_{vapor} = nk_BT$
  - $F_{vapor} = n\bar{C}/4$
- $Flux_{(vapor,evaporation,flat)} < Flux_{(vapor,evaporation,sphere)}$
- At equilibrium, $Flux_{(vapor,evaporation)} = Flux_{(vapor,condensation)}$

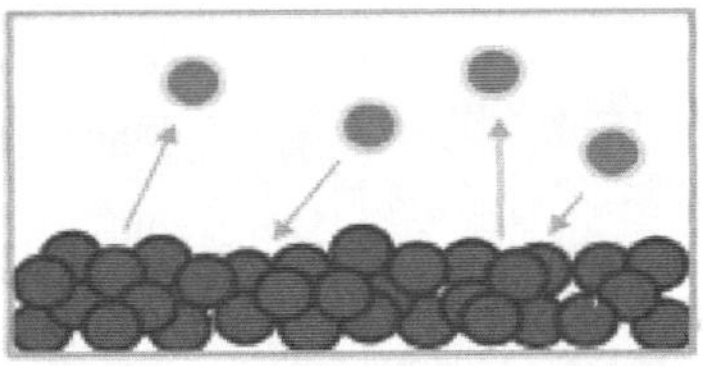

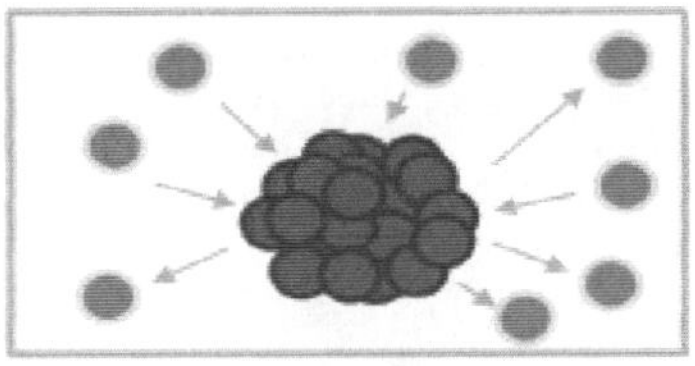

$$P_{(vapor,\ equilibrium,\ flat)} = P_{(vapor,\ saturation)} < P_{(vapor,\ equilibrium,\ sphere)}$$

Page 20

When does vapor condense onto particles?

Third: Introduce the Kelvin Equation

$$radius_{equilibrium} = \frac{-2\sigma V_{vap}}{k_B T} \ln\left(\frac{P_{sat}}{P_{vapor}}\right)$$

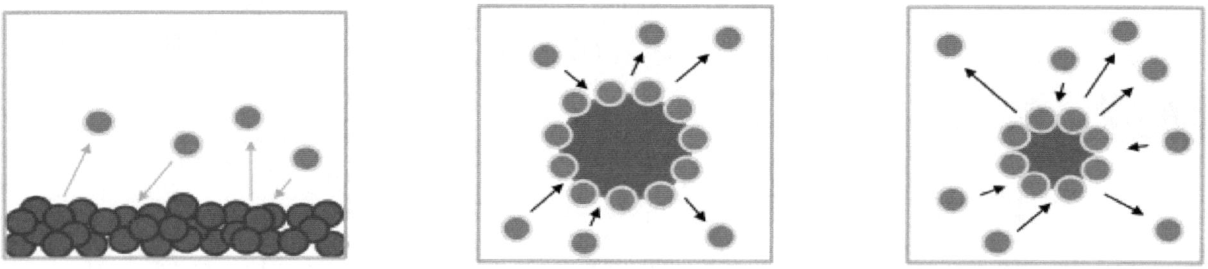

$P_{(vapor,sat)}$ < $P_{(vapor,\ large)}$ < $P_{(vapor,small)}$

Answer: When $Radius_{(particle)}$ > $Radius_{(equilibrium)}$

Page 21

Contamination Monitoring on Photomask Substrates with TXRF

Logic Technology Development, Intel Corporation
Semiconductor Metrology Division, Rigaku Corporation

Intel Mask Operation (IMO): Sergey Mezhenny, TD Engineer and Maxwell Slaton, TD Engineer

Rigaku: Ben Buford, Principal Engineer; Meredith Beebe, Application Scientist;

Hiroshi Kono, General Manager; Motohiro Nakanishi, Application Scientist;

Shinya Kikuta, Application Scientist; Markus Kuhn, Vice President

Outline

- Photomask Contamination
- Conventional TXRF Overview
- TXRF Photomask Capability Demonstration and Hardware
- TXRF Contamination Monitoring on Photomask Blanks
- Beyond Photomask Blanks
- Conclusion

Outline

- Photomask Contamination
- Conventional TXRF Overview
- TXRF Photomask Capability Demonstration and Hardware
- TXRF Contamination Monitoring on Photomask Blanks
- Beyond Photomask Blanks
- Conclusion

Contaminants on Different Photomask Types

- Optical
  - Transmission through substrate (quartz)
    - On quartz: Absorbs scanner illumination
    - On light absorbing material: Excessive, clusters or mobile
- EUV
  - SLM mirror reflection
    - On mirror: Almost all materials
    - On absorber: Excessive, clusters or mobile

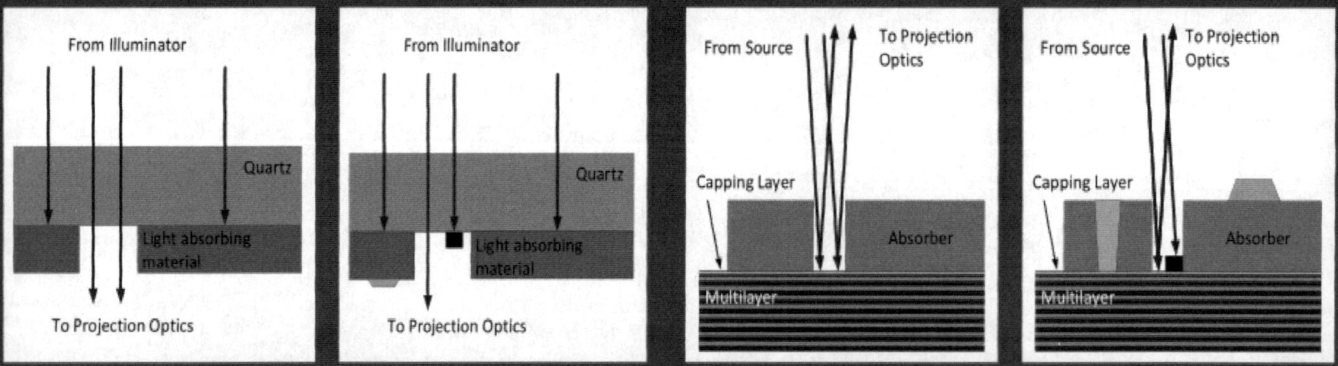

Defect Free Mask | Mask with Defects | Defect Free Mask | Mask with Defects

Contaminants on Photomasks – A Chemical Perspective

- Optical
  - Quartz
    - Clean interface on both sides of quartz
  - Light absorbing material
    - Embedded contaminants removed during etch
    - Limited non-printing defects without clusters or mobile contaminants
- EUV
  - Capping Layer
    - Post etch residue completely removed for maximum reflectivity
  - Absorber
    - Embedded contaminants removed during etch
    - Limited non-printing defects without clusters or mobile contaminants

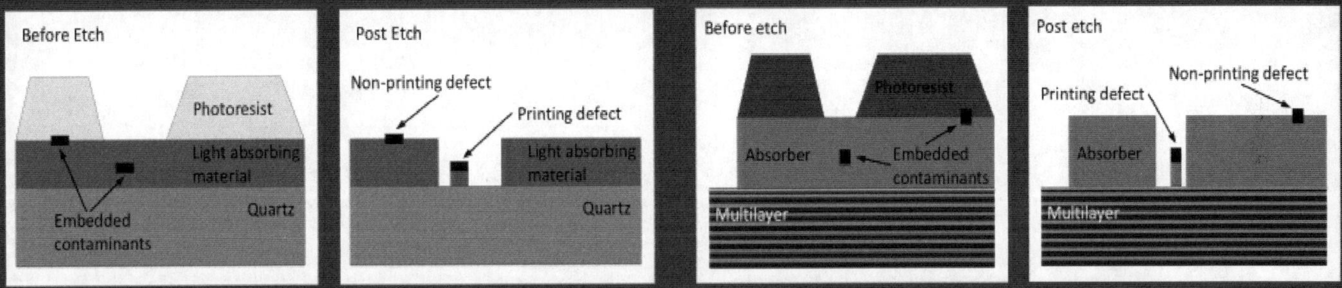

Photomask TXRF Motivation

- Existing Techniques
  - Non-Destructive
    - Actinic and blank inspections
    - SEM
    - Scanning Probes
    - Vibrational Spectroscopies
  - Destructive Analysis
    - Cross-sectional TEM
    - Extractive analysis
  - Current and Future Needs
    - Affordable non-destructive surface-sensitive methods capable of detecting both metallic and non-metallic inorganic contaminants

Failure Analysis Ecosystem

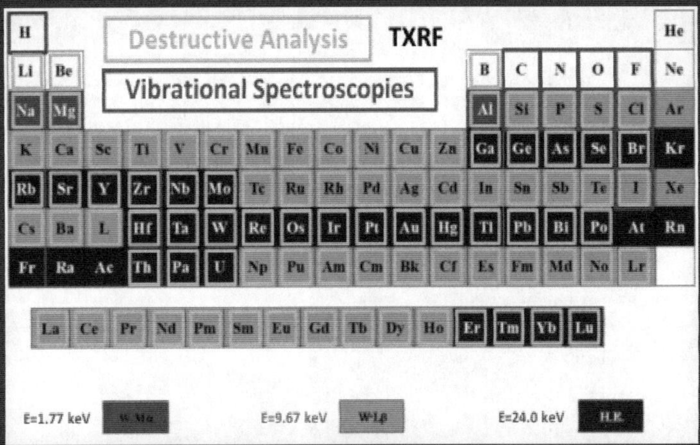

Outline

- Photomask Contamination
- Conventional TXRF Overview
- TXRF Photomask Capability Demonstration and Hardware
- TXRF Contamination Monitoring on Photomask Blanks
- Beyond Photomask Blanks
- Conclusion

Total Reflective X-ray Fluorescence

- Advantages: non-destructive wafer surface measurement, multi-element, 10^9 atoms/cm^2 detection limits for transition metals, sensitive to particles, easy to operate
- Disadvantages: Light element sensitivity, spectral overlaps

Instrumentation:

- X-Ray source
- Monochromator
- X/Y/Z and θ/ω sample stage
- Silicon drift detector (SDD)

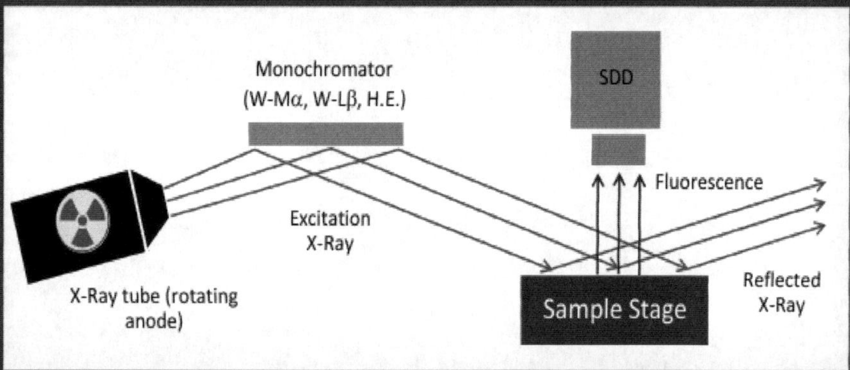

Fitting Algorithm

- Spectra fitting is done using a tool simulation including all known tool response artifacts and relative XRF peak intensities.
- The tool model is optimized over a large database of measured spectra, then fine tuned for each tool
- Bayesian sampling is used to integrate peaks from fit distribution, to avoid errors from imperfect peak fits

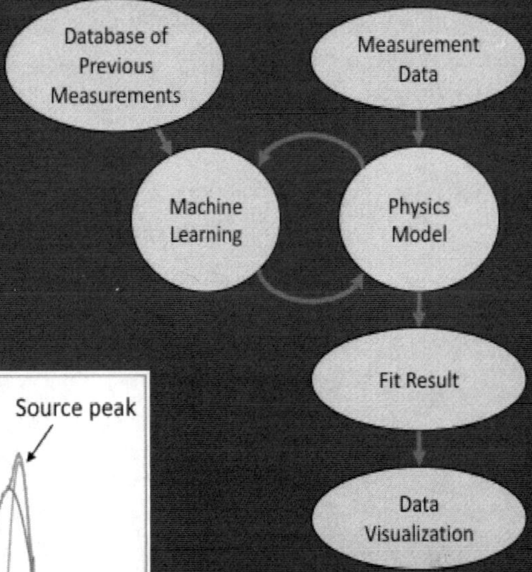

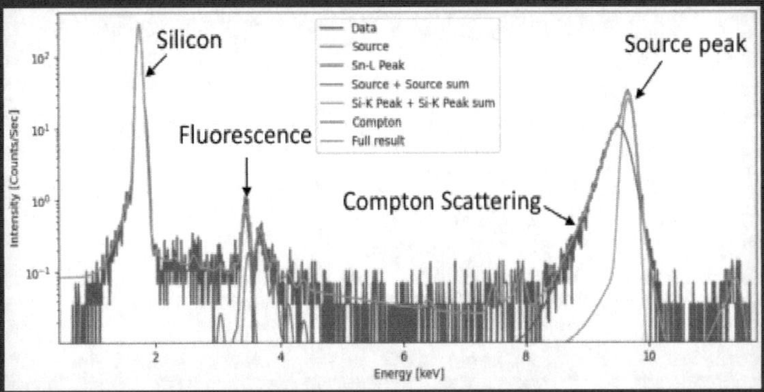

Reporting

VPD:	False
Beam:	82
Angle:	0.1
Time:	30
Elements:	Al, P, S, Cl, Ar, K, Ca, Sc, Ti, V, Cr, Mn, Fe, Co, Ni, Cu, Zn, Ga, Y, Zr, Nb, Mo, Ru, Pd, Ag, In, Sn, Sb, I, Ba, La, Hf, Ta, W, Pt, Pb, Bi

Wafer Maps:		Si Map	Al Map	Cl Map	Sc Map	V Map	Fe Map	Ni Map	I Map	La Map	
Units:		cps	e10 at/cm^2	e10 at/cm^2	e10 at/cm^2	e10 at/cm^2	e10 at/cm^2	e10 at/cm^2	e10 at/cm^2	e10 at/cm^2	
Detection Limit		NA	865.2	47.8	6.7	3.0	1.4	1.1	9.9	3.2	
X	Y	Si	Al	Cl	Sc	V	Fe	Ni	I	La	Plot
Wafer Average		2,128.0	ND	ND	Trace	Trace	ND	ND	224.3	ND	I_105U - Average
1.0	1.0	1,939.8	ND	ND	ND	ND	ND	ND	9.8	ND	I_105U - (1, 1)
21.0	21.0	2,158.6	ND	ND	ND	ND	ND	ND	19.8	ND	I_105U - (21, 21)
21.0	-21.0	2,253.1	ND	ND	ND	ND	ND	ND	19.1	ND	I_105U - (21, -21)
-21.0	-21.0	2,555.6	ND	ND	ND	ND	ND	Trace	23.5	ND	I_105U - (-21, -21)
-21.0	21.0	2,218.9	ND	ND	ND	ND	ND	ND	19.7	ND	I_105U - (-21, 21)
10.0	59.0	1,569.5	ND	ND	ND	ND	ND	ND	52.2	ND	I_105U - (10, 59)
46.0	39.0	1,559.7	ND	ND	ND	ND	ND	ND	50.4	ND	I_105U - (46, 39)
60.0	1.0	1,568.8	ND	ND	ND	ND	ND	ND	46.5	ND	I_105U - (60, 1)
46.0	-39.0	2,011.6	ND	ND	ND	ND	ND	ND	51.5	ND	I_105U - (46, -39)
10.0	-59.0	2,050.6	ND	ND	ND	ND	ND	ND	53.1	ND	I_105U - (10, -59)
-30.0	-52.0	1,917.0	ND	ND	ND	ND	ND	ND	54.3	ND	I_105U - (-30, -52)
-56.0	-21.0	1,844.6	ND	ND	ND	ND	ND	ND	49.1	ND	I_105U - (-56, -21)
-56.0	21.0	1,502.8	ND	ND	ND	ND	ND	ND	45.1	ND	I_105U - (-56, 21)

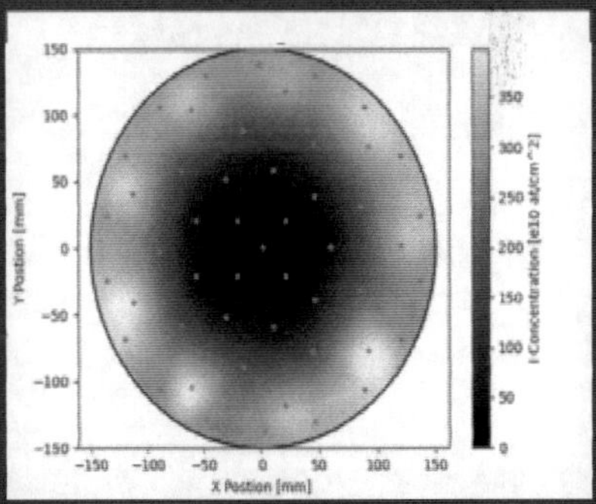

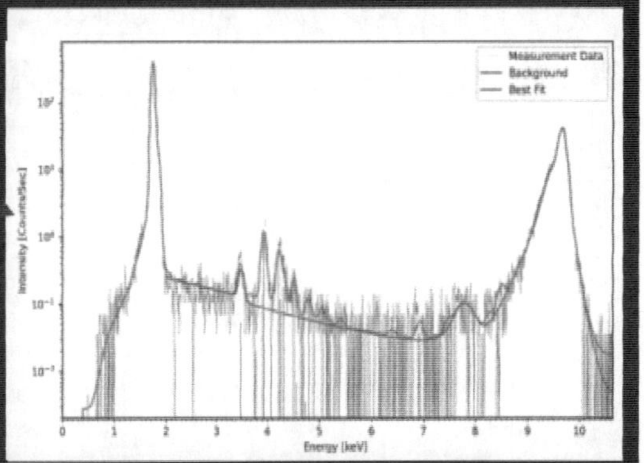

- We report hits (red background) when signal > 2x calculated noise.
- If the signal is greater than the noise, but less than 2x noise, we report a "Trace" amount (yellow background) to flag for possible manual review.

Excitation Source E = 9.67 keV

Outline

- Photomask Contamination
- Conventional TXRF Overview
- TXRF Photomask Capability Demonstration and Hardware
- TXRF Contamination Monitoring on Photomask Blanks
- Beyond Photomask Blanks
- Conclusion

Photomask TXRF

- Rigaku Proposal
  - Evaluate the capability on the existing wafer platform and if successful proceed with the photomask tool development
- Solution
  - Mount thin photomask samples on Si

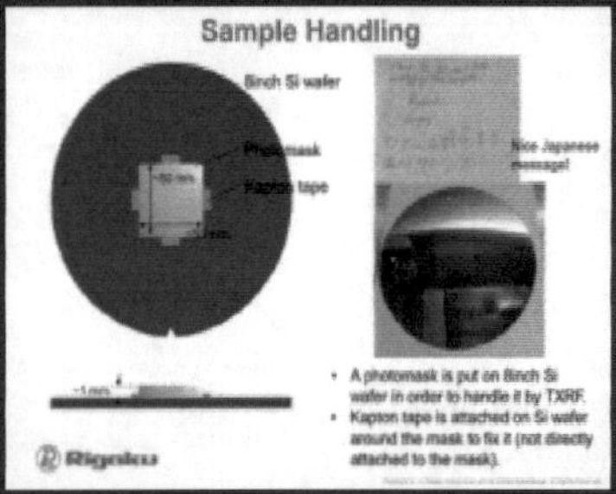

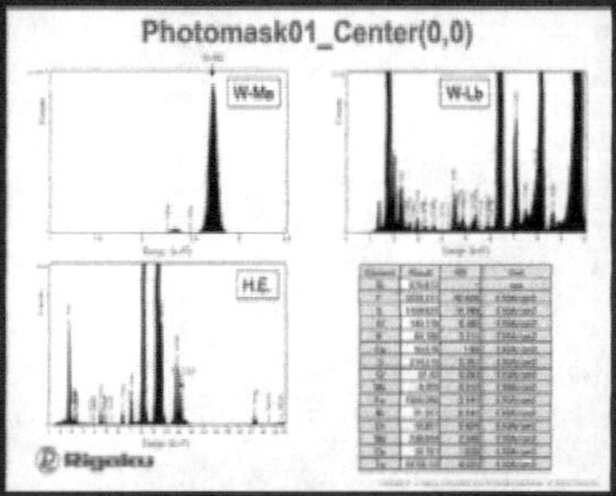

- 310Fab Photomask TXRF
  - Based on 300 mm wafer platform

https://www.rigaku.com/products/semi/txrf310#specs

Outline

- Photomask Contamination
- Conventional TXRF Overview
- TXRF Photomask Capability Demonstration and Hardware
- TXRF Contamination Monitoring on Photomask Blanks
- Beyond Photomask Blanks
- Conclusion

TXRF on Quartz Blanks

- Quartz Blank as Received and Post Chemical Processing

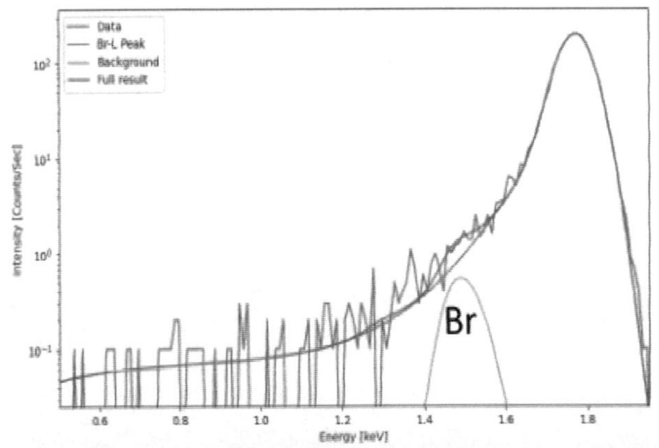

As Received

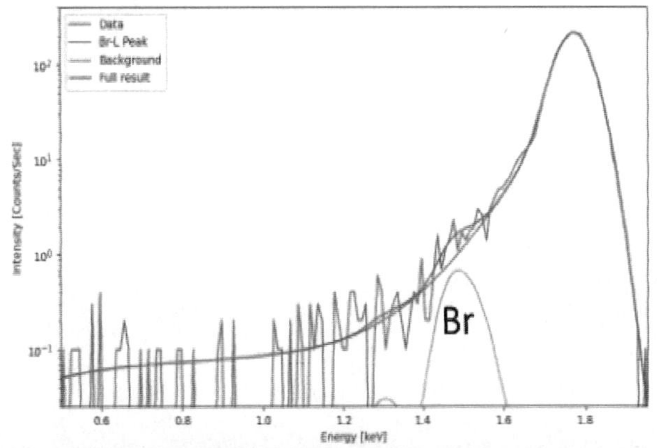

Post Processing

Excitation Source E = 1.77 keV

TXRF on Quartz Blanks

- Quartz Blank as Received and Post Chemical Processing

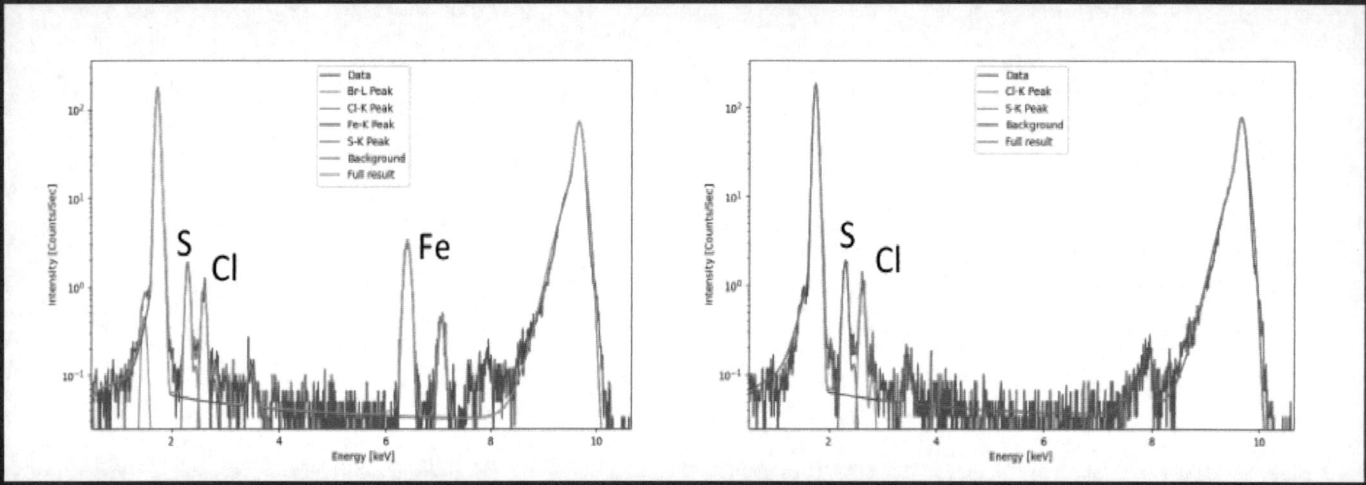

As Received

Post Processing

Excitation Source E = 9.67 keV

TXRF on Quartz Blanks

- Quartz Blank as Received and Post Chemical Processing

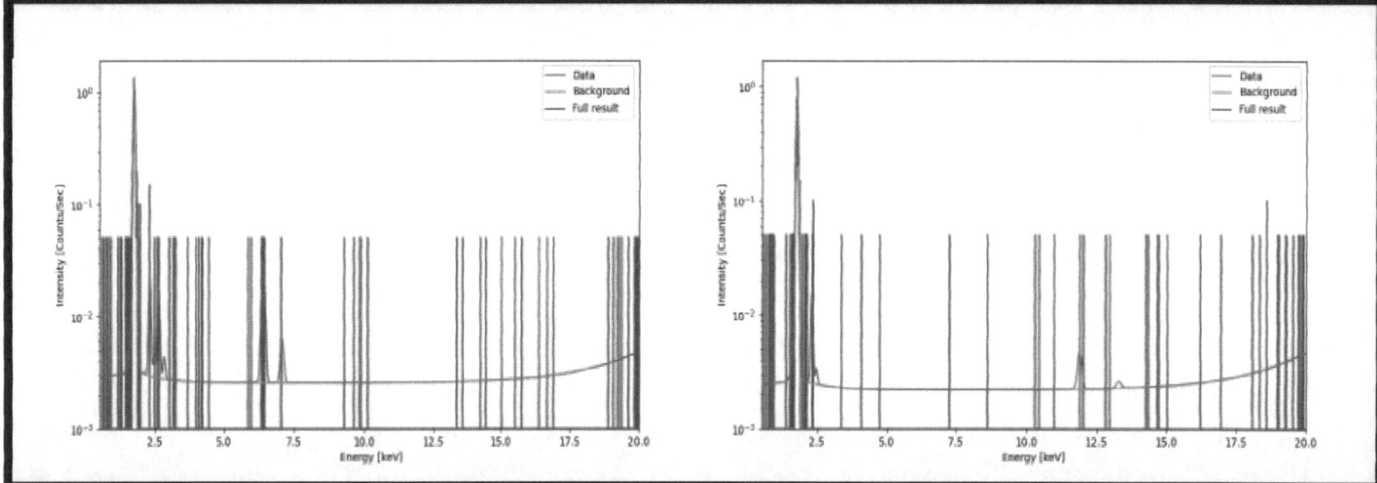

As Received

Post Processing

Excitation Source E > 24.0 keV

TXRF on Quartz Blanks

- Post Chemical Processing with increased acquisition time

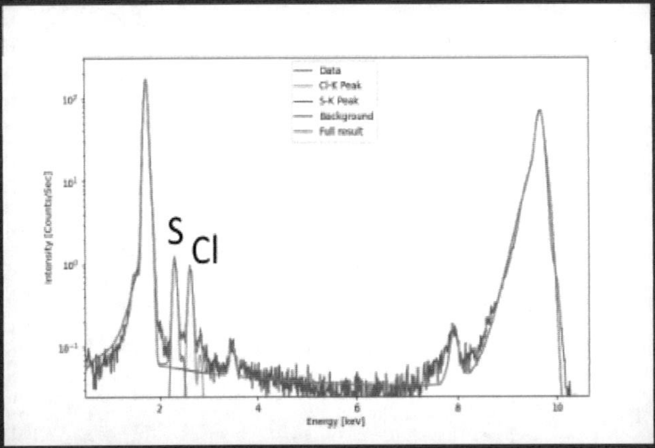

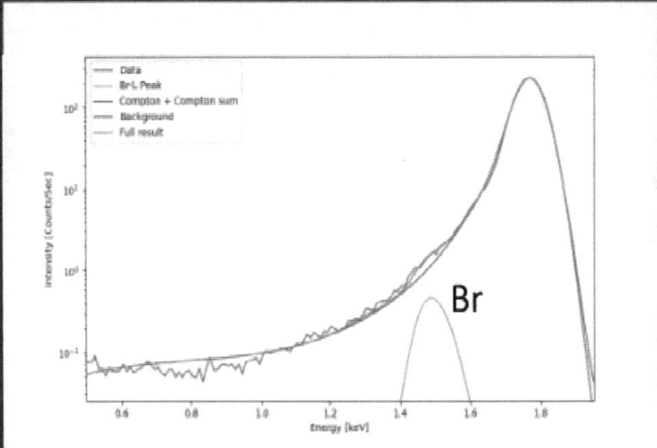

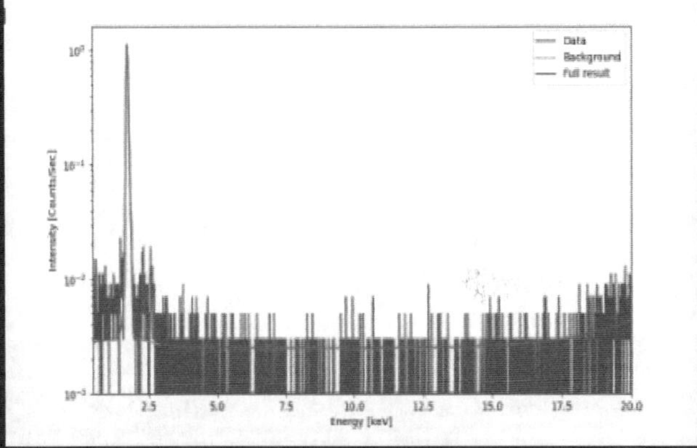

Distributed Contamination on Quartz

- Post Chemical Processing

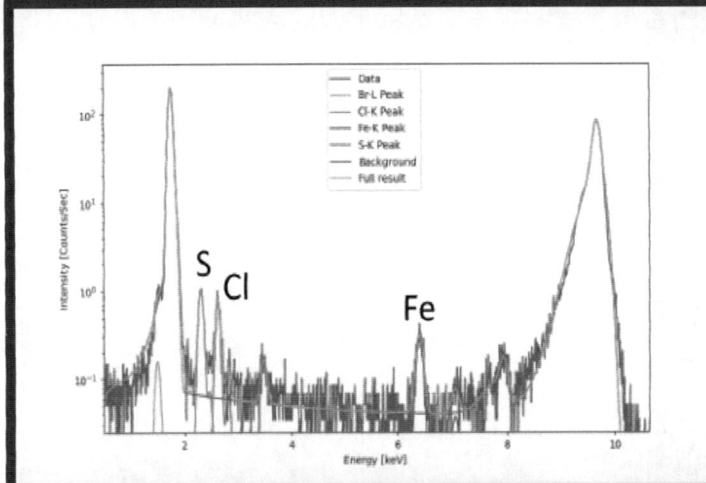

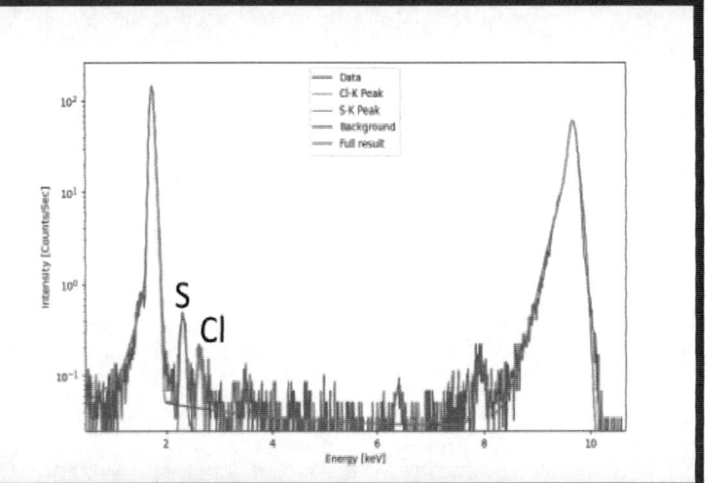

Excitation Source E = 9.67 keV

Distributed Contamination on Quartz

- Post Chemical Processing

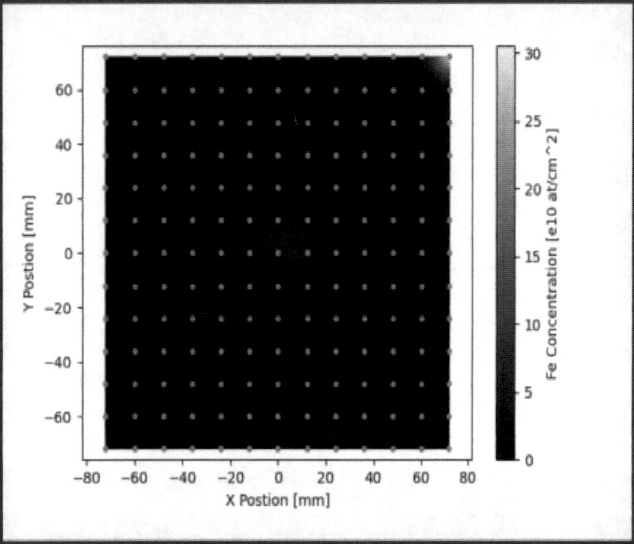

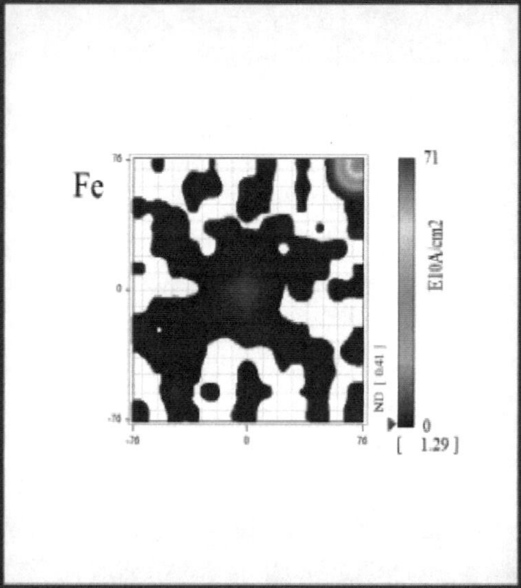

Excitation Source E = 9.67 keV

TXRF on Chrome Blanks

- Chrome Blank as Received

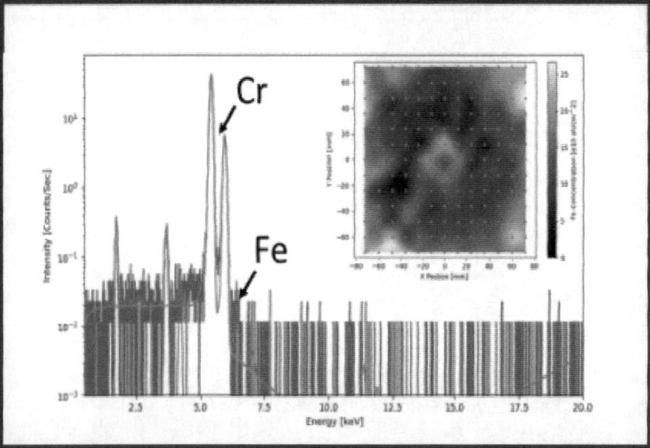

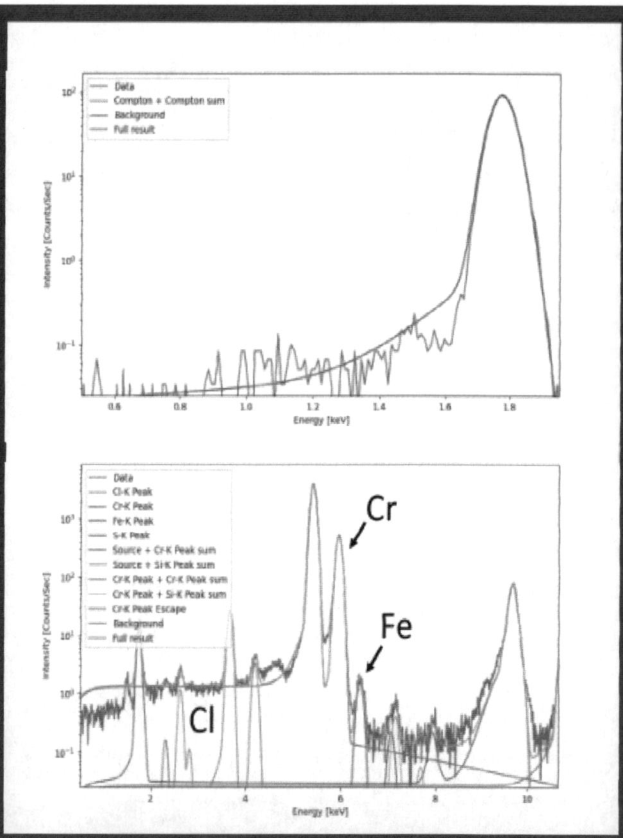

Outline

- Photomask Contamination
- Conventional TXRF Overview
- TXRF Photomask Capability Demonstration and Hardware
- TXRF Contamination Monitoring on Photomask Blanks
- Beyond Photomask Blanks
- Conclusion

Beyond Photomask Blanks

- EUV Blank Example

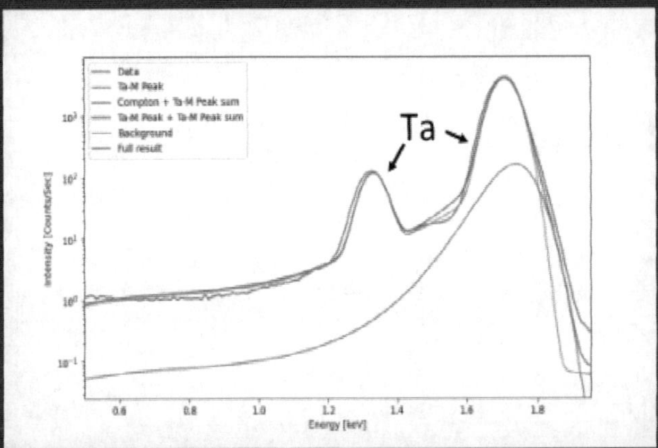

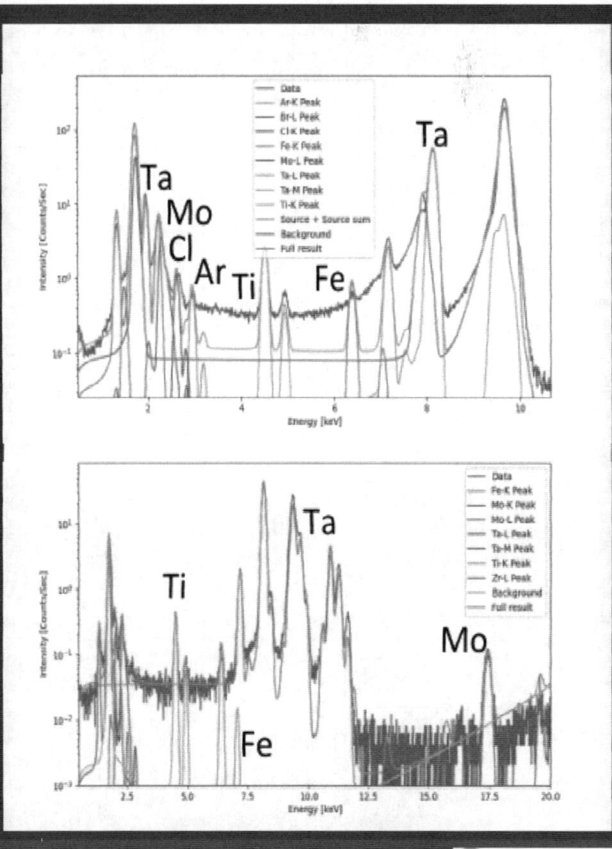

Outline

- Photomask Contamination
- Conventional TXRF Overview
- TXRF Photomask Capability Demonstration and Hardware
- TXRF Contamination Monitoring on Photomask Blanks
- Beyond Photomask Blanks
- Conclusion

• Conclusion

- Developed industry's first photomask TXRF.
- Successfully demonstrated the contamination monitoring capability on blank photomask substrates.
- Defined the path for platform development.

Acknowledgements

- Intel: Mitchell First, Malahat Tavassoli, Travis Hebden
- Rigaku: Chris Bayne

intel®

THE SURFACE PREPARATION AND CLEANING CONFERENCE (SPCC)

SPCC 2024 | April 24-25, 2024

Evolution and Characterization of PVA brush in post CMP cleaning for 10 nm below metal CMP process

APRIL 25, 2024

Prof. Jin-Goo Park
Dept of Materials and Chemical Engineering
Hanyang University ERICA, South Korea
jgpark@hanyang.ac.kr

The Engine of Korea

Contents

1. Introduction

- CMP Technology Evolution
- Post CMP Cleanings
- PVA Brush; its understanding

2. Brush study at HYU

- Quantification of brush impurities
- PVA Brush Particle Loading and Conditioning
- Contact area analysis

3. Summary

2

1. Introduction

- CMP Technology Evolution
- Post CMP Cleanings
- PVA Brush; its understanding

Short History Review on CMP Technology

1983 Proprietary development initiated at IBM
(Pb-silicate glass for STI)
1988 *Oxide CMP* **in production at IBM**
1989 *Tungsten CMP* **in production at IBM**
Aluminum-copper CMP **in production at IBM**
1992 *Copper CMP* **process announces by IBM**

- **A consequence of proprietary CMP development**
- **All production firsts were implemented using existing polishing consumables;** ***Cabot, Rodel, IPEC***

Planarization and Damascene Technology

CMOS Transistor Scaling; Materials and Structure

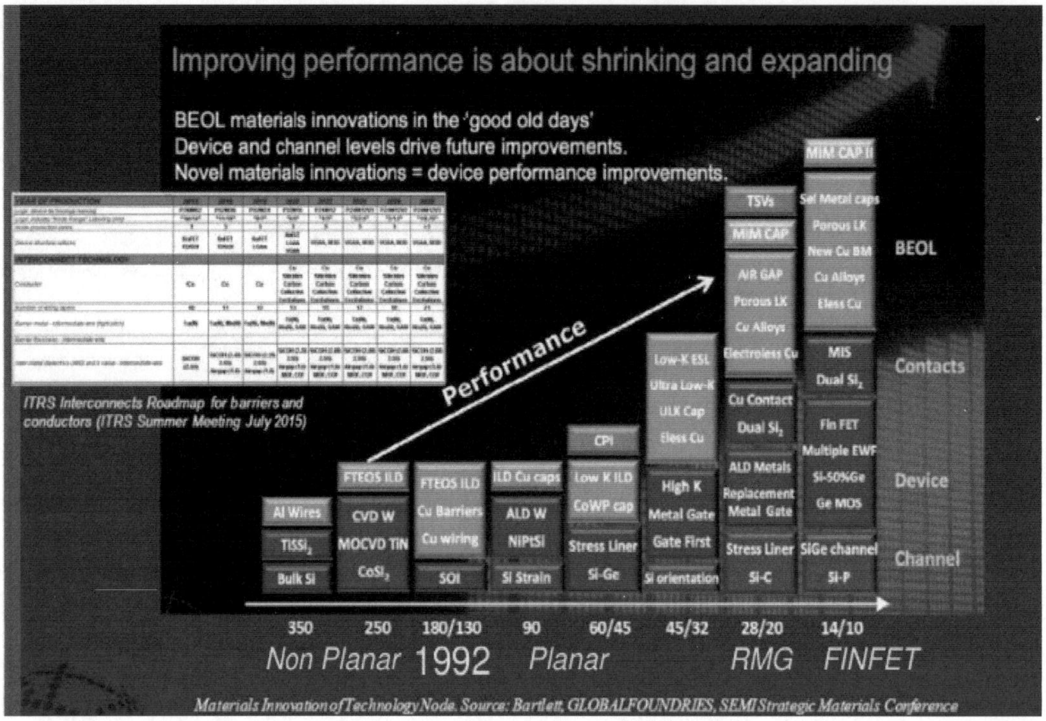

CMP Steps Shift to FEOL/MOL

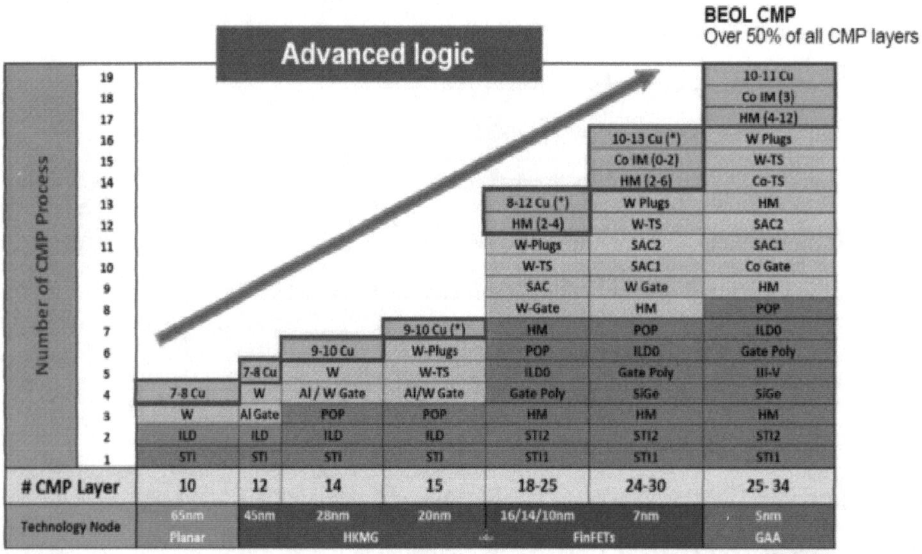

- Number of CMP steps doubled from 28nm to 10nm node in order to enable new integration schemes such as replacement metal gate and self-aligned contact.
  - ~ 75% of the increase is in MOL due to the complexity of contact module.
- New CMP processes such as Co, Cap SiN and SiOC CMP in MOL

Effect of FINFET/RMG Structure on CMP

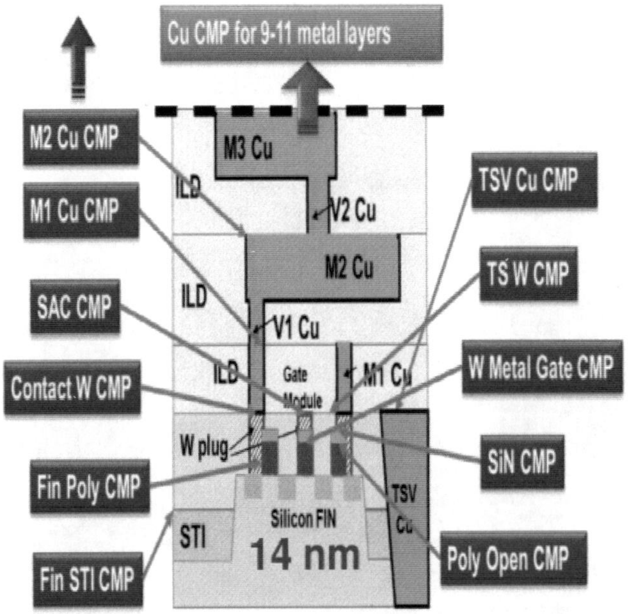

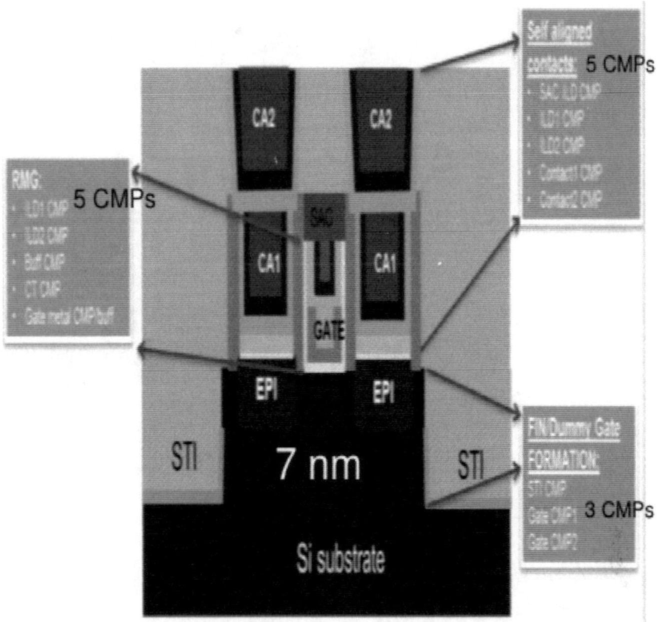

Effect of FINFET/RMG Structure on CMP

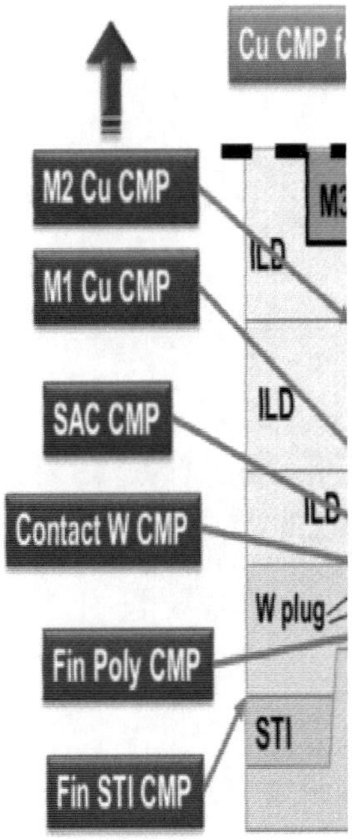

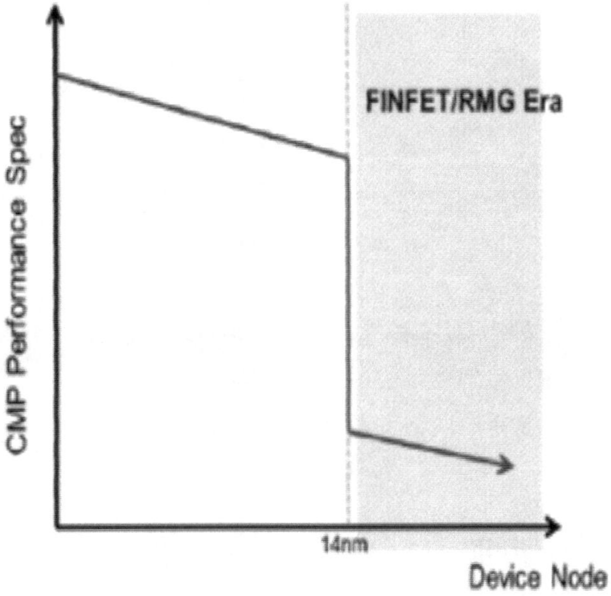

Multi-Materials Exposure at RMG

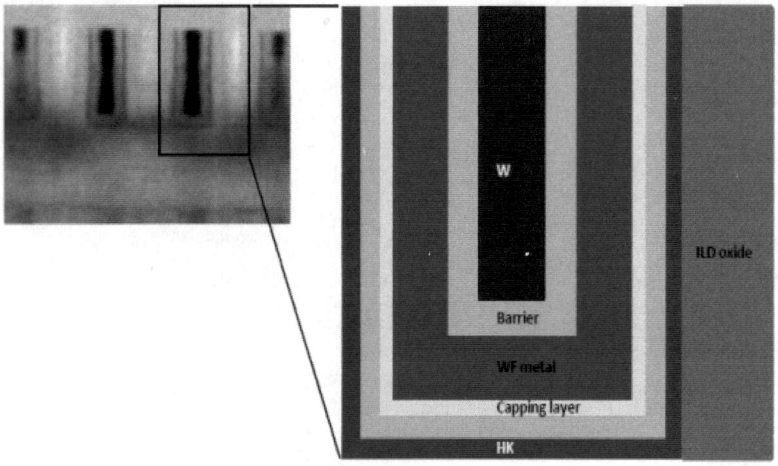

Bulk Slurry Requirements:

1. Stop on ILD Oxide (or Stop on BM)
2. Enough OP Margin of W/WF Recess Perspective
3. Minimize W Corrosion
4. No Scratch

Buffing Slurry Requirements:

1. Non-Selective for Multi-Materials
2. Minimize W Corrosion
3. No Scratch
4. Superior Planarity

Clean Chemistry Requirements:

1. No damage on various metals
2. No W corrosion
3. Minimal material loss (etching)

Post Metal Gate CMP: 1)Gate Metal, 2)Barrier Metal, 3)WF Metal, 4)Capping Metal, 5)High-k Dielectric, 6)ILD Oxide → Various materials exposed to polishing slurry/clean chemical

Paradigm Shift in CMP

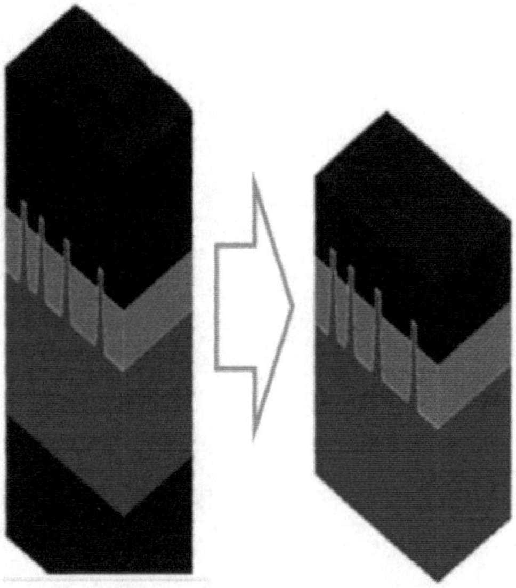

Planarization & Polishing

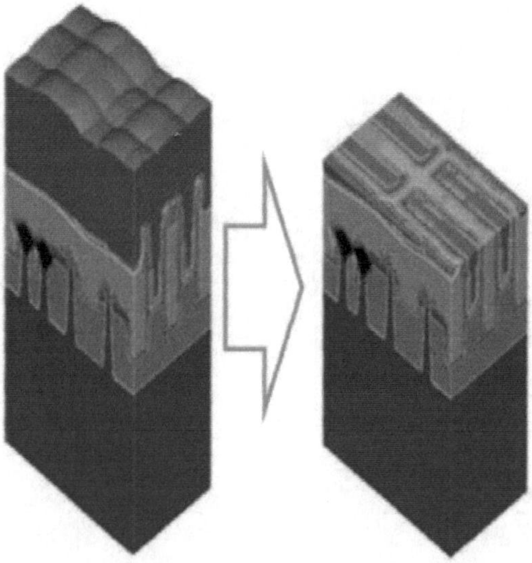

Material Removal & Separation

<u>Defects at FEOL and MOL surfaces – Yield killer</u>

Impacts of CMP Cleaning on Yield Detractors

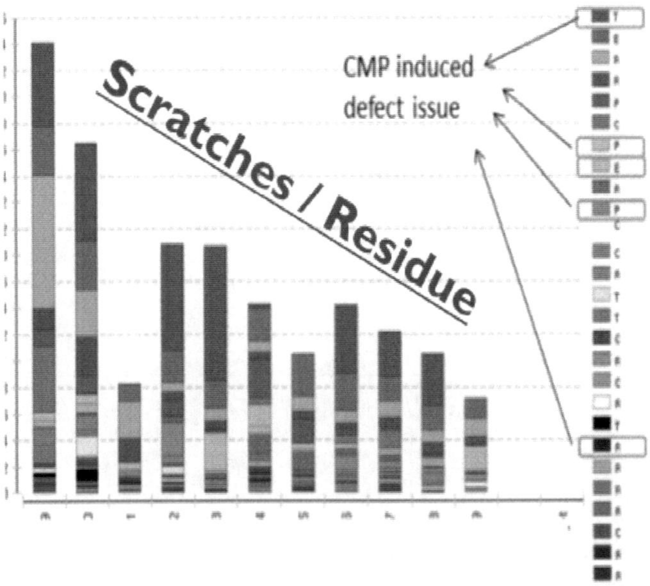

Defects	Failure Mode
Metallic Particle	Leakage Failure
Non-metallic Particle	Open Contact (Block Etch)
Organic Residue	Leakage Path
Microscratch	Pattern Damage
Metal Corrosion	Resistance

Post CMP cleaning becomes one of the key yield impact detractors

How have we cleaned polished wafers last 30 years?

(*FOUP = Front Opening Unified Pod)

PVA Brushes!!

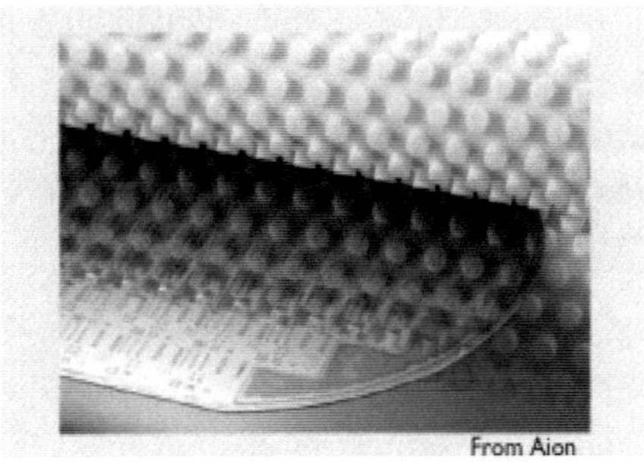

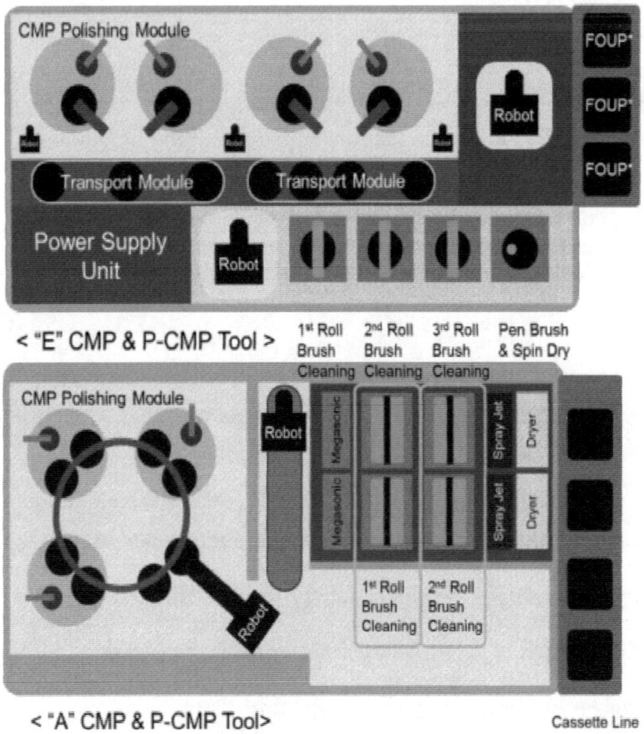

Brush; Defect Adders in sub 10 nm Devices

Defect Signature by Brush Cleaning

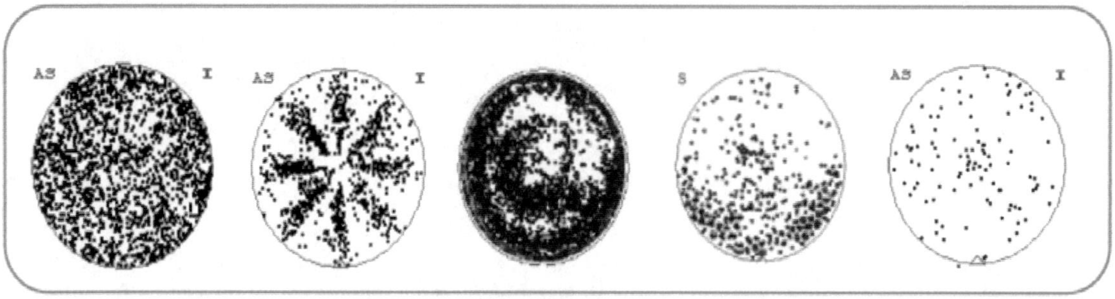

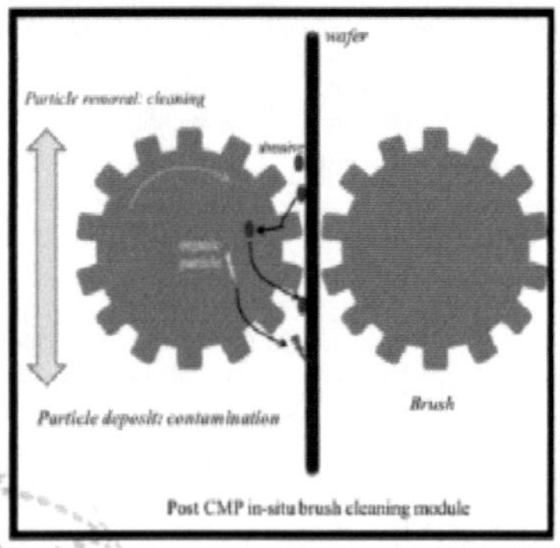

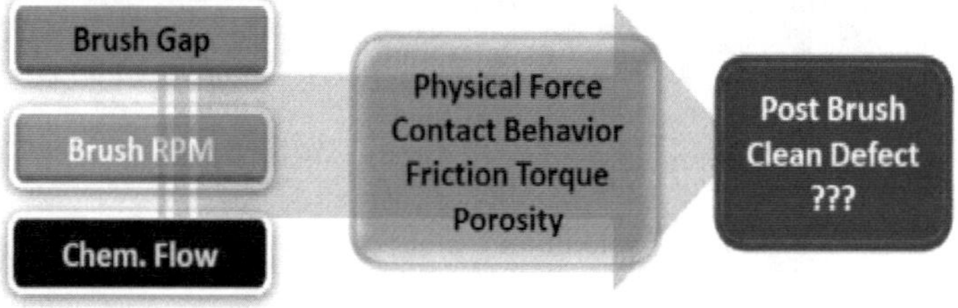

Brush Cleaning is NOT ONLY clean the wafer BUT ALSO contaminate wafer → **PRE vs C.C.**

Current status of brush in HVM

❖ Expected Defect Level

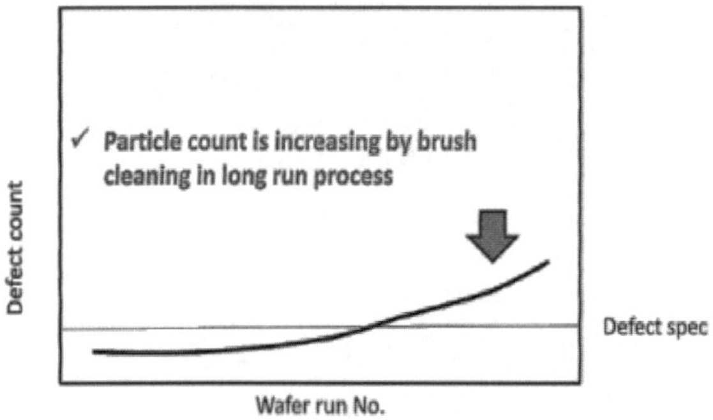

❖ End-user's Comment

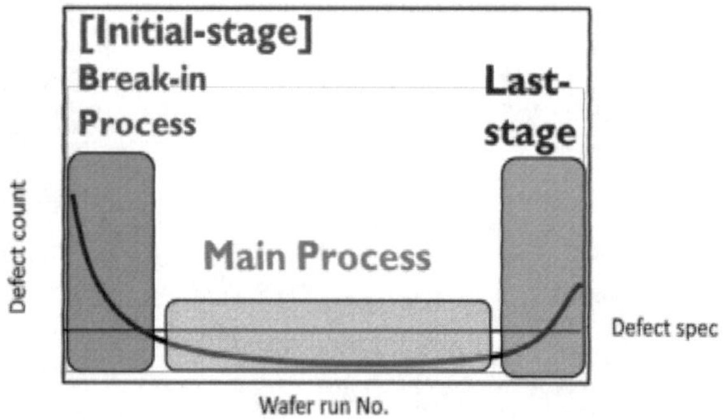

- **Defect Level: - Initial-stage > Last-stage >> Main Process**
- **Break-in process is required for new brush due to higher defect level at initial-stage.**

- **Optimization is required for break-in process**

- Break-in Method / Detection method / Time

Problems in New Brush (1)

❖ **Effect of Brush Soaking**

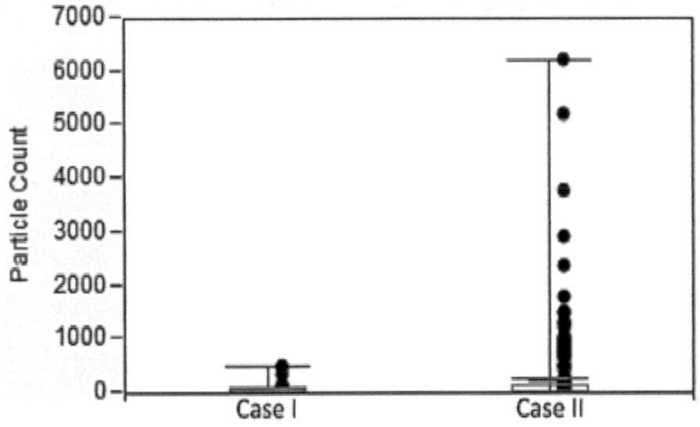

Case 1: New Brush implement -> **DI flow through the brush core**-> Break-in -> Dummy run

Case 2: New Brush implement -> Break-in -> Dummy run

Problems in new brush : 1) insufficient soaking of brush, 2) variation in nodule's height, 3) surface property of the nodules, 4) **contamination**, 5) brush batch variation

Problems in New Brush (2)

❖ **Brush Break-in Effect**

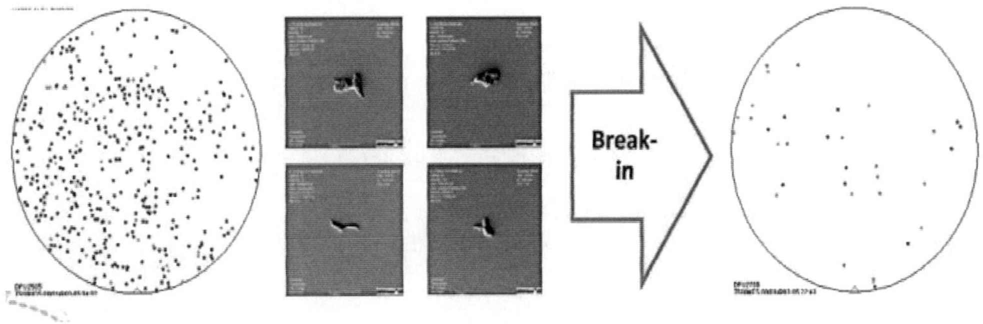

➢ **Optimized break-in process is essential before using a new brush.**

Brush Life Time vs. Organic Defects

- No public Information and public research
- Even not much understanding even in endusers until recently
- Too small market size; very poor quality control
- Unpredictable

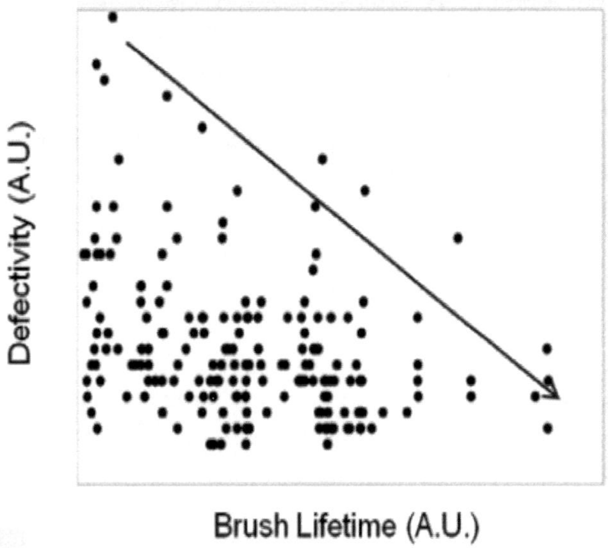

2. Brush study at HYU

- Quantification of brush impurities
- PVA Brush Particle Loading and Conditioning
- Contact area analysis

Typical Formulation Process for Making PVA Brush

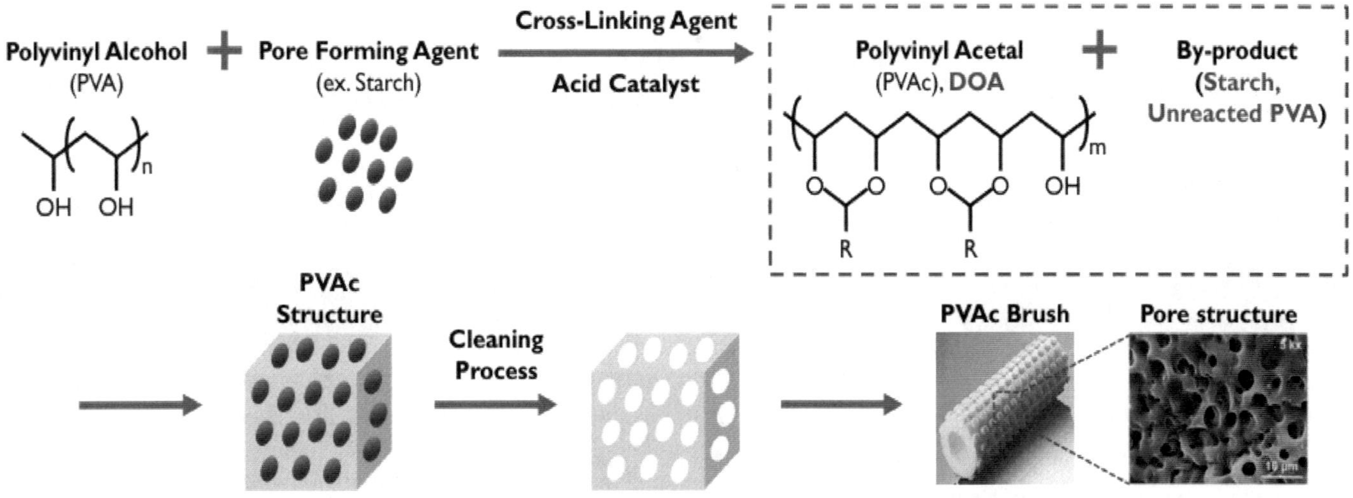

- The PVAc is formulated by the acetalization reaction between PVA and cross-linking agent
- After the cleaning process, the by-product and pore forming agent are removed from the PVAc brush
- **The development of characterization methods was necessary to identify the by-products**

PVA-Brush Research at HYU

PVA-brush synthesis process

- **Optimization of manufacturing process parameter to develop residue free brush**
  - **Quantification of brush impurities from synthesized brush**
  - **Removal of starches and unreacted PVA to produce defect free brush**
  - **Effect of brush skin layer during PVA brush scrubbing process**
  - **Root causes of defects from PVA**

- **Optimization of scrubbing parameters to control defects on wafer**
  - **Development of brush cleaning solution**
  - **Understanding slurry loading on brush**
  - **Development of physical and chemical brush conditioning system**
  - **Metal ion removal and identification of PVA induced scratches**
  - **Brush induced metal corrosion**

PVA-brush scrubbing process

Current and Future Defect Free Brush based PCMP Cleaning

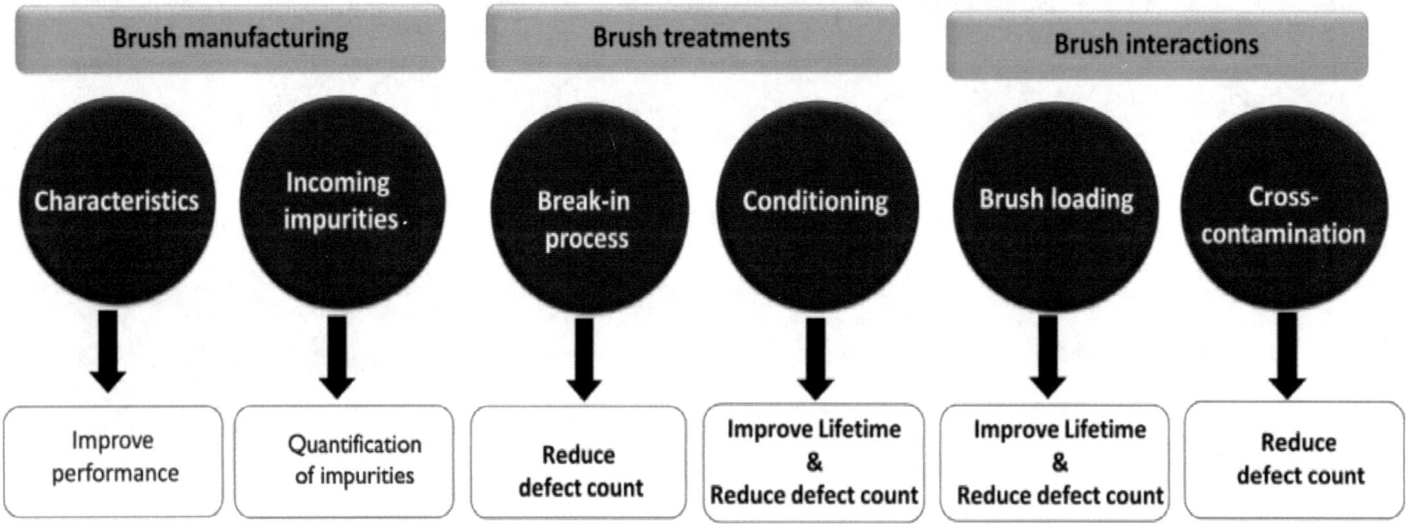

- **Defect free PVA brush**
- **Brush in-situ conditioning**
- **Brush induced defects reduction: scratches and corrosion**

Quantitative Analysis of Starch Content using UV-visible Spectra

- **UV-visible spectra of indicator solution by pure starch concentration**
- **Absorbance values of starch-indicator peak as a function of starch concentration**

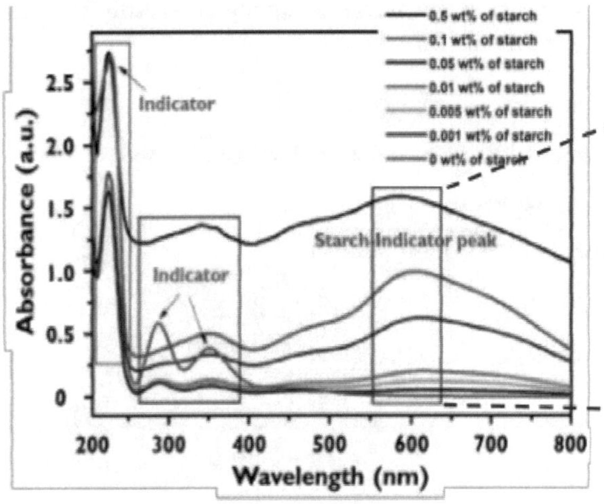

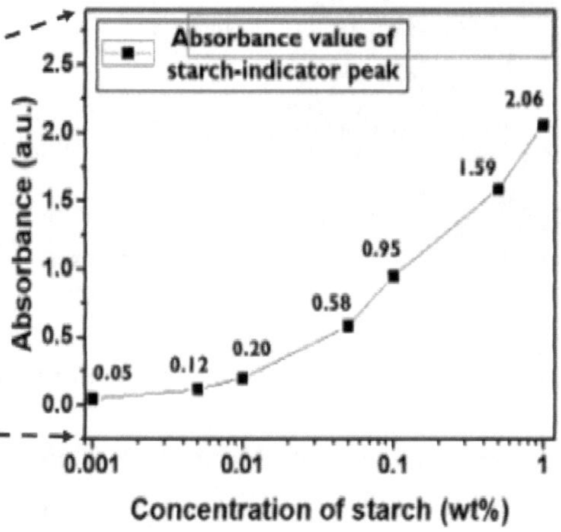

- To identify the starch amount quantitatively, UV-visible analysis was performed using indicator solution
- The intensity of absorbance peak increased with increasing the starch concentration
- The absorbance value at the starch-indicator peak increased as a function of starch concentration

^{1}H NMR Analysis of unreacted PVA

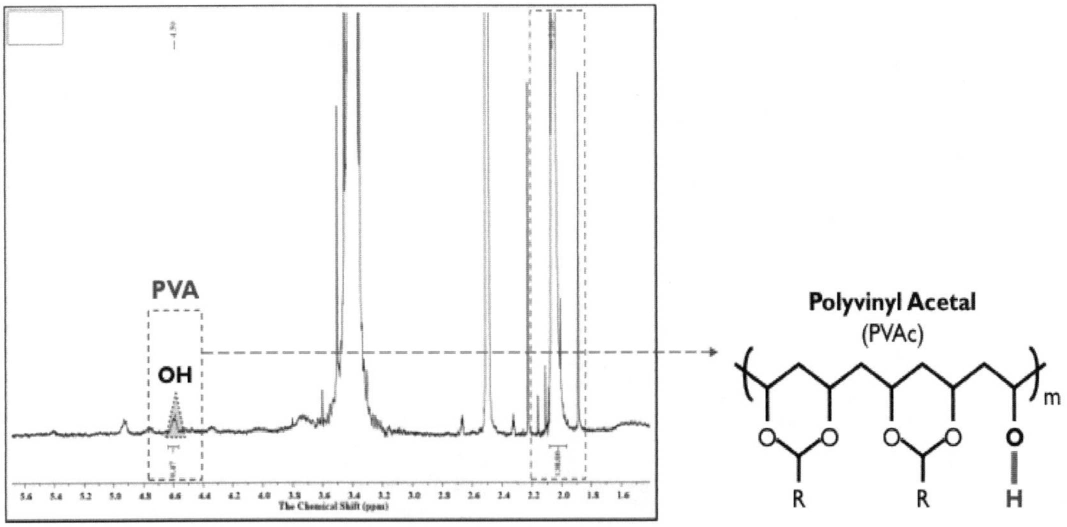

- The proton peak of OH group appeared at around 4.5 ppm in the measured ^{1}H NMR spectra
- The proton peak is the unreacted PVA in the formulated PVAc brush structure

23

^{13}C NMR DOA Analysis of PVA Brush

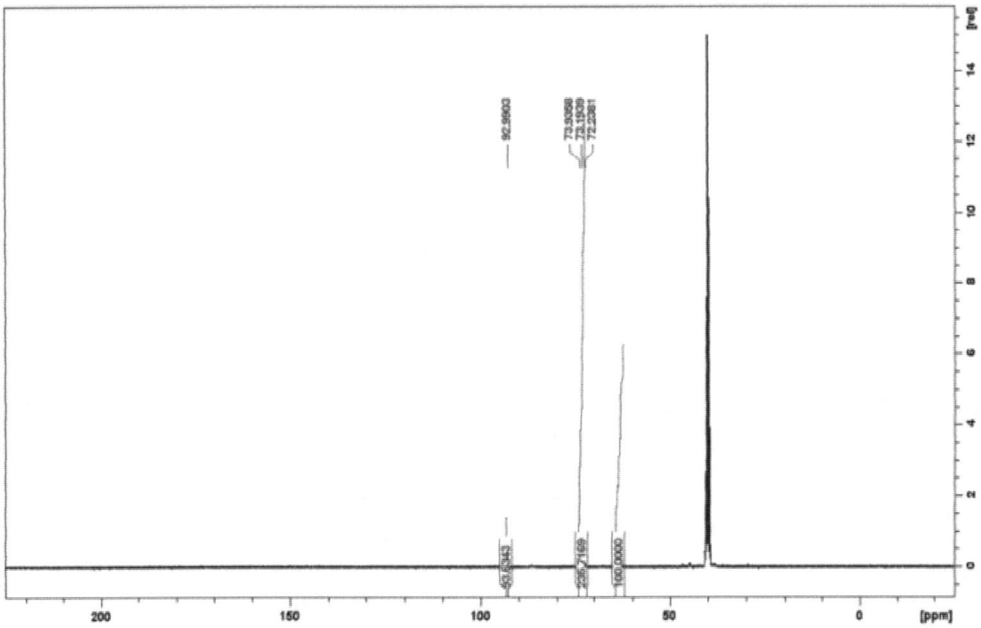

- The degree of acetalization for PVA was analyzed using ^{13}C NMR spectra

24

Scrubbing test for wafer defects from brushes

As-received wafer

Surface scan before scrubbing

Optical Surface Analyzer
(Candela CS10, KLA-TENCOR, USA)

Scrubbing test with EBARA scrubber

300 mm Brush scrubber
(Special-made, EBARA, Japan)

Surface scan after scrubbing

Analysis of particle number

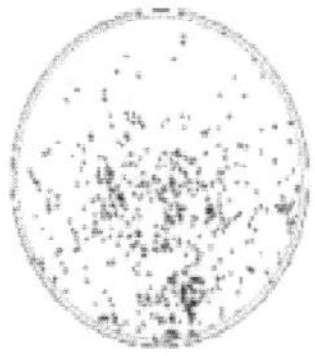

As-received Brush

- AMAT type (O.D. 70 mm) with EBARA core
- Roll inner rinse with *DIW for 30 min

*5 nm particle filter is used for DIW filtration

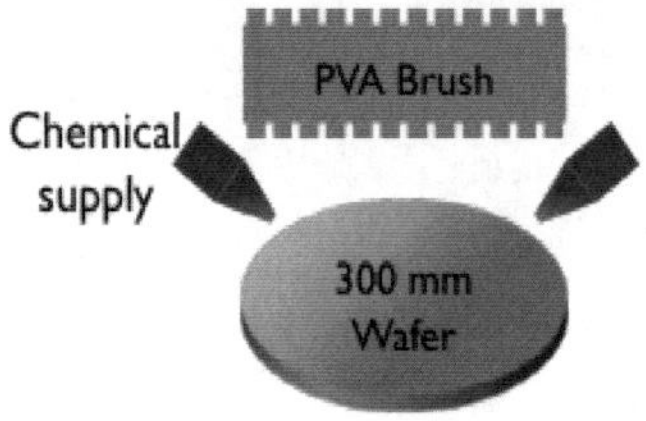

Brush	Commercial brushes
Chemical	DIW
Wafer	300 mm Bare-Si

Brush rotation : 100 rpm
Wafer rotation : 100 rpm
Gap distance : -1.5 mm
Flow rate : 500 ml/min

Observation of Wafer Contamination during Break-in Process

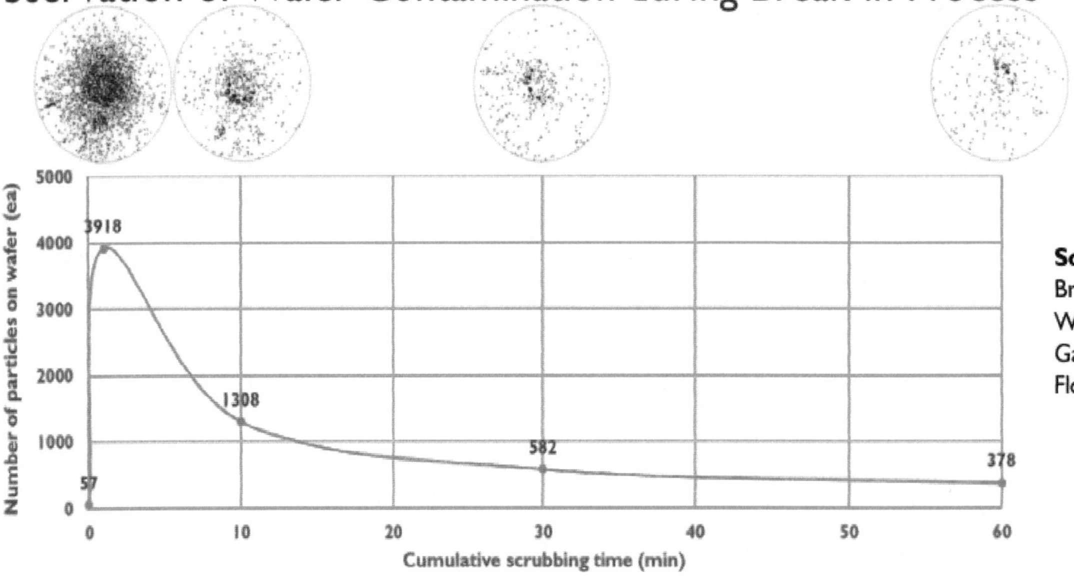

Scrubbing Condition
Brush rotation : 100 rpm
Wafer rotation : 100 rpm
Gap distance : -1.5 mm
Flow rate : 500 mL/min

- In the initial break-in process, many particles were contaminated on the wafer surface
- As the scrubbing time increased, the number of particles on the wafer decreased

26

Evaluation of Initial Brush Quality and Integrity

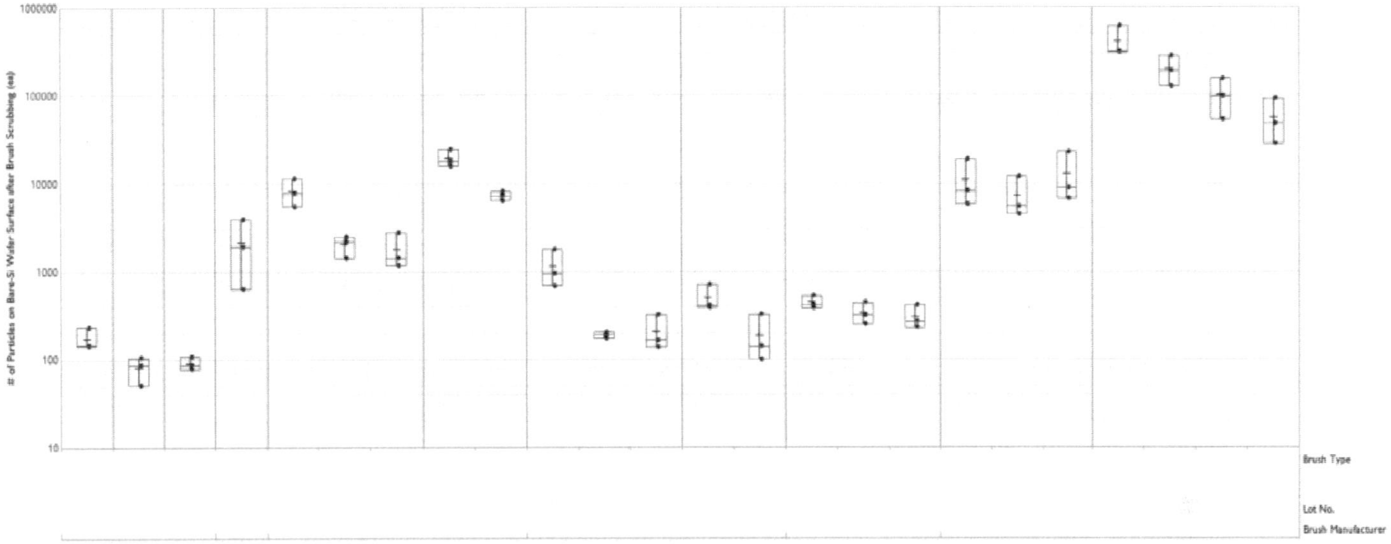

Observation of Impurities on Wafer Surface after Drop Deposition

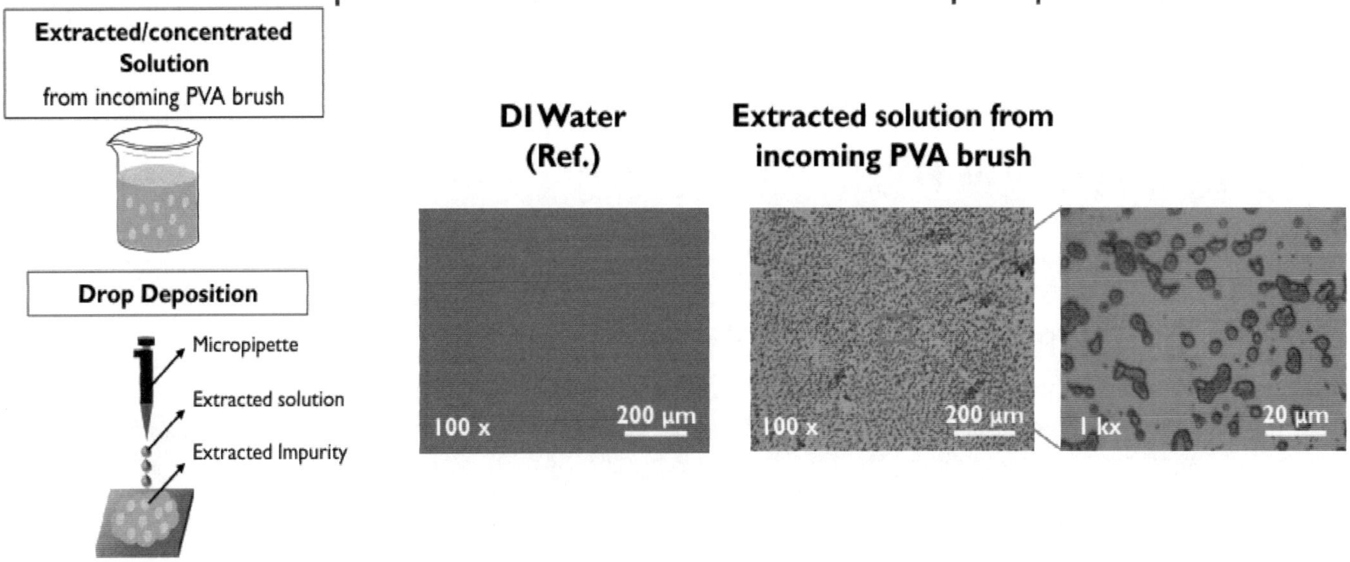

- To investigate the impurity shape in the extracted solution, drop deposition method was performed
- Chemical analysis will be followed to find root causes of defects

Quantification of Brushes

- Starches
- Unreacted PVA concentration
- Degree of Acetalization
- Particles on wafer during scrubbing

Defect free PVA polymerization
Defect free brush cleaning process
Longer shelf life time
Brush quality control

2. Brush study at HYU

- Quantification of brush impurities

- PVA Brush Particle Loading and Conditioning

- Contact area analysis

Loading and cross contamination

- **Schematic of wafer contamination after brush scrubbing**

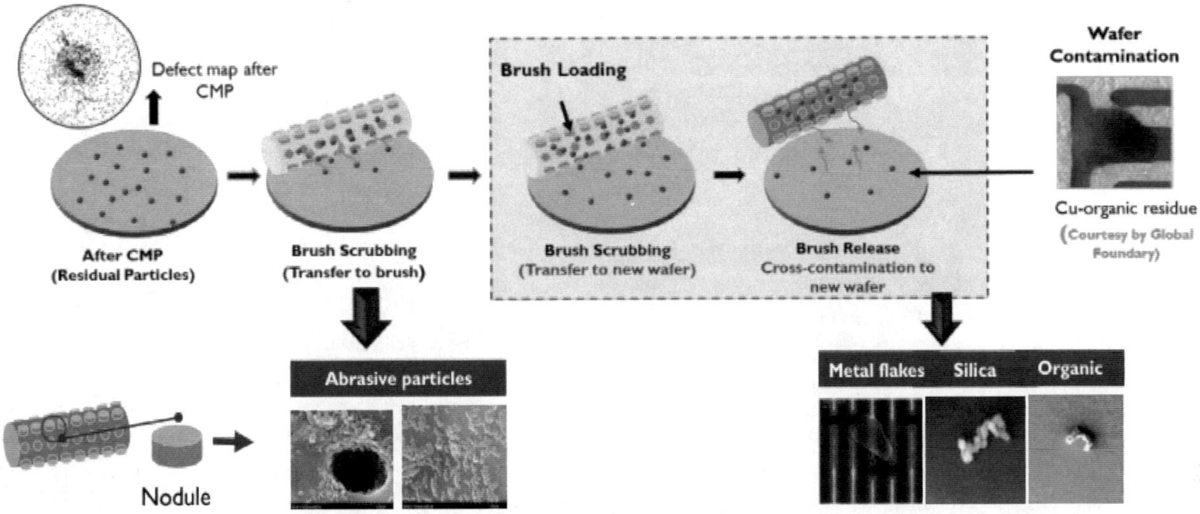

- **Brush loading affects the brush lifetime**
- **Cross-contamination causes scratch and wafer contamination**
- **Small particles (< 10 nm) are difficult to detect, and their removal is more difficult**

PVA brush cleaning | Effect of slurry abrasives on brush loading

Brush loading - Silica slurry

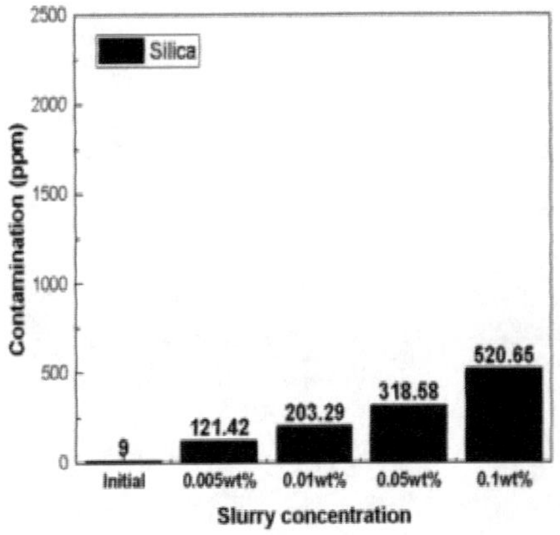

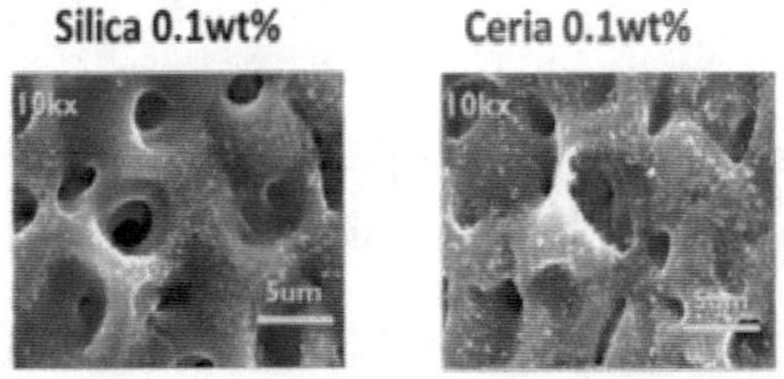

Brush loading: Ceria >> Silica

Brush loading-Ceria slurry

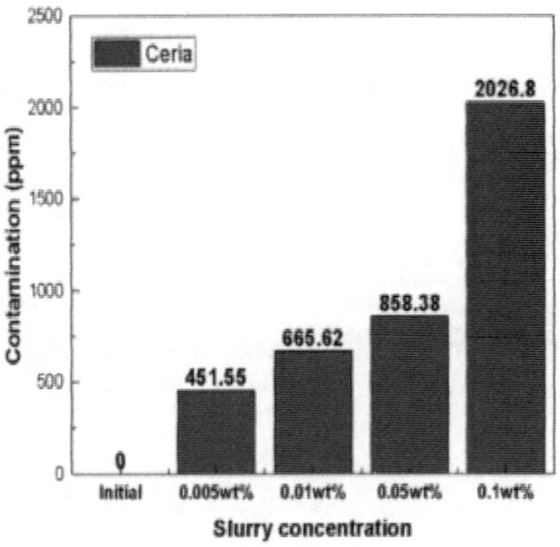

- **Brush loading is severe with ceria compared to silica**
- **Strong electrostatic interactions and Ce^{3+} concentration play dominant role in high ceria contamination**

Brush loading with pH and its effect on cross contamination (Silica)

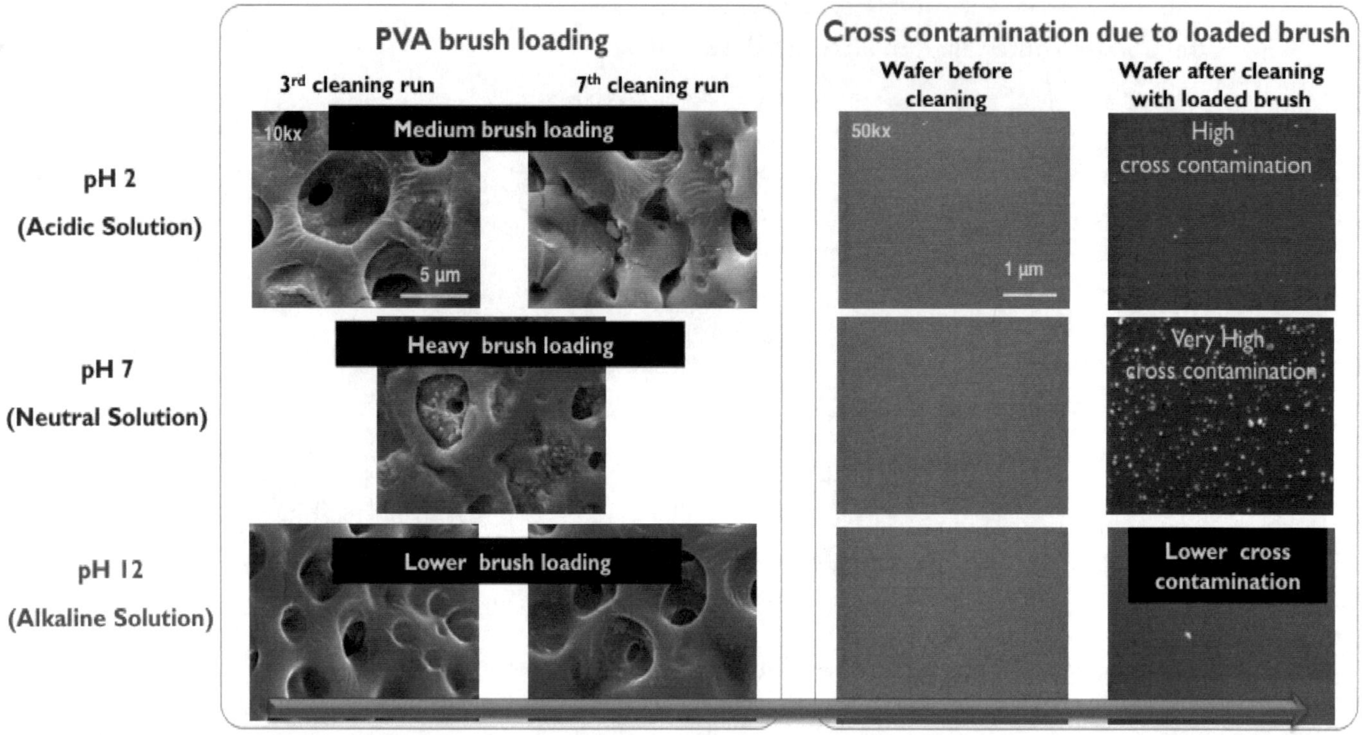

Effect of metal ions on the brush loading with particle at different pH

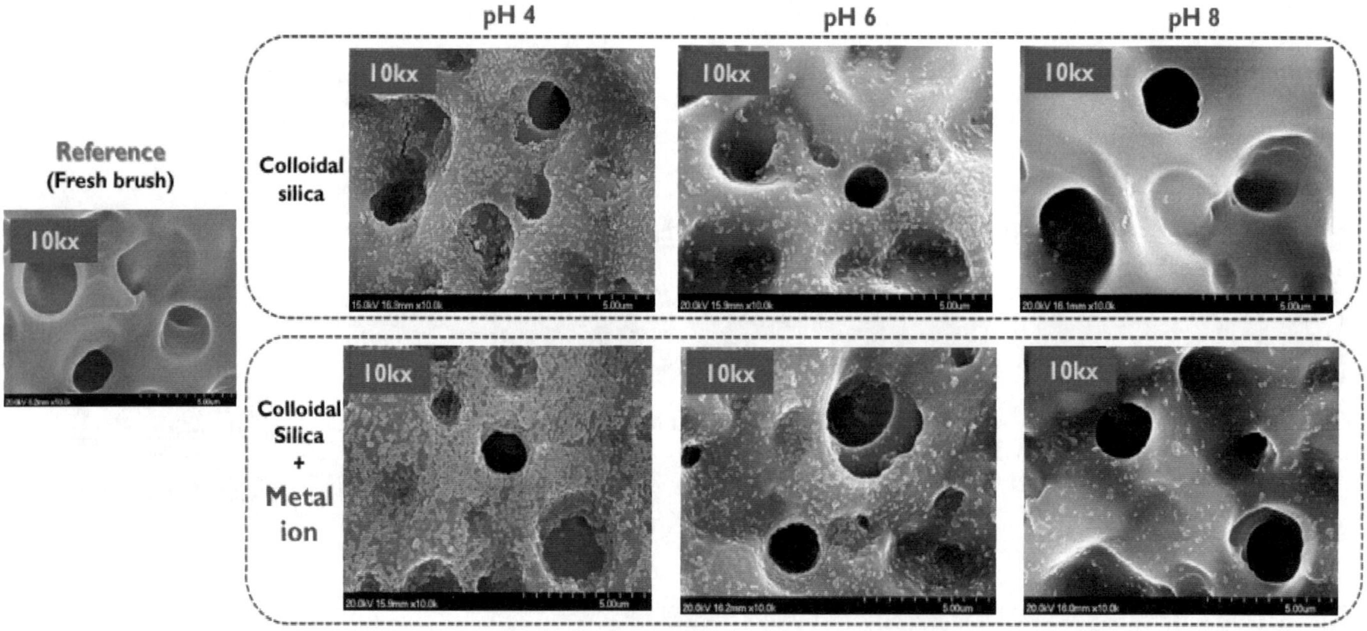

- **Very high particle loading at pH 4 and 6: with and without metal ions**
- **High particle loading at pH 8 with metal ions**

Effect of chemicals on the brush loading with metal ions at different pH

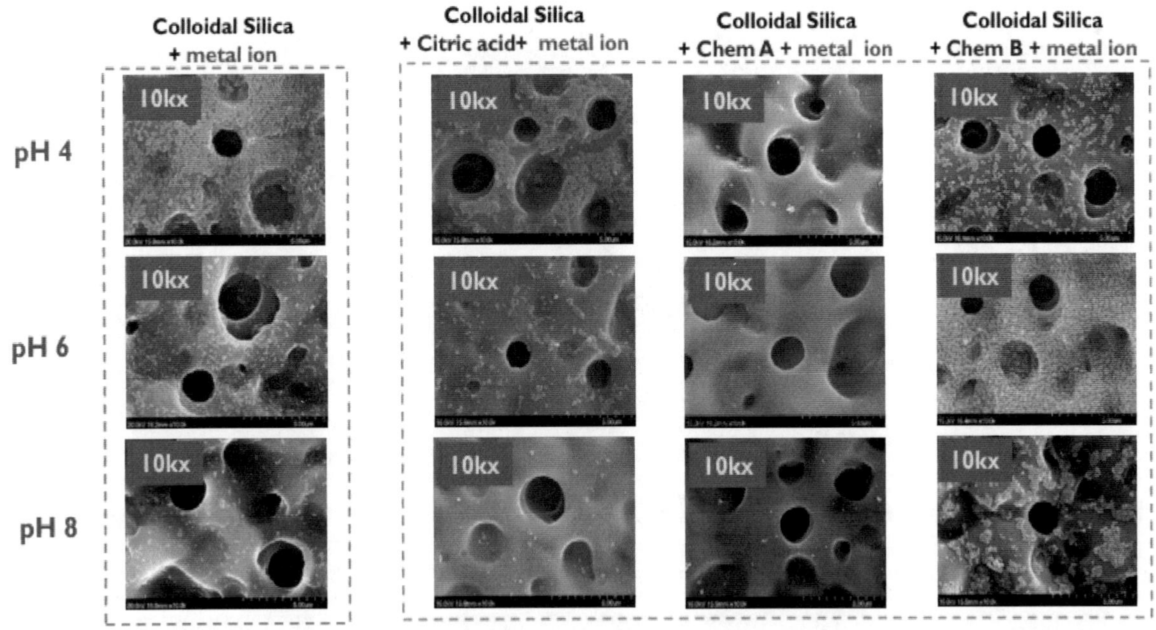

- Citric acid did not show significant difference on contamination at any pH condition
- Chem A showed the lowest particle contamination on brushes

Brush contamination by colloidal ceria slurries (FE-SEM)

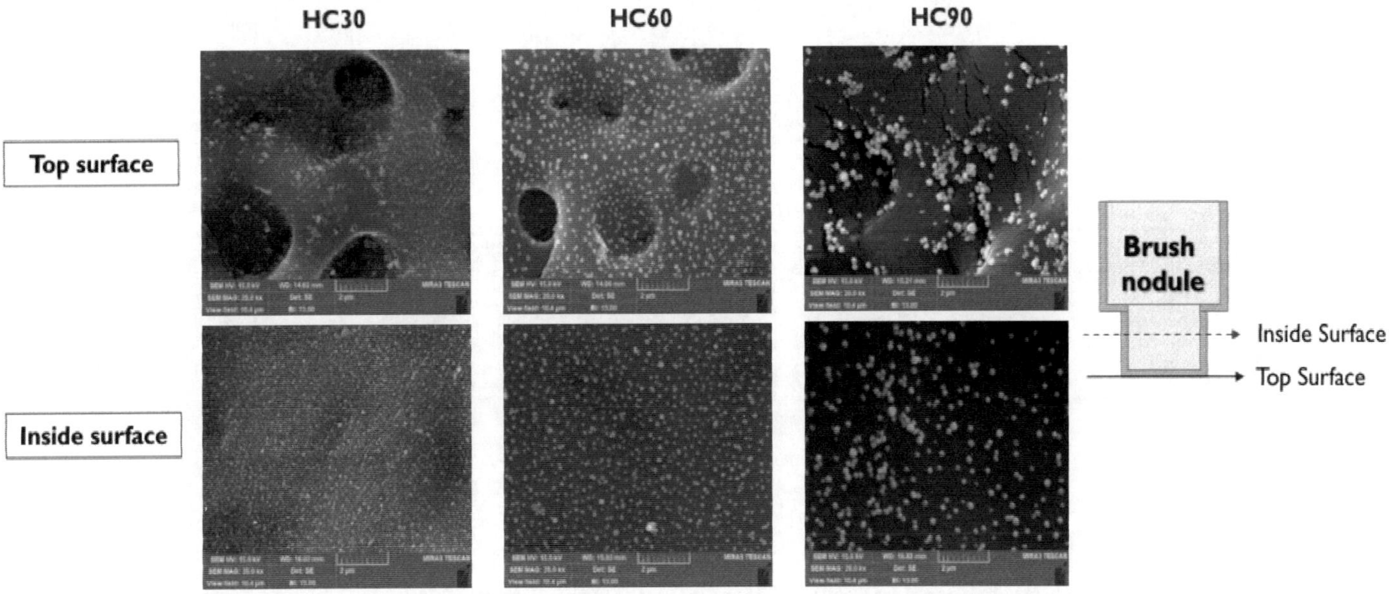

- High ceria loading at brush top surface along with high ceria penetration inside brush was observed for HC30 and HC60, whereas slightly reduced brush loading at top surface for HC90 along with high penetration.

Conditioning of brush after dipping

I. Brush Contamination

II. Conditioning

a) Ultrasonication

b) Scrubbing

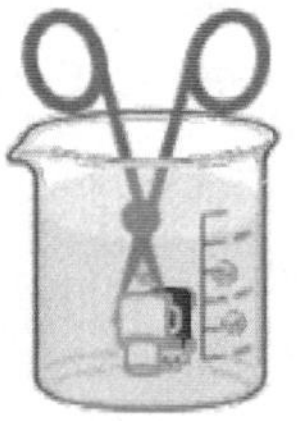

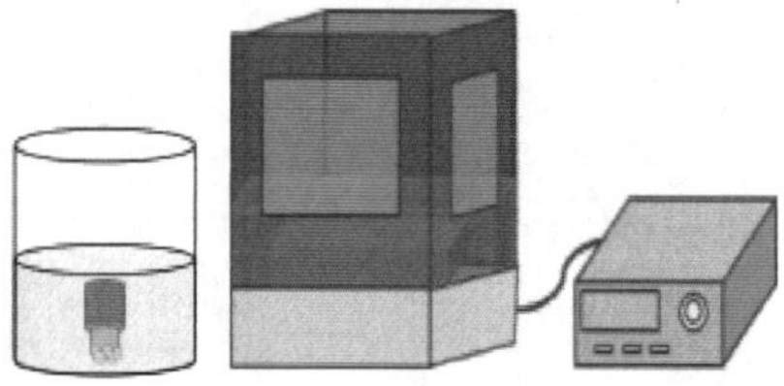

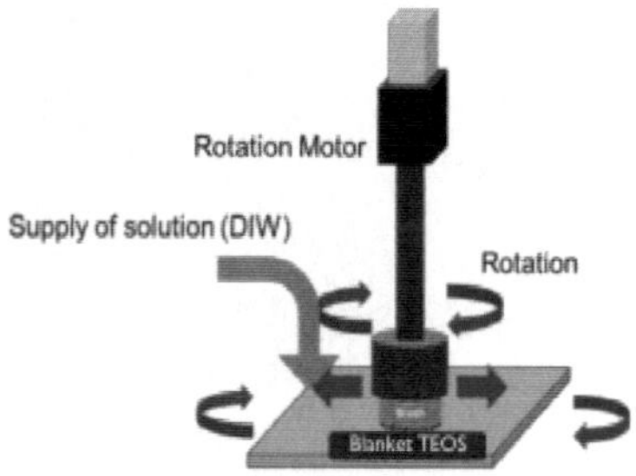

- **Calcined ceria slurry (pH 2 & 7)**
- **Dipping Time :** 1 min
- **DIW Rinsing :** 30 sec

- **Sample:** brush after dipping
- **Solution:** 100ml

  2 wt% NH_4OH
- **Conditioning:** ultrasonication at 600W for 10 & 30 min

- **Solution:**

  2 wt% NH_4OH
- **Flow rate:** 100 ml/min
- **Wafer speed :** 60 rpm
- **Nodule speed :** 60 rpm
- **Gap distance :** -2 mm
- **Time :** 10 & 30 min

37

Conditioning of brush after dipping contamination @ pH 7

Brush loading by dipping of brush nodule in ceria slurry prepared at pH 7

1) Ultrasonication in 2 wt% NH_4OH

2) Scrubbing in 2 wt% NH_4OH

Before Conditioning

a) 10 min **b) 30 min**

a) 10 min **b) 30 min**

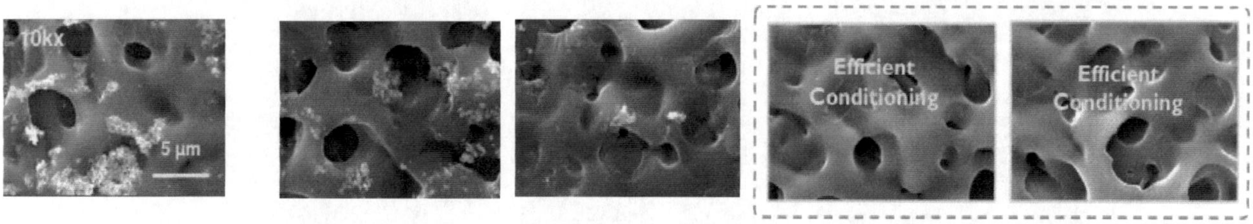

- **Increasing time during ultrasonication resulted in decreased ceria particles present on brush (not efficient conditioning)**
- **Scrubbing of contaminated brush nodules in 2 wt% NH_4OH resulted in efficient conditioning due to strong shear forces provided during scrubbing along with electrostatic repulsion**

38

The effect of skin layer of brush on

- Brush loading
- Penetration depth
- Particle removal efficiency
- Cross contamination

PVA brush skin layer formation due to molding process

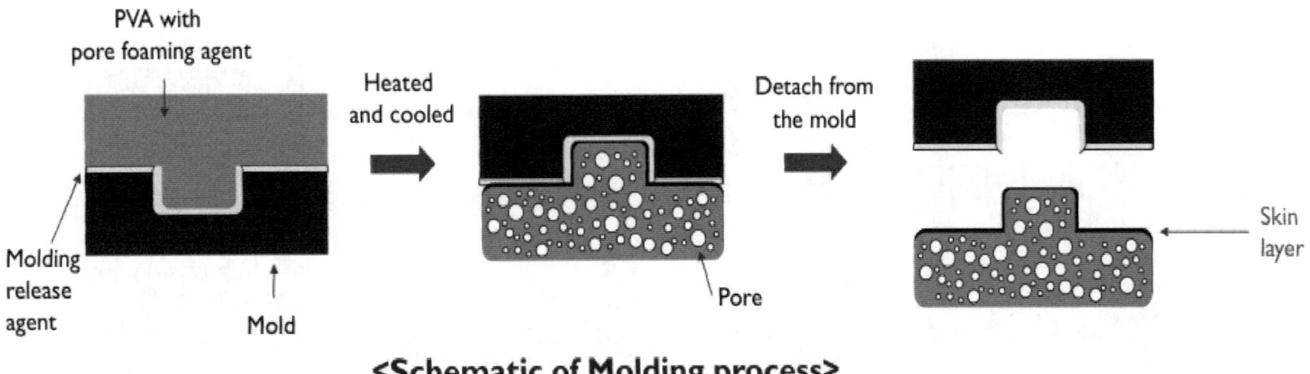

<Schematic of Molding process>

- **Manufacturers use a <u>molding process</u> to make PVA brush.**
- **The top surface of brush contacts the mold to form a 'skin layer'.**

FE-SEM images of PVA brush: w/ and w/o skin layer

Cross-section

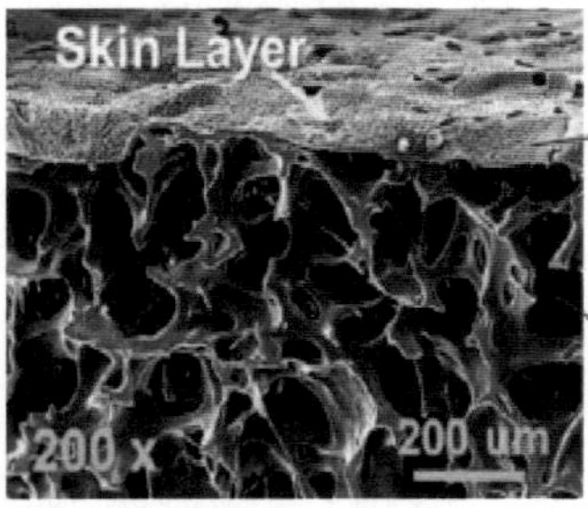

Top view

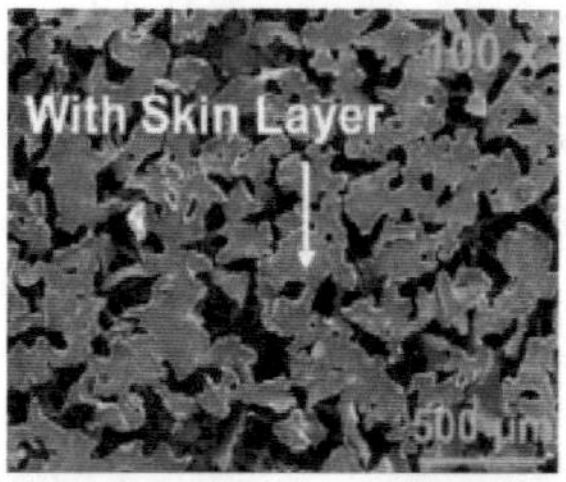

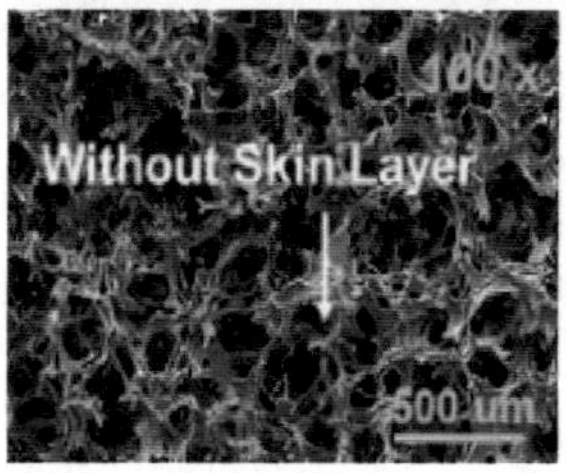

Role of skin layer on contamination & cross-contamination

??

- **The thickness of skin layer is about 1 μm.**
- **The morphology of brush top surface w/ and w/o skin layer is different.**
- **Effect of brush skin layer on contamination is needed.**

FESEM analysis of brushes : w/ and w/o skin layer

<Brush with skin layer>

<Brush without skin layer>

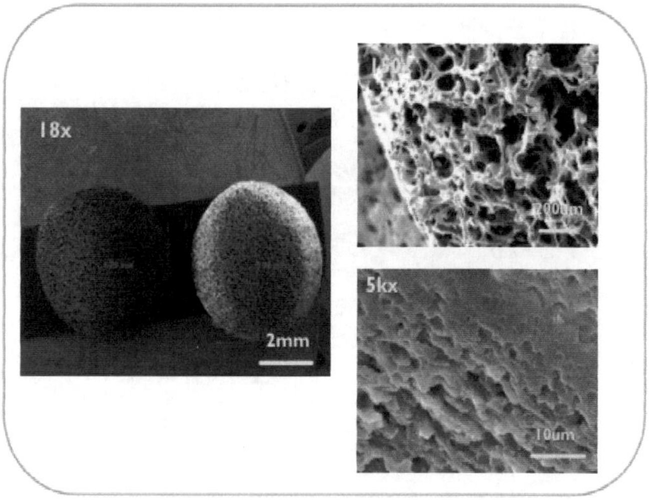

- **Brushes w/ and w/o skin layer were received from Rippey.**
- **Top surface seems to be mechanically worn (observed from 18x and 150x images).**
- **The contact area might be different in brushes w/ and w/o skin layer.**

Ceria brush loading comparison: w/ and w/o skin layer

FE-SEM

w/ skin layer　　w/o skin layer

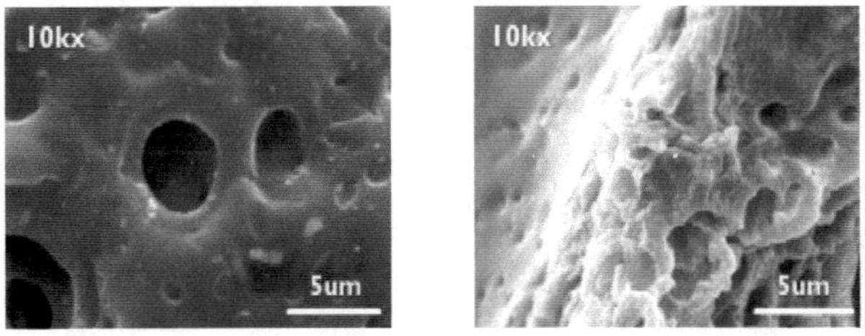

Ceria 0.1wt% pH 4

Hard to compare because structure is different

ICP-OES

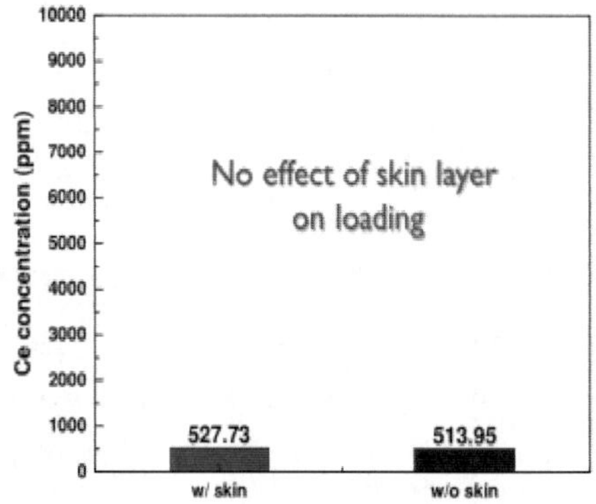

- **It is hard to compare the loading using FESEM as top surface of w/o skin layer was damaged**
- **ICP-OES results showed no difference in the loading level between w/ and w/ skin layer.**

Skin layer effect on cleaning performance: FE-SEM wafer results

<FE-SEM wafer results>

Before scrubbing

w/ skin layer

w/o skin layer

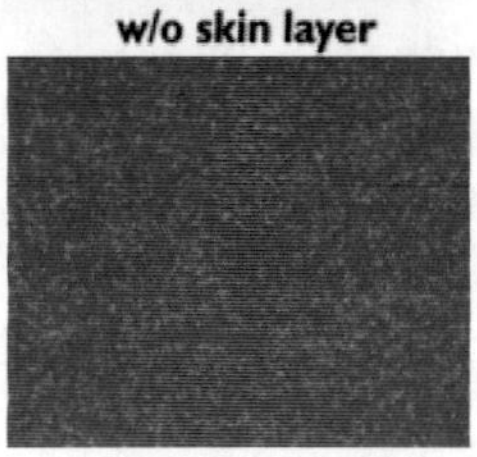

Check White area

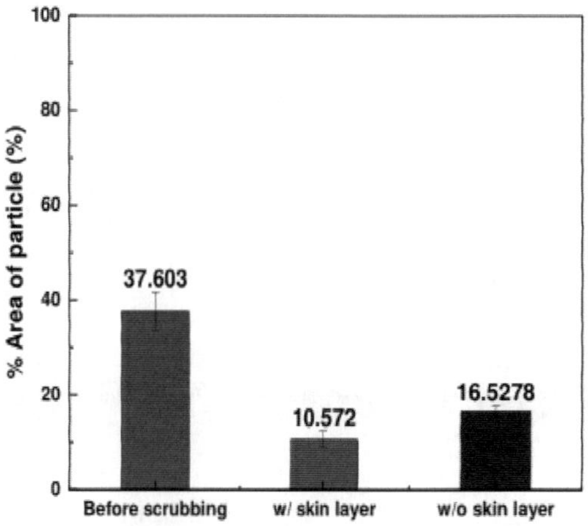

- **DIW cleaning with brush did not remove all particles and hence skin layer effect can be studied.**
- **A slightly higher particle removal observed with the brush w/ skin layer than brush w/o skin layer.**

  → **Brush w/ skin layer can remove particle more than brush w/o skin layer.**

Wafer cross contamination results: OM analysis

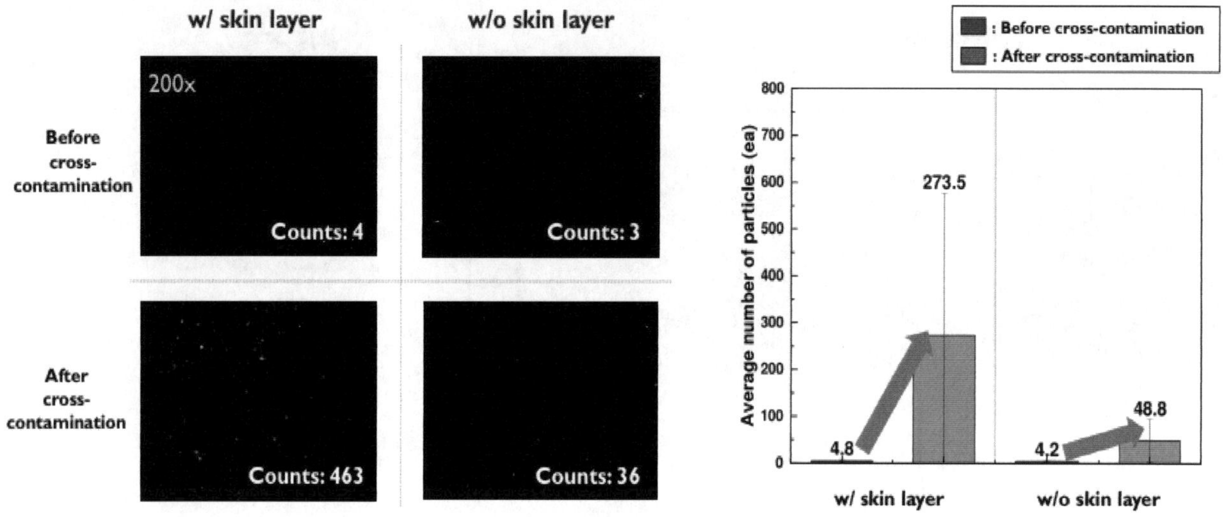

- **Some of particles are transferred to the wafer confirming cross-contamination.**
- **Amount of cross-contaminated particles in w/ skin layer is higher than that in w/o skin layer.**

→ Skin layer seems to occur higher cross-contamination.

2. Brush study at HYU

- Quantification of brush impurities
- PVA Brush Particle Loading and Conditioning
- Contact area analysis

PVA-brush wear | Brush contact area analysis

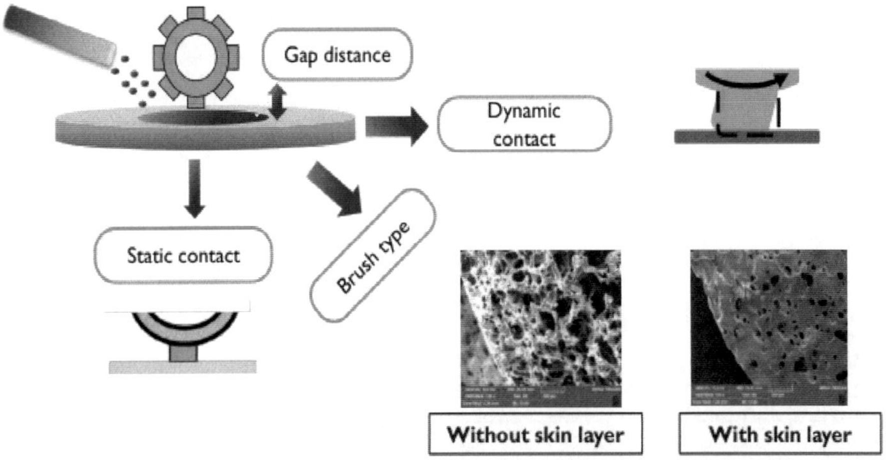

Working principle of contact area

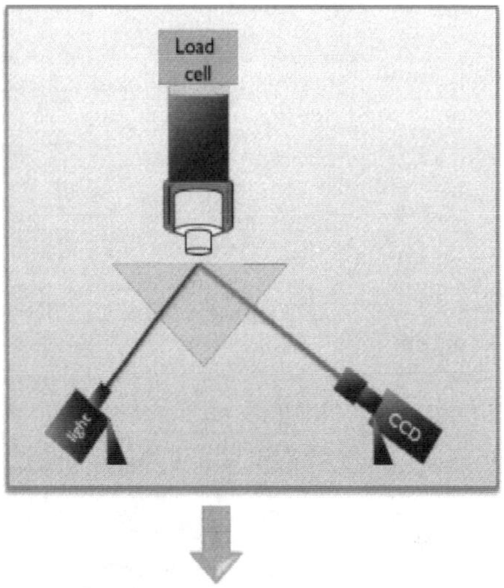

1. Effect of gap distance on contact area
2. Effect of brush type in static condition on contact area
3. Effect of brush type in dynamic condition on contact area

Analyse contact area

- **ImageJ**
- **MATLAB**
- **FE-SEM**

Brush contact area analysis system

- **Schematic of the experimental setup for contact area measurements**

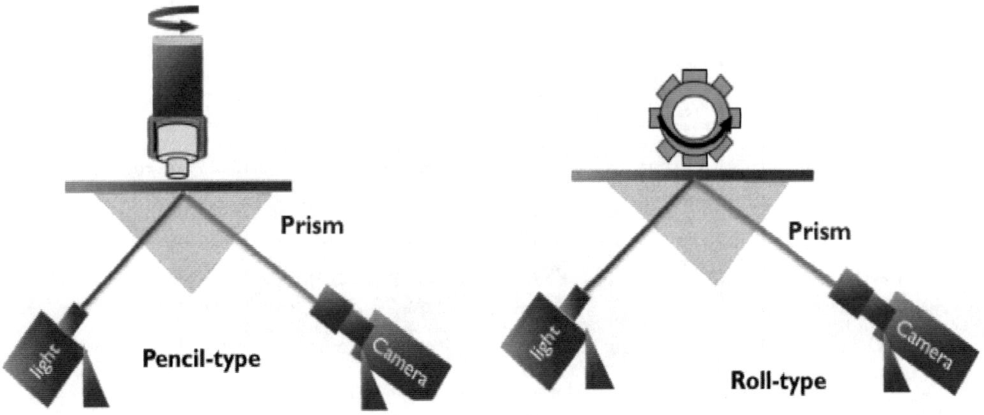

Contact area has been measured using the experimental setup

① **Parts**

- Prism
- CCD camera
- LED light
- Motors
- Load cells

② **Working Principle**

- Prism
- Reflection
  - Critical Angle
- Refraction
  - Bends towards the contacted area

FE-SEM analysis of different brush types

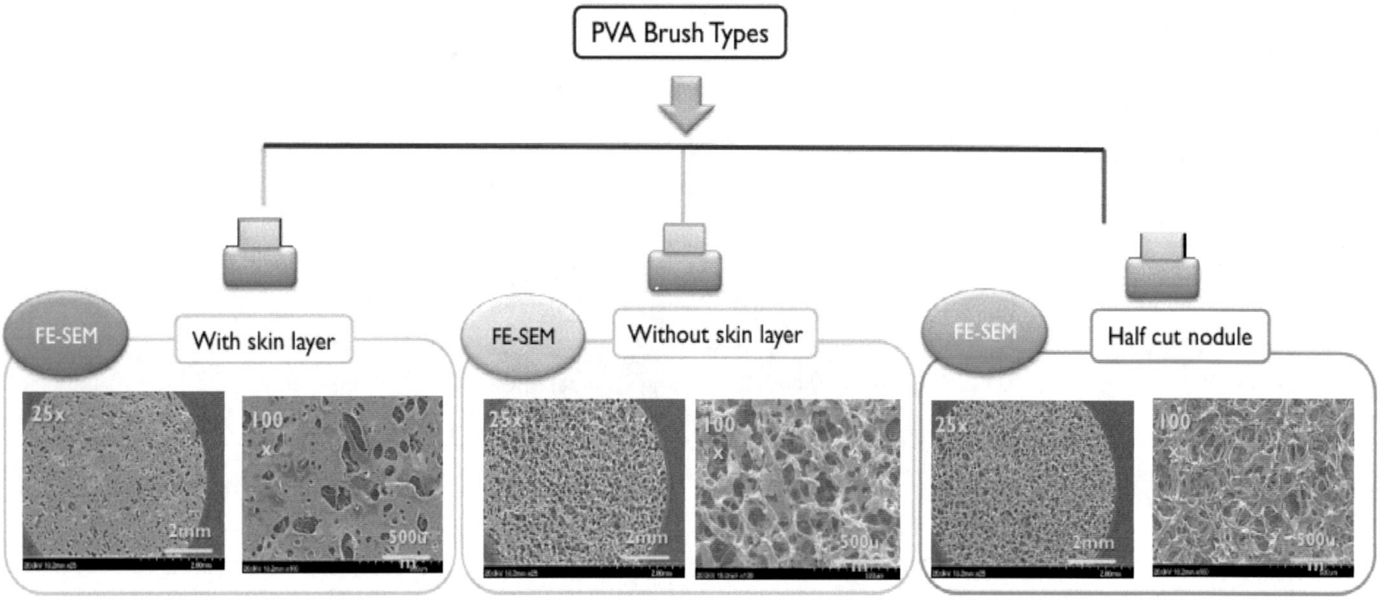

- PVA brush surface condition for with, with-out skin layer as compare to half cut nodule (self removed top skin layer)
- Similar structure was observed for brush without skin layer and self removed skin layer

Contact area at static and dynamic condition after 0.5 sec (-3mm gap distance)

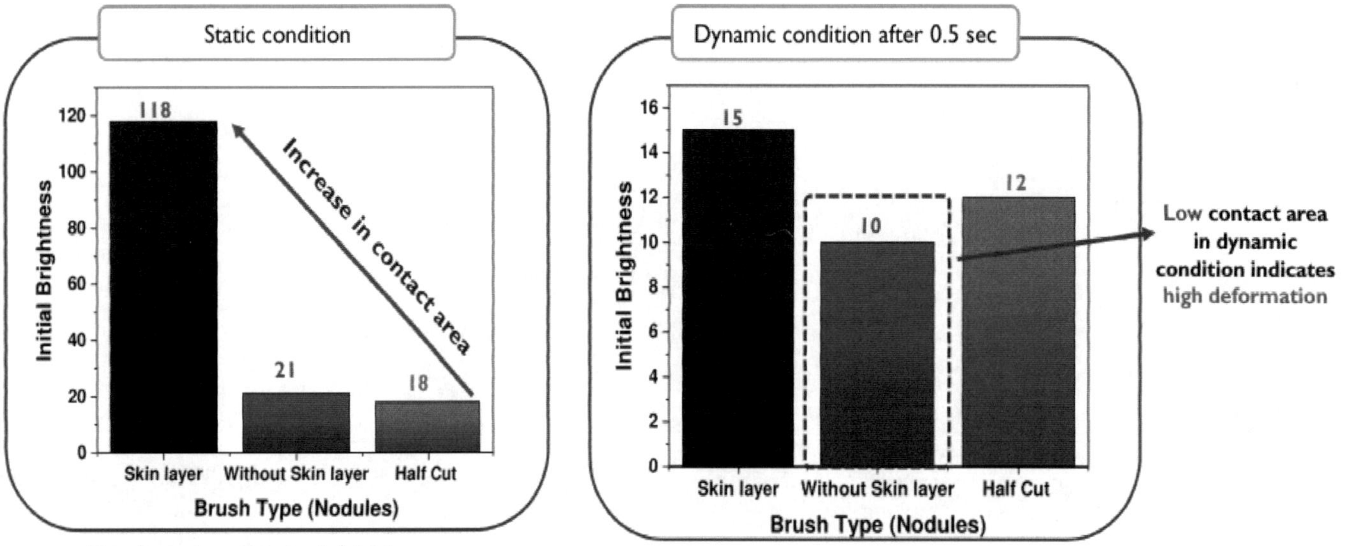

- Skin layer brush is having high contact area
- Without skin layer brush is having high contact area as compared to half cut nodules due to round edge which become in contact area by decreasing gap distance
- Low contact area observed in without skin layer brush during dynamic condition which indicates brush deformation

Effect of wearing on contact area

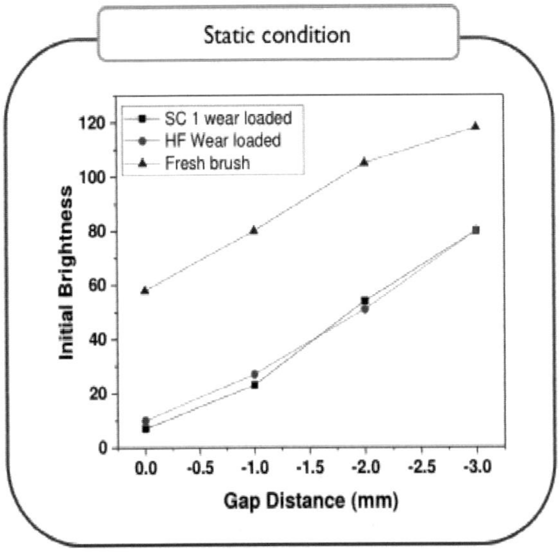

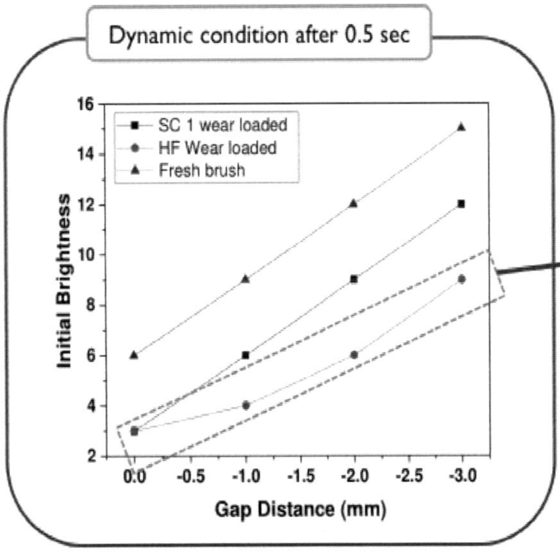

Low contact area in dynamic condition shows high wear loads

- HF treated brush having high wear when compared to SC1 treated brush
- Contact area increases with decreasing gap distance
- During scrubbing after 0.5 seconds the contact area also increased

Summary

- Understanding on brush just started
- HYU is working on all aspects of brush technology
- Full infrastructure and analysis tools are in house
- Polymerization, analysis and evaluation are possible

Sustainable Pyrrolidinone-Based Solvents Refinement for Wet Cleaning Process

Chih-Hung Ko, Jeff Powell, and Ken Hampton
100 Eastman Road Kingsport, TN 37660, and 300 Kodak Blvd. Longview, TX 75602
chihhung.ko@eastman.com | +1-919-590-9008

Eastman EastaPure™
electronic chemicals

Challenges and Recommendation

Specialty chemicals formulated with dipolar aprotic solvents such as NMP, NEP, DMF, DMAc, and 2-Pyrrolidinone are widely utilized for stripping and cleaning processes due to their high polarity and solvation Yet, these solvents raise environmental concerns and pose health risks. For instance, NMP is classified as a substance of very high concern (SVHC) by the European Chemicals Agency (ECHA) due to reproductive toxicity, initiating Notice of Proposed Rulemaking (NPRM) under Toxic Substances Control Act (TSCA) by Environmental Protection Agency (EPA). In response, major technology firms have committed to managing hazardous substances by reducing or eliminating undesired solvents.

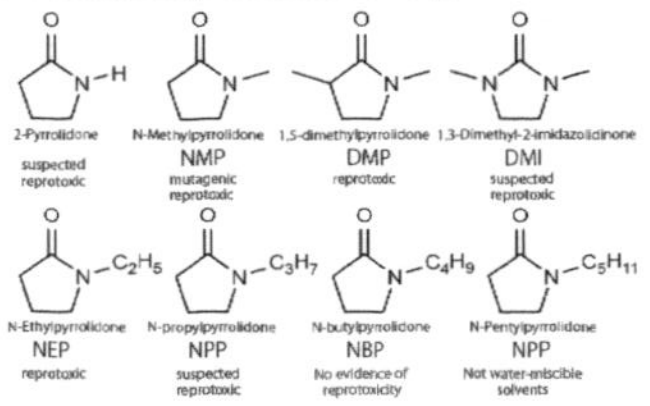

A significant challenge in the searching of sustainable solvents resulting from time-consuming reproductive toxicity studies in which NEP and DMP were initially defined as green solvents, and eventually categorized as reprotoxic category 2 by ECHA. To solve the issues, the targeted solvent at least fulfill three requirements:

- *Pyrrolidinone (amide) functional group*
- *Non-reproductive toxicity compound*
- *Water miscible*

N-butylpyrrolidinone (NBP) emerges as a promising candidate which identified as non-reproductively toxic (OECD 414 and 421), non-mutagenic (OECD 471), and inherently biodegradable (OECD 302B). The structural similarity, wetting properties, and water miscibility facilitate a smoother process in product and process development.

Properties and Parameters

Solvent	b.p / °C	m.p / °C	viscosity / cp	E_T	lowlight
NBP	241	< -75	4.0	0.323	
NMP	202	-24	1.9	0.355	H360
DMF	153	-60	0.8	0.386	H360
DMAc	163	-20	1.9	0.377	H360
Sulfolan	285	26	10.3	0.410	H360, high cp
DMSO	189	18	2.0	0.444	skin penetration
Acetonitrile	82	-44	0.3	0.460	class 3 flammable
PolarClean	280	-60	9.8	-	high cp

The longer alkyl chain of NBP leads to slightly higher boiling point and viscosity at room temperature while reducing polarity. Hansen Solubility Parameters (HSP) and Kamlet-Taft (K-T) parameters are introduced to estimate the type of interactive forces responsible for solvent–polymer compatibility and synthesis process. In sum, rather than being a drop-in replacement, the adaptation of NBP may require refinement of raw materials, reformulation of specialty chemicals, and redesigned process flow to accommodate new solvent selection.

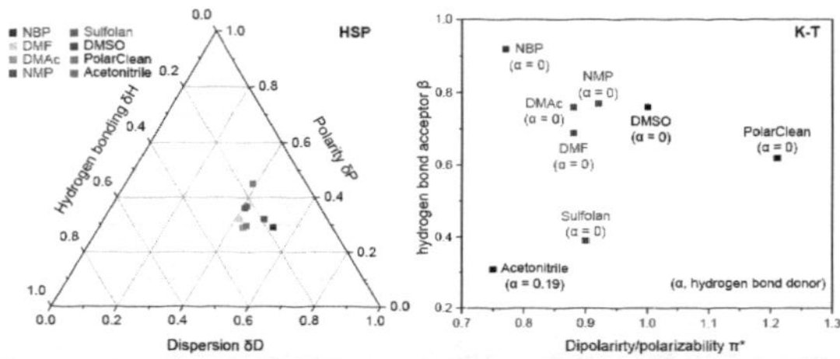

Refinement Processes and Stability Test

- Since the required purity or controlled impurity level is determined by FEOL/BEOL processes, packaging or synthesis, various NBP polishing methods are selected to effectively reduce water/acidity/basicity (titration), ions (ICP-MS), and anions (IC), minimize the surface contamination or prevent unwanted side reactions of reactants.

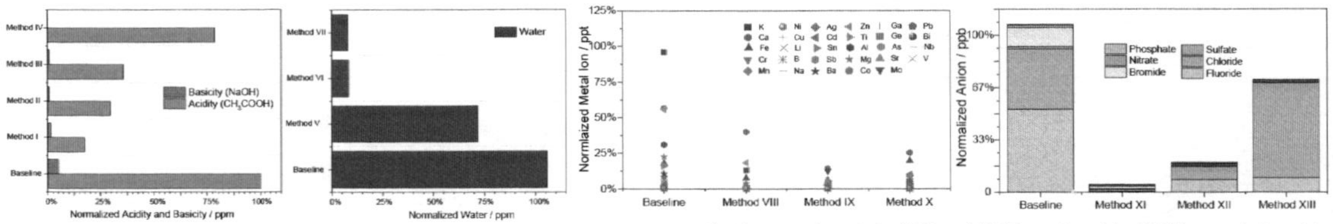

- Etching rate (ER) refers to the rate at which a thin film is removed when exposed to an etchant, and critical is to ensure that the solvent used in a formulation does not damage critical layers. NBP demonstrates a minimal but slightly higher relative ER than NMP on Cu and a lower relative ER on W. The rest of them are at a comparable level. Linear Polarization Resistance (LPR) Technique is employed under a 0.05M electrolyte concentration.

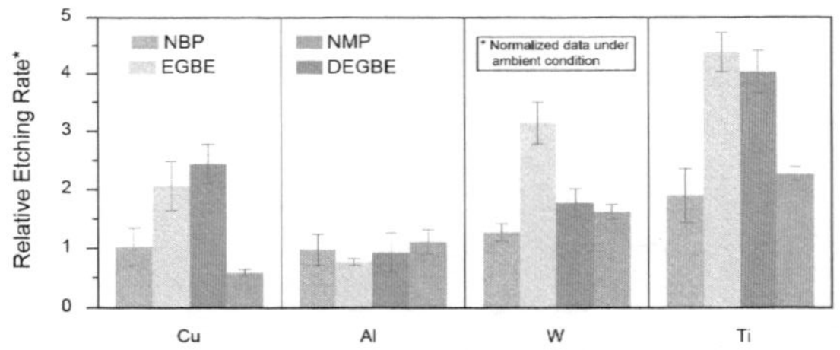

- Heating experiments for NBP and NMP, monitored by UV-Vis, are designed to be conducted at 70 degrees Celsius for up to 120 hours. Lower absorption at 280 nm of NBP indicates higher heat stability compared to NMP. The degraded intermediate and succinic acid results from high temperature and oxygen-rich environment were verified through ^{1}H NMR analysis or titration.

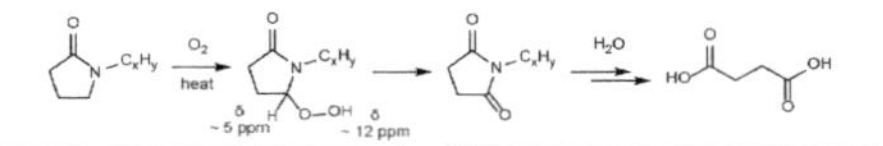

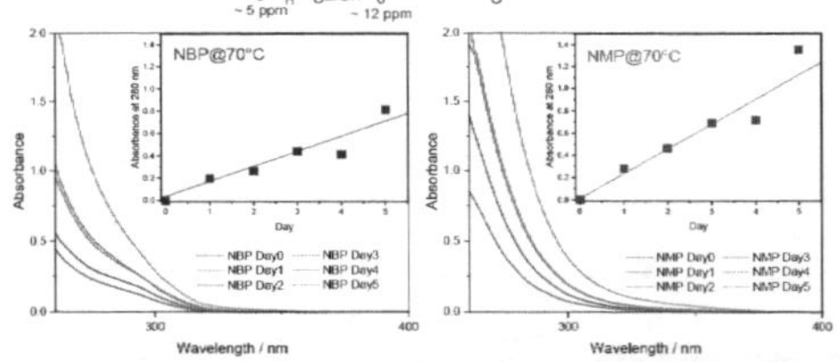

Polymer Removal and Compatibility

Polymer	Solubility	Polymer	Solubility	Polymer	Solubility
PE	Not soluble	PTFE	Not soluble	PC	Soluble
PP	Not soluble	PFA	Not soluble	PET	Partially soluble
PS	Soluble	PMMA	Soluble	PVDF	Not solube
PVC	Solube	CPVC	Solube	PI	solube

- NBP exhibits exceptional solubility in resists containing PMMA and PI structures, while maintaining good compatibility with standard HDPE containers and PTFE/PFA piping. Taking photosensitive polyimide (PSPI) as an example, NBP was heated to evaluate its capability as a PSPI stripper. As the temperature increased, the polymer residues on a 1.2 x 1.2mm area were significantly reduced. The removal capability, process temperature, and time can be further adjusted with common additives.

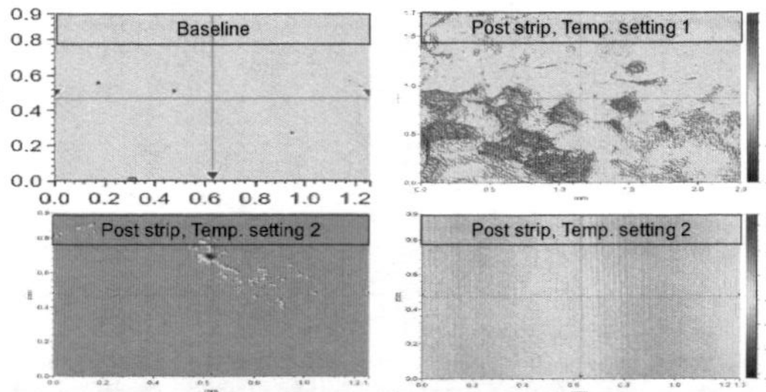

Waste Stream Reclamation

- A proposed solvent recovery flow, as a representative scheme, may contain stripping, distillation, and post-treatment to demonstrate the customizable feasibility of reclaiming a mixture of NBP with selected resin in which spiked yellowish polyimide (PI) is removed through the process verified by GPC. The recovery ecosystem will help end users on:
  - *Reducing incineration cost*
  - *Creating sustainable manufacturing*
  - *Delivering environmental, social, and governance (ESG) value*

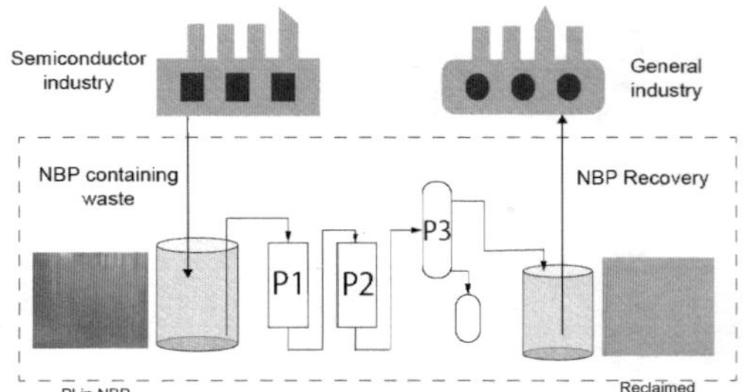

Proposed NBP Recovery Flow

EASTMAN

HUNTSMAN

Enriching lives through innovation

Study of Quaternary Ammonium Hydroxide (QAH) for Photoresist Applications

E-GRADE® AMINES

Hui Zhou[1,a], Victoria White[1], Ke Zhang[1]

Graham Gibson[2]

[1]: Huntsman Corporation, 8600 Gosling Road, The Woodlands, TX 77381, USA; [a]: hui_zhou@huntsman.com

[2]: NanoFabrication Kingston, Queen's University, 945 Princess Street, Kingston, ON K7L 0E9, CANADA

Introduction

Tetramethylammonium hydroxide (TMAH) is broadly used as a key component in photoresist developing formulations, however, it is known to have toxicity concerns. Huntsman has been developing a family of quaternary ammonium hydroxides (QAHs) to replace TMAH, such as E-GRADE® Choline Hydroxide, E-GRADE® THEMAH, and XHE experimental products. These QAHs offer comparable performance to TMAH, while being safer to handle and can be commercially available at higher concentrations. In addition, the ability to adopt new molecular structures, which can impact the performance, allows the user flexibility in finding the right formulations to a specific application.

The three key criteria for a QAH in a formulation are development rate, selectivity, and compatibility, which can be used to evaluate the performance of a photoresist developer. In this systematic study on selected QAHs we have benchmarked performance against 2.38 wt% TMAH, Ethyltrimethylammonium Hydroxide and Tetraethylammonium Hydroxide.

In the presentation, both positive and negative resists studied are commercially available Novolac-based photoresists.

Positive photoresist development

Development of positive PR

- All three QAHs cleared the resist at the given conditions
- XHE-125 at higher concentration, 7 wt%, can clear the resist in less than 30 seconds

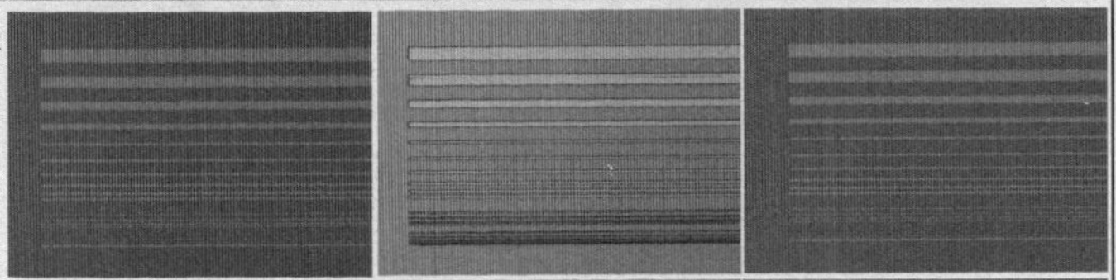

TMAH (2.38%, 30s) | XHE-125 (6%, 30s) | Ethyltrimethylammonium Hydroxide (3%, 120s)

Linear relation of resist removal and developing time

- For all QAHs tested, resist removal was linear with time at all concentrations measured
- Linear relationship indicates a stable developing rate giving good process control

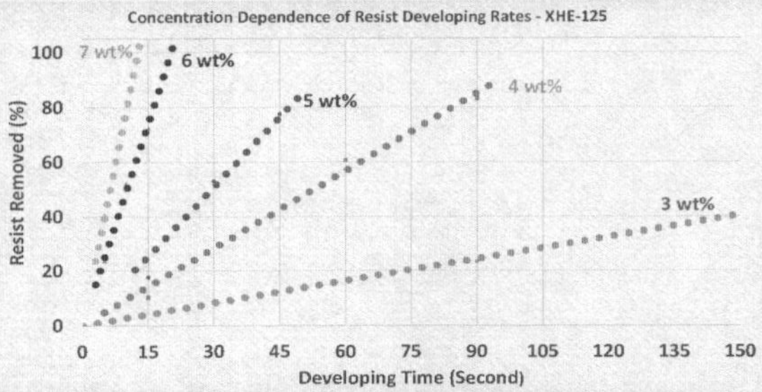

Resin Loss Rate

at equal developing rates (59 nm/s)

- At the target rate, XHE-125 demonstrated low dark erosion
- The ER/DR ratio indicates XHE-125 has similar selectivity to TMAH at the given concentration
- Huntsman QAHs are available in higher concentration and offer improved EHS profiles allowing greater flexibility in formulating for rate or selectivity across a broader working window

QAH Developers for Positive PR	Concentration*	Unexposed Resin Loss	
	wt%	nm/s	ER/DR**
TMAH	2.38	0.18	0.003
XHE-125	5.70	0.40	0.007
Ethyltrimethylammonium Hydroxide	3.25	0.40	0.007
THEMAH	8.10	1.00	0.017
Tetraethylammonium Hydroxide	7.50	1.00	0.017
XHE-150	7.50	1.40	0.024
Choline Hydroxide	4.30	1.80	0.031
XHE-148	8.00	3.60	0.061
XHE-128	4.80	4.00	0.068

*concentration giving comparable developing rate to 2.38% TMAH (59 nm/s) for comparison
**ER: Erosion rate; DR: developing rate

Negative photoresist development

Film Loss Rate at equal developing rate (78 nm/s)

- XHE-125 demonstrated loss rate <0.2 nm/s with good selectivity
- Choline Hydroxide has slightly higher Film Loss at comparable concentration.

QAH Developers for negative PR	Concentration*	Exposed Film Loss	
	wt%	nm/s	LR/DR**
TMAH	2.38	0.07	0.001
Ethyltrimethylammonium Hydroxide	2.90	0.10	0.001
XHE-125	6.30	0.18	0.002
Tetraethylammonium Hydroxide	6.90	0.19	0.002
THEMAH	16.60	0.82	0.011
XHE-148	11.20	1.60	0.021
Choline Hydroxide	3.10	1.70	0.022
XHE-146	9.80	1.70	0.022
XHE-150	11.20	3.40	0.044

* where the developing rate was similar to 2.38% TMAH (78 nm/s) for comparison
** LR: film lost rate; DR: developing rate

Concentration Dependence Developing Rate

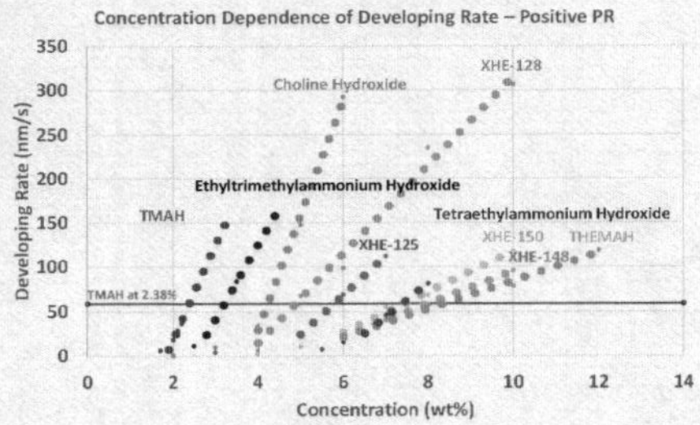

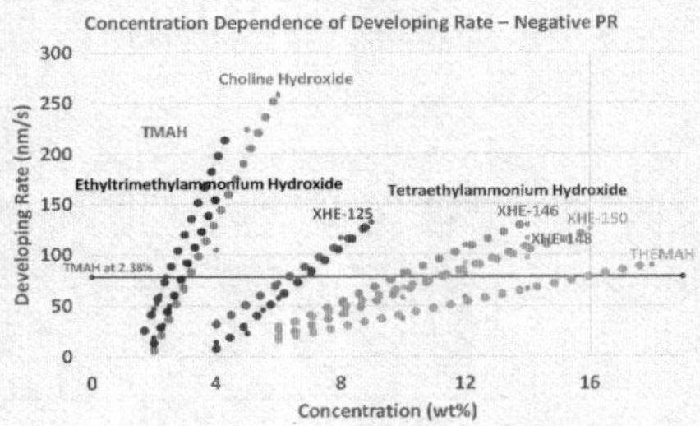

- Choline Hydroxide, XHE-125, and XHE-128 have comparable developing rates relative to 2.38 wt% TMAH for positive PR
- Developing rates can be optimized across a broader working window with concentration adjustments to modify performance for specific applications
- Developing rates have a linear relationship with concentration, above certain minimum concentrations

- Choline Hydroxide demonstrated good performance for negative PR
- XHE-125 and Tetraethylammonium Hydroxide have similar performance, their developing rates can be improved by increasing the concentration to 5 wt% and match 2.38 wt% TMAH
- Developing rates have a linear relationship with concentration, even at low concentrations

Summary

This study provides detailed information about optimizing developing rate, dark erosion rate and concentrations based on a family of aqueous quaternary ammonium hydroxides, such as Choline Hydroxide, THEMAH, XHE-series we developed, which offer a variety of structures, and lower toxicity options to enable flexibility in pursuing improved EHS profiles for photoresist developing formulations. QAHs help performance improvement with process flexibility, lower corrosion rates on metal substrate, faster photoresist developing speed at higher concentrations and wider process window for critical layer at lower concentrations, which gives benefit for cost saving, yield improvement and enabling leading-edge technology.

A Copper (Cu) Post-CMP Cleaning Formulation For Both Wafer and Pad Clean

Jacky Cheng, Peter Sun, Wayne Lin, Wei-En Huang, Ping Hsu

Advanced Cleans & Slurry Technologies (ACST)

DuPont Electronics & Industrial

1. Introduction

New barrier metals such as Cobalt (Co) applied to interconnects in advanced chips and impose challenges on compatibility of post-CMP cleaners. Also, the demand for clean performance is increasing, especially with respect to small particle residues. Furthermore, stains on pads are also a proven source of defects and micro-scratches. Therefore, a cleaner that can clean both wafer and consumables is highly desired.

PCMP5688 is a cleaner that gives excellent residue removal performance for both wafers and pads. It contains an advanced chelating component for dissolving/dispersing metal oxides, organometallic residue, and abrasives. It also includes an effective inhibitor that provides good Cu/Co compatibility for a smooth metal surface. It combines wafer and pad cleaning into a single cleaner that not only reduces defect level but also extends pad life.

Component	PCMP5675	PCMP5688
Metal Compatibility	+	+++
Roughness	+	+++
Undercut Efficiency	+++	++
Redeposition Prevention	+	+++
Organic Removal	+++	+++

Table 1. Performance comparison of PCMP5675 (POR) and PCMP5688 (new formulation for wafer and pad clean)

Objectives of this research:

A Post-CMP cleaner to provide:

- Good metal compatibility
- Good particle / organic residue clean

2. Dominated Clean Mechanism

PCMP5675 (Aggressive etching)	PCMP5688 (Clean & compatibility)

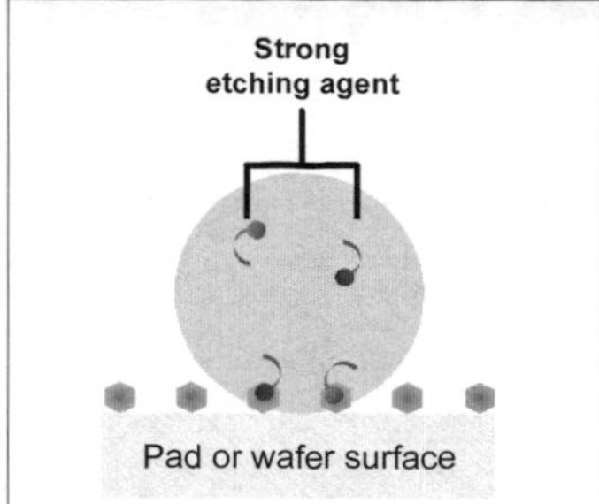

Fig 1.
(a.) Good clean performance derived from strong dissolution and undercut.
(b.) Not suitable for advanced node due to high etching to Cu and barrier metals.

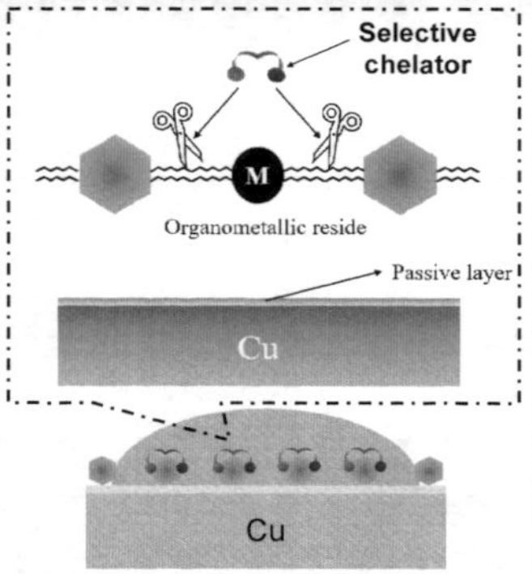

Fig 2.
(a.) Improved affinity and dissolution to organic residue; less etching to Cu and Co.
(b.) Hydrophilic passivation layer provides good wetting and metal protection.

3. Performance Comparison

1. Metal Compatibility

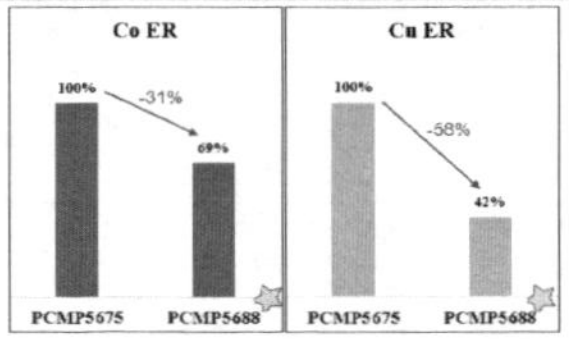

Fig 3. Normalized Cu/Co etching rate

2. Metal Oxide Solubility

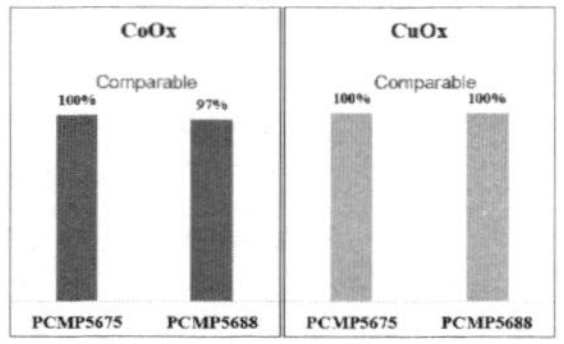

Fig 4. Normalized metal oxide solubility

3. Organic Residue Clean

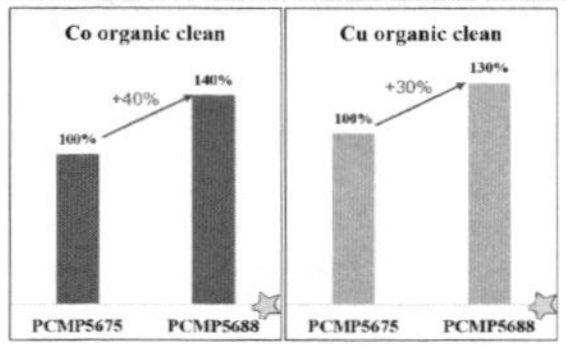

Fig 5. Normalized organometallic residue solubility.

4. Abrasive Removal Test (Worse condition)

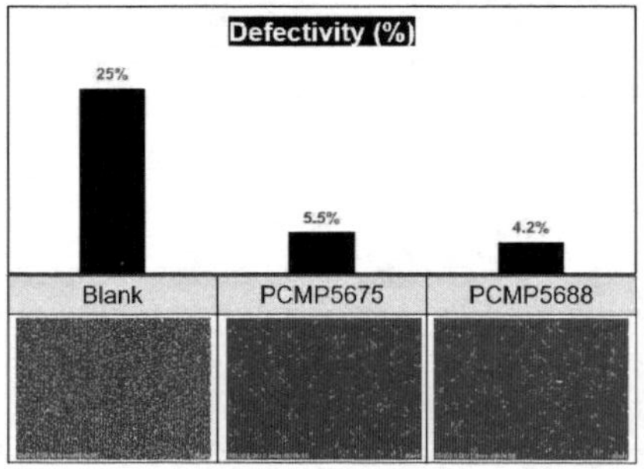

Fig 6. Defect coverage test on Cu surface by table-top polisher.

5. Pad Clean Test (Worse condition)

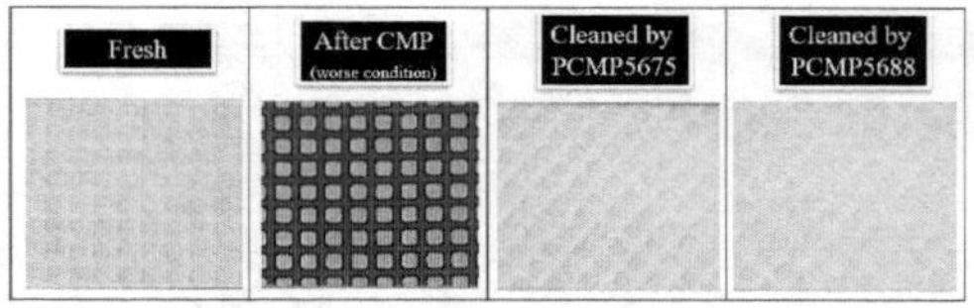

Fig 7. Clean demo on pad heavily stained with CuOx residues. Both PCMP5675 & PCMP5688 showed excellent pad clean performance.

4. Conclusion

By introducing selective chelating agent and hydrophilic corrosion inhibitor, DuPont PCMP5688 improved metal compatibility over existing product PCMP5675 while maintaining good particle and organic clean capability on both wafer and pad. It shows the potential to simplify chemistry used during CMP & PCMP processes and cost saving by extending pad life.

PCMP5688 Formulation	
Metal compatibility & clean performance	• Improved metal compatibility through 30%~60% reduction in Cu/Co etching rate. • Comparable metal oxide residue removal on wafer/pad and improved organic clean.
Feature	• Non TMAH • Highly concentrated alkaline formulation with robust quality and stability. • A single formulation for both wafer and pad clean.

ACM RESEARCH

ACM Research

Application of TMAH In Semiconductor Wet Etching

Kai.Liu, Zexian.Ma, Yao.wu, Xiaoyan.Zhang, Jian.Wang, Sally-Ann Henry

Email: kai.liu@acmrcsh.com

Introduction

Semiconductor wet etching process is a method that uses chemical solutions to remove part or all of the material on the surface of semiconductors, which can be used to form circuit patterns, clean the surface or adjust the material properties. TMAH (tetramethylammonium hydroxide) is a kind of wet etching chemical solution, which has strong alkalinity and easily absorbs CO_2 in the air. TMAH can perform anisotropic or isotropic etching on the silicon surface at certain temperatures and concentrations, to form different patterns or achieve the purpose of thinning the wafer. TMAH has a low selectivity ratio for silicon crystal planes 100 and 110 ,therefore TMAH is often used as the source-drain epitaxy etching solution in germanium-silicon source-drain epitaxy technology.

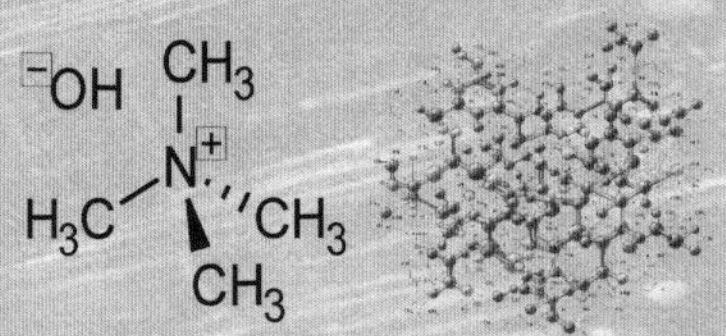

Fig. 1 Tetramethylammonium hydroxide

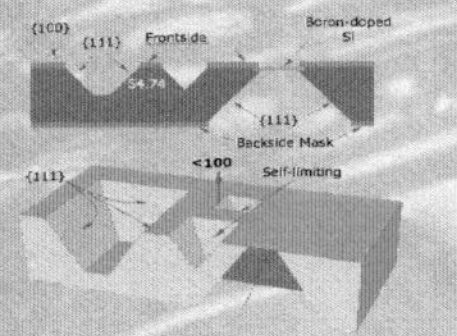

Fig. 2 Silicon anisotropic etching

Research Content

SiGe Etch
(For Low Temperature TMAH 25 °C)

The devices with Σ-shaped pattern devices obtained by anisotropic etching have a 23% higher maximum transconductance than normal shape .

In order to stably and accurately etch out the shape of SiGe Recess in large-scale production, it is necessary to study the effects of TMAH temperature, concentration, service life, oxygen concentration and doping on the silicon etching rate.

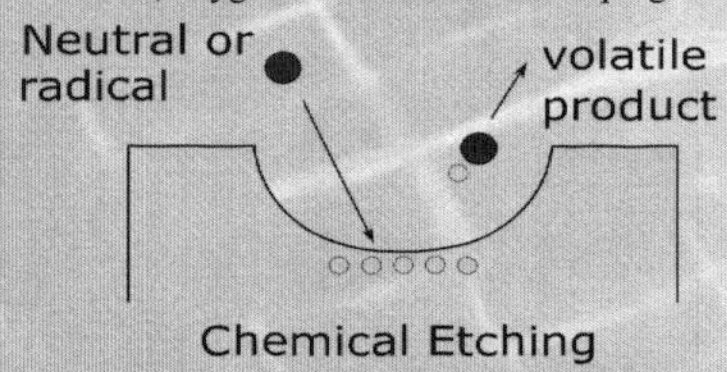

Fig3 . Schematic diagram of wet silicon etching

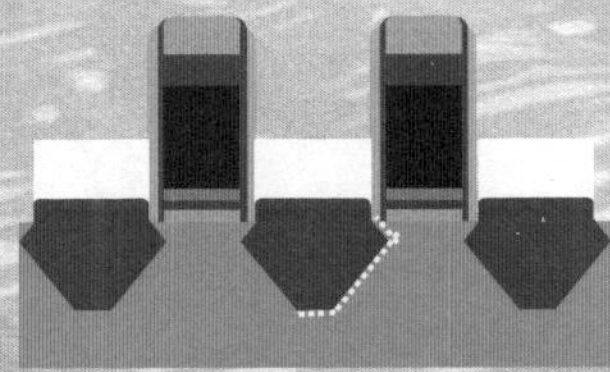

Fig. 4 Σ-shaped patterns

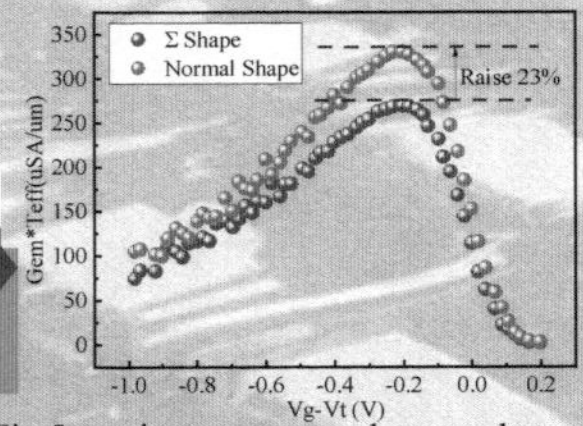

Fig.5 maximum transconductance between Σ-shaped and normal shape devices

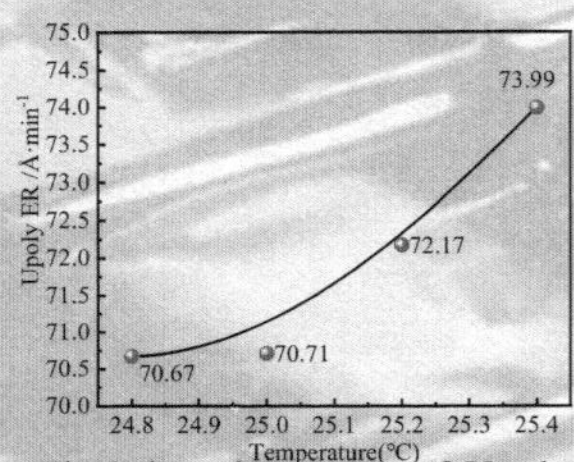

Fig 6. the etch rate curve of U poly at room temperature TMAH

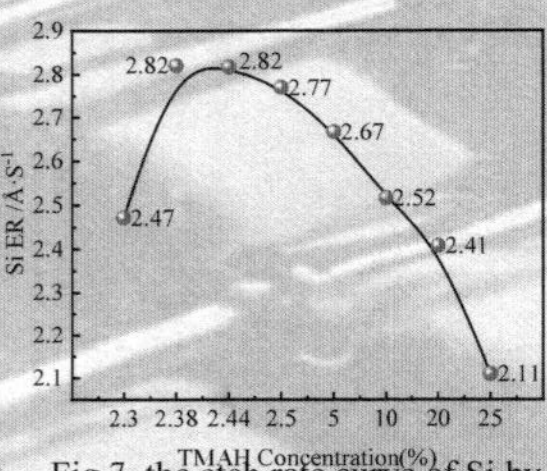

Fig 7. the etch rate curve of Si by different TMAH concentration

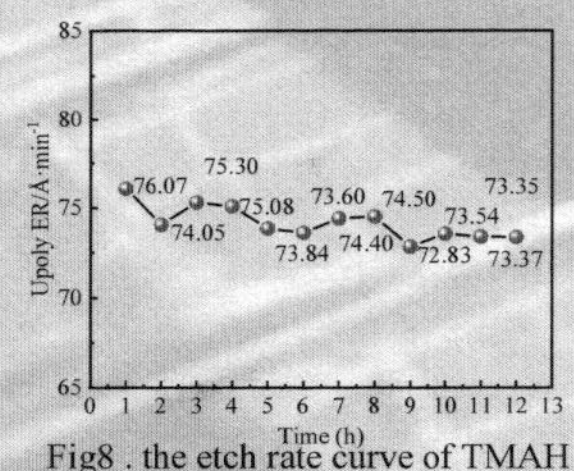

Fig8 . the etch rate curve of TMAH within 12H lifetime

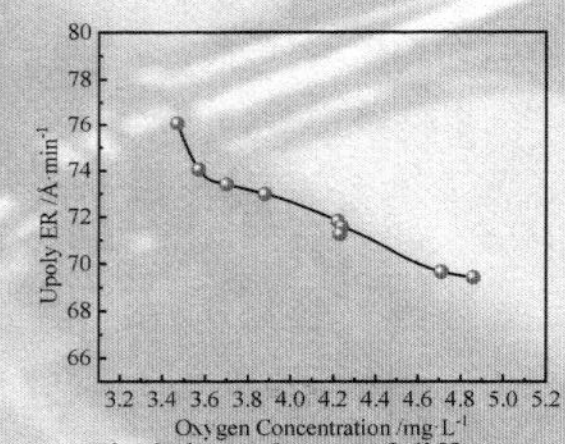

Fig.9 the etch rate of different oxygen concentrations in TMAH

Dummy Poly Remove
(For High Temperature TMAH)

Polysilicon is used as a dummy gate before filling the metal gate, and the traditional silicon gate process is still followed in the ion doping steps. After the source/drain formation is completed by high temperature annealing, the polysilicon gate is etched away and filled with suitable metal to form MG.

High-temperature (70°C) TMAH solution can be used as a choice for dummy poly gate removal.

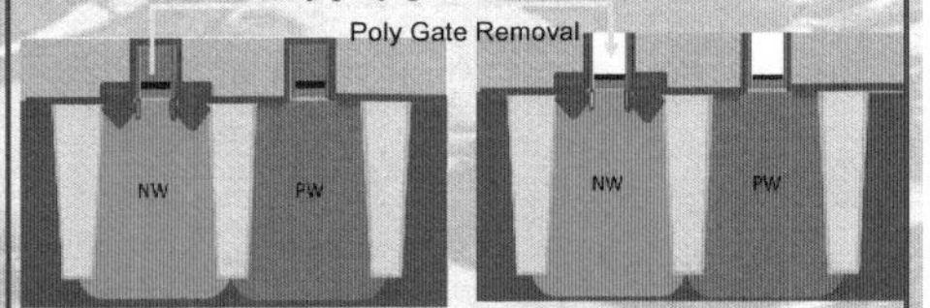

Fig 11 Dummy Poly Remove

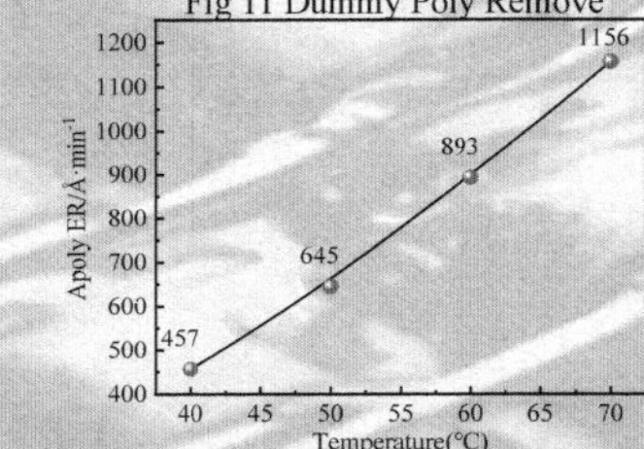

Fig 12. Different high temperature TMAH etch rate

Summary

1.The ER of TMAH increases approximately linearly with increasing temperature; decreases with increasing oxygen content; reaches the peak value in the concentration range of 2.38% to 2.44%; and fluctuates slightly within the lifetime of 12 hours.

2. Under high temperature conditions, the etch rate of TMAH increased linearly with temperature。

Development of a Wafer Heating Stage for Thermal Desorption Spectroscopy

Xiaoyu Zou, Matthew Fisher, Hugh Gotts
Air Liquide Electronics – Balazs NanoAnalysis
46409 Landing Parkway, Fremont, CA 94538

Introduction

Controlling the outgassing of contaminants is critical to maintaining the cleanliness of semiconductor components and materials exposed to vacuum. While many outgassing analyses are done at ambient temperature, this is not a realistic representation of thermal conditions such as those encountered in ALD and CVD processes.

Thermal desorption spectroscopy (TDS) is a mass spectrometry technique that heats a sample under ultrahigh vacuum (UHV) and detects the outgassed species with a residual gas analyzer (RGA). This allows for full characterization of the thermal outgassing products. A limitation in sensitivity is the outgassing of wafer stage materials at high temperatures, which may hide trace outgassing of the sample. Here, we present the development and application of an ultra-clean, UHV compatible TDS wafer stage for trace outgassing analysis of solid samples.

Experimental Design

- SS304, all metal sealed, oil free turbopumps, load-lock chamber
- Mass spectrometer: 1-200 amu RGA, $<1\times10^{-12}$ mbar LOD

- Laser machined 100x100x5 mm Al stage, 50x50 mm 250W AlN heating element, 250°C max T, cleaned with Balazs process
- UHV grade, $< 1\times10^{-9}$ mbar L/(s cm^2) outgassing at 150°C
- Temperature uniformity is +/- 2°C within scan time of RGA

Results

Thermal and vacuum outgassing simulations are conducted to ensure design specifications are met prior to fabrication and installation. Experiments are then performed to validate the design, cleanliness and functional performance in real conditions.

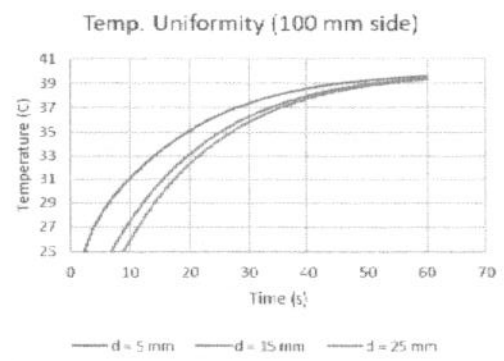

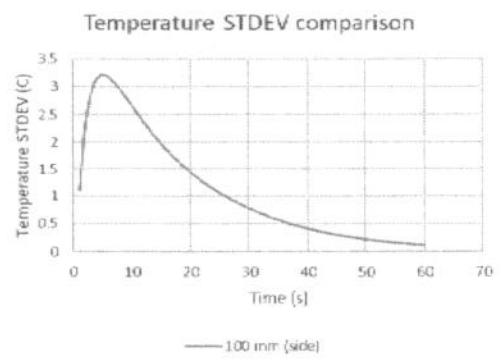

- Energy2D is used to simulate the time response of the thin heater plate to a 15°C step as function of distance from the heating element.
- <1°C STDEV temperature vs. position within 30 sec

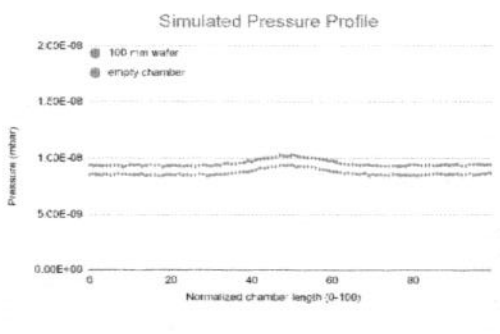

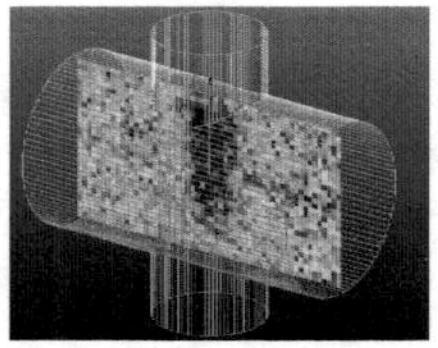

- molFlow is used to simulate the equilibrium pressure of the vacuum chamber under various conditions.
- Comparing an empty chamber vs. a "clean" 100 mm wafer sample with assumed 1×10^{-9} mbar L/cm^2 s outgassing shows a detectable signal.

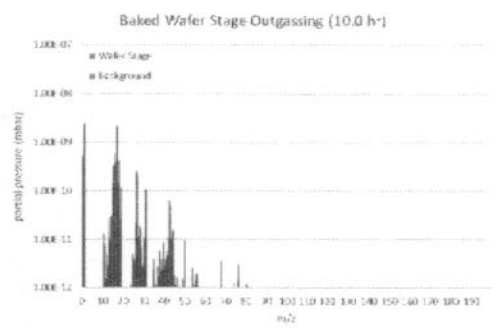

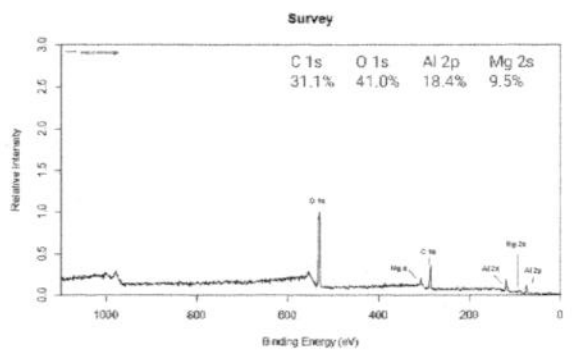

- External room temperature RGA of uninstalled stage shows it has room temperature (RT) outgassing levels indistinguishable from background.
- X-ray photoelectron spectroscopy (XPS) shows the cleaning process is effective in reducing surface carbon to ~30% and removing ionic impurities

Results

Experiments comparing the empty wafer holder and clean 100 mm wafers briefly exposed to lab atmosphere (~5 min) show detectable differences in outgassing rates for common species at 150°C.

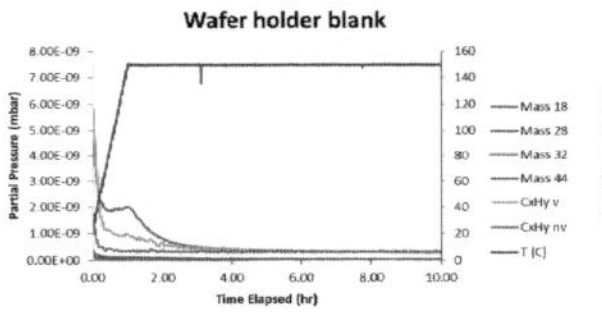

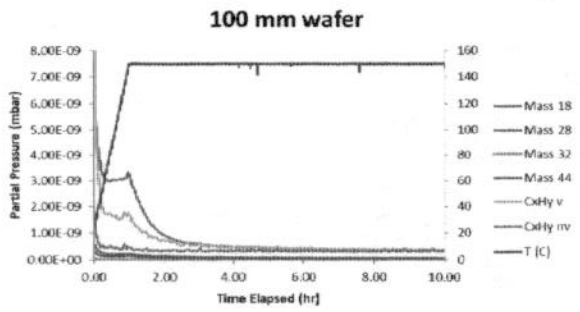

TDS results show the complete instrument is capable of detecting $\sim1\times10^{-10}$ mbar level partial pressure difference between an empty wafer holder and a clean 100 mm wafer at elevated temperature.

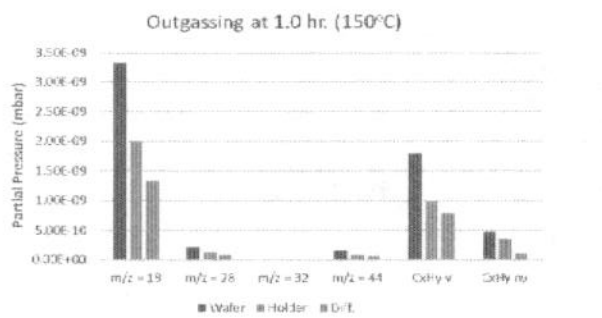

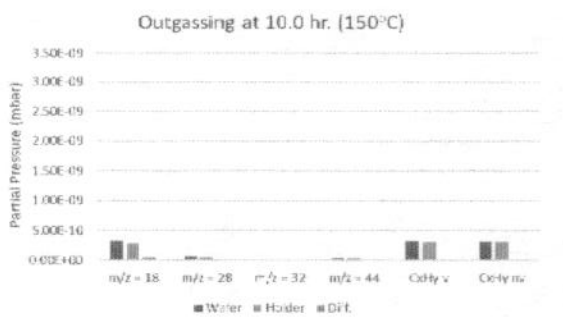

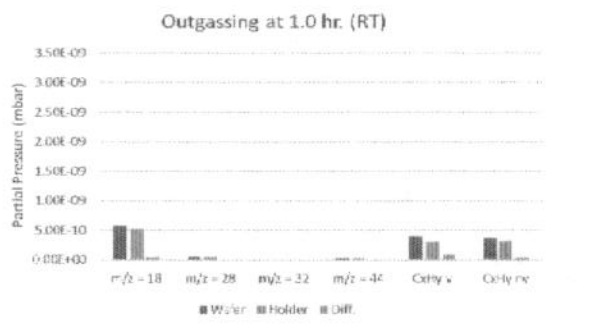

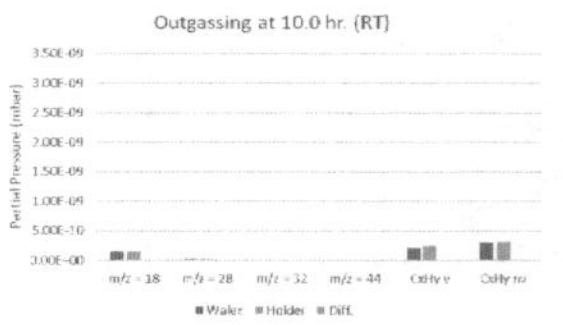

TDS shows improved distinction between clean samples compared to RT RGA analysis, with low outgassing contributed by the wafer stage itself.

Conclusion

In this work, we present the development of a UHV wafer heater stage suitable for TDS analysis through design, construction and validation. We then demonstrate the capability of the complete instrument to carry out high sensitivity outgassing experiments using TDS. Based on this experience, we believe that in addition to direct use in analysis, this stage serves as a prototype for development of larger sample stages useful for process simulation and TDS analysis of commercial sized wafers.

Conjointly address stringent device specifications, long bath life considerations and HSE/ESG standards for specific copper seed wet etching applications

TECHNIC

Dr. Jérôme DAVIOT,
Anaïs PASSELANDE & al

TechniEtch Cu-Select for selective metal etching

• OBJECTIVES

- Alternative to acidic-H_2O_2 based mixtures with:
  - Better Cu PVD/ECD selectivity
  - High Metal compatibility
  - Stable Etch Rate (ER) and low roughening
  - Higher metal loading & stability
  - CMR, bioaccumulating agent free

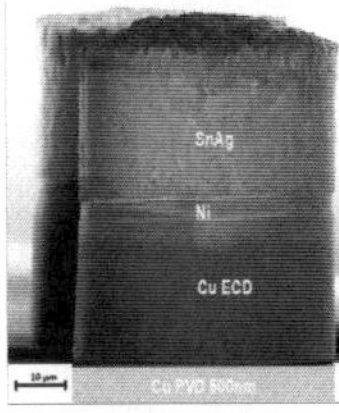

• FUNCTIONALITIES & INGREDIENTS

Functionality	Ingredient
Oxidizer	H_2O_2 (3-5wt%)
H_2O_2 stabilizer /bubble nucleation quencher	Poly-carboxylic acid $RxC(OOH)_n$
Selective Cu chelating agent	Carboxylic acid
Surface energy modulator	Anionic surfactant
Solvent	Water

• STABILITY

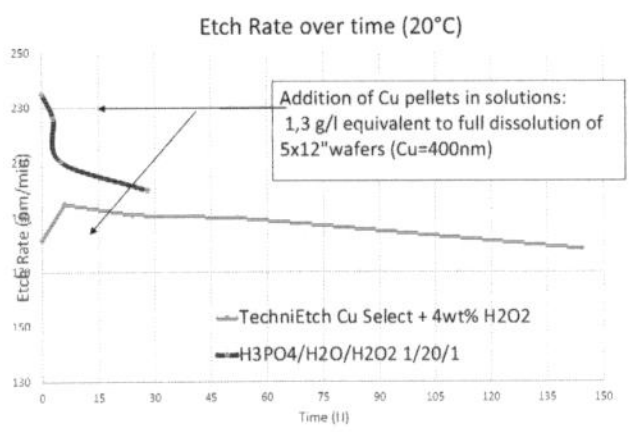

ER & H2O2 monitoring overtime

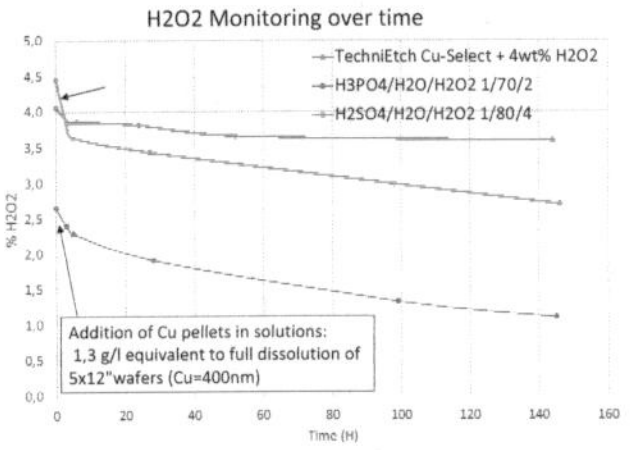

• MECHANISMS

$Cu + H_2O_2 \xrightarrow{\text{Target reaction}} Cu^{2+} + H_2O$

$Cu^{2+} + R_xC(OOH)_n \xrightarrow{\text{Stabilization}} R_xC(OO)_{n-2}Cu + 2H^+$

$Cu^{2+} + H_2O_2 \xrightarrow{\text{Undesired reaction}} Cu^+ + {}^\circ O^{2-} + 2H^+$ (Superoxide)

$M^{n+} + H_2O_2 \xrightarrow{\text{Fenton reaction}} M^{(n+1)+} + {}^\circ OH + OH^-$ (Hydroxyl radicaux)

${}^\circ O^{2-} + H_2O_2 \xrightarrow{\text{Haber Weiss reaction}} O_2 + {}^\circ OH + OH^-$

${}^\circ O^{2-} + M^{(n+1)+} \xrightarrow{\text{Metal attack}} O_2 + M^{n+}$

Cu etching, H_2O_2 decomposition & metal attack reactions

• SELECTIVITY / COMPATIBILITY

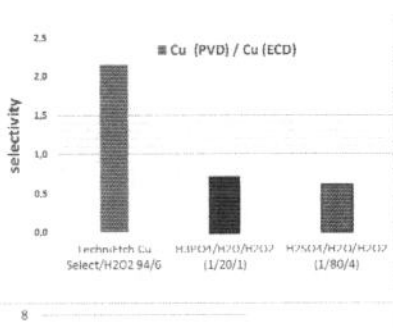

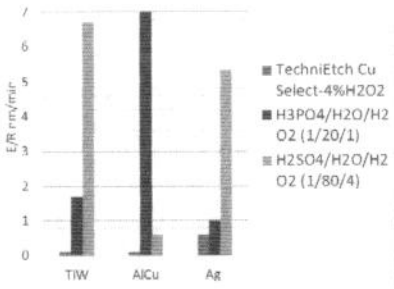

ER in nm/min @ 25C	Immersion 200RPM	Selectivity
Cu (PVD)	>180	1
Cu ECD	90-100	2
TiW	<10	>18
Sn	<2	90
Ag	<0,6	250
Al(5%Cu)	<0,2	750
Ni, Bi, In	<0,2	750
Ti/TiN/Au	<0,2	750
Ta/TaN	<0,2	750
Undoped Si, SiO2	<0>	

ER and metal selectivity of TechniEtch Cu-Select

Silver coupons immerged 2 mins with agitation.

Choosing particular chemical functionalities allow fine-tuning of:

- Surface selectivity
- Metal ion affinity
- Quenching catalytic H_2O_2 decomposition reaction

Ultimately facilitating the selective removal of Cu PVD in presence of different metals.

• INDUSTRIAL VALIDATION on PATTERNED WAFER

- Cu PVD/ECD selectivity and metal compatibility confirmed on patterned wafers.

Feature CD and undercut change in nm with an overetch of 30%

	U_r	U_l	U_h	U_l/U_r	Ni
	ECD	PVD	PVD & ECD	ECD/PVD	EVD
REFERENCE					
TechniEtch Cu Select 120s @ 20C	300	900	800	3	OK
H3PO4/H2O/H2O2 (1/20/1) 120s @ 20C	2100	500*	1500	0,2	Attack
H2SO4/H2O/H2O2 (1/80/4) 120s @ 20C	1500	500*	1700	0,3	Attack

- Examining the under-bump recess (U_r), as well as lateral and height undercut (U_l, U_h) measurements, let foresee the possibility of enhancing U_l further by reducing process time and achieving an even lower CD and U_l with TechniEtch Cu-Select.

- Through online end point detection (EPD), the process etching time can be automatically adjusted, allowing for simultaneous improvement and better control of feature undercut, formulation loading capacity and bath life.

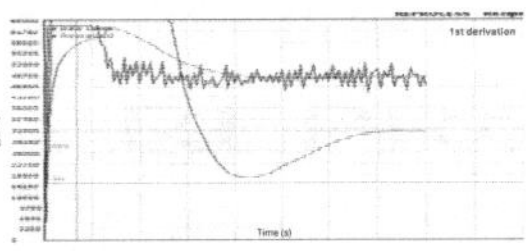

Standard EPD graph collected on Siconnex batch spray.

- TechniEtch Cu-Select successfully processed hundreds of wafers with minimal CD loss, feature undercut, Ni, Ag and Sn attack over a few days trial period.

The novel TechnicEtch Cu-Select underscored significant performance attributes including materials compatibility, Cu PVD/ECD selectivity, process stability and low CD loss. Additionally, the study provided valuable insights into extending the solution bath life through online monitoring. Finally, the team is constantly refining the processes, actively seeking further enhancements to respond to your process challenges.

SCAN ME

* Exclude the simultaneous sidewall ECD recess when determining the relative lateral undercut.

siconnex

Reduced chemical consumption and metal loss for polymer clean by batch spray

Patented by Siconnex

Contact: Siconnex, Austria
Bernhard Hammerl
bernhard.hammerl@siconnex.com
www.siconnex.com

Introduction

Every semiconductor manufacturing facility has the issue of polymers caused by dry etching processes. This poster describes how wafers can be cleaned from polymers in a batch spray cleaner using highly diluted chemicals called perc™. Optimized control of chemical concentrations enables a realiable polymer clean and significantly reduced chemical consumption. **Polymer formation:**

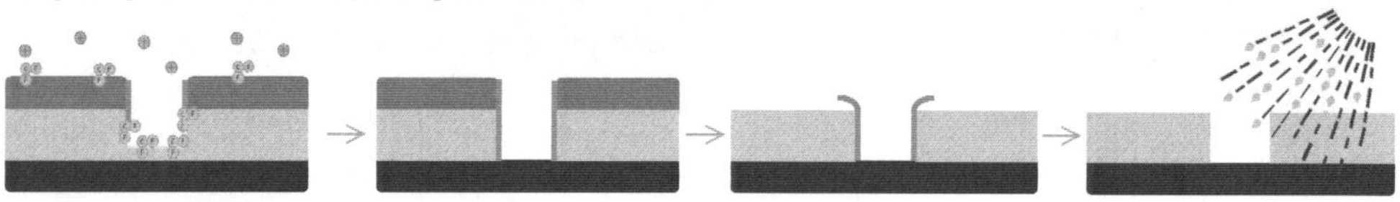

Fluorocarbons are used in plasma etching to form a polymer passivation layer, which adheres to every surface. The ion bombardement will remove this layer from the horizontal planes.

The directional ion movement will leave the polymers on the sidewalls untouched.

The polymer will hang over the structure edges after the photoresist has been removed.

Polymer removal is traditionally performed with a solvent based chemical.

Experiment

Batch spray is a technology for performing wet processes on a full lot of wafers. Either (25) 300 mm wafers, or (50) 200 mm or smaller wafers are processed within a single process step. Chemical nozzle bars are fitted with 19 spray nozzles that cover all wafers within the process chamber.

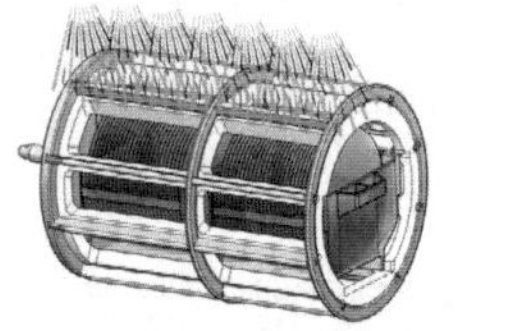

Three spiking systems are configured to hold H_2SO_4, H_2O_2 and HF. To utilize these systems, a consistent DI water flow of 10 liters per minute into the process chamber is required. Small amounts of these chemicals (5 – 120 mL per minute) are available to be injected on demand into the DI water stream.

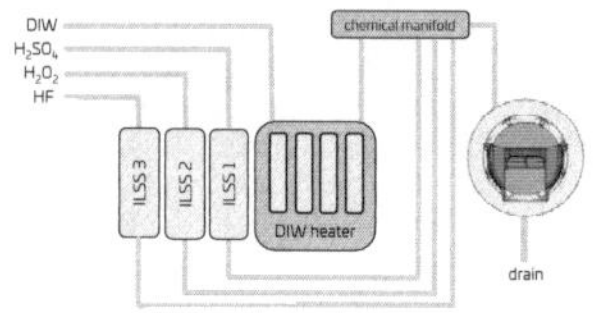

The mixture of DI water and the chemicals is then sprayed onto the wafer surface and the polymers will be removed.

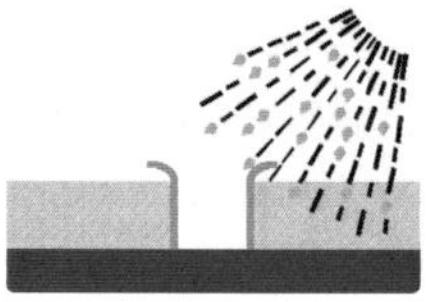

Results

Visual inspection

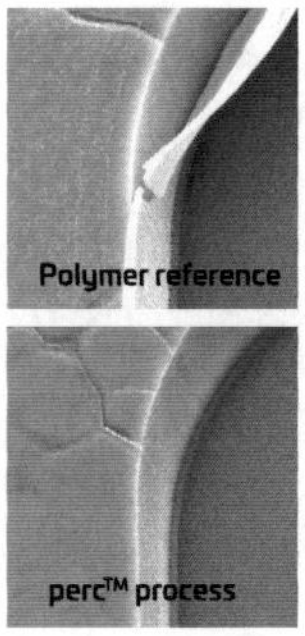

Metal loss

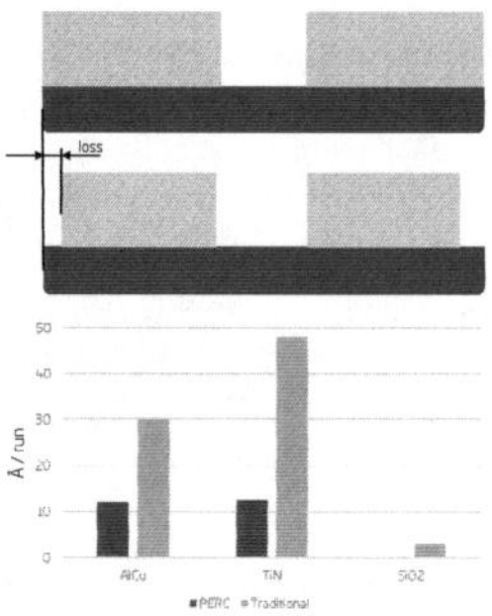

Chemical consumption

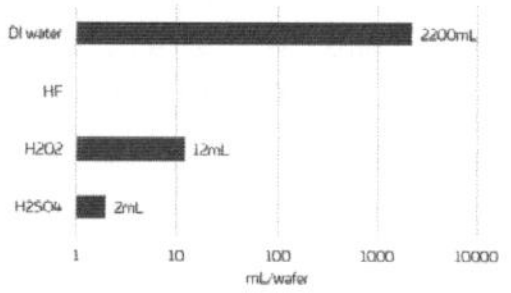

The chemical consumption values refer to a process sequence of 5minutes of perc™, followed by 5 min of rinse and 10 min of drying.

Throughput

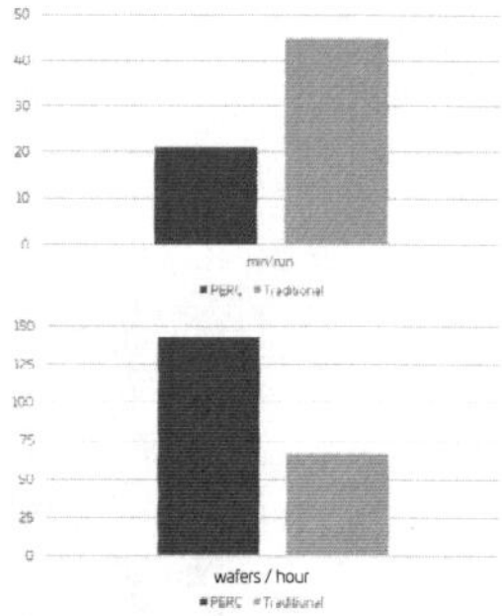

Conclusion

100% of polymers are removed.

Throughput is increased by **140%**.

Reduction of chemical consumption by **79,8%**.

Reduction of waste treatment by **99,95%**. Solvents are eliminated by **100%**.

Cost per wafer is significantly reduced by **58,3%**.

AUTHOR INDEX

Altamirano-Sanchez, E. 11
Altun, A.O. 1, 121
Alvarez, A. 198
Audouin, M. 147
Baek, S. 173
Barnett, J. 162
Bassett, D. 162
Basu, I. 81
Beebe, M. 233
Beyne, E. 51
Bhabhe, A. 81
Buford, B. 233
Chae, H. 73
Chan, A. 121
Cheng, J. 274
Chin, K.S. 90
Cho, W. 73, 127
Cho, Y. 73
Cho, Y.H. 127
Cleveland, G. 33
Coulthard, R. 1
Daviot, J. 147, 277
De Zeria, A.D. 11
Dong, R. 162
Dufek, B. 22, 113
Fisher, M. 276
Flores, N. 154
Fujita, A. 162
Gabette, L. 147
Gotts, H. 276
Ha, H. 73
Hammerl, B. 278
Hampton, K. 272
Harumoto, A. 41
Hazarika, J. 127
Henry, S.A. 275
Herek, T. 22
Hinode, T. 154
Hsu, P. 274
Huang, S. 113
Huang, W.E. 274
Jourdain, A. 51
Jung, T. 73
Kawarazaki, H. 41, 51, 186
Ke, J. 90
Ketkar, S. 22, 113
Khan, M.J. 127
Kikuta, S. 233
Kim, J. 33, 90
Kim, M. 141
Kim, T.G. 98
Kim, T.U. 98
Kim, Y. 179
Klipp, A. 62, 73, 127
Ko, C.H. 272
Ko, Y. 73
Kono, H. 233
Kuhn, M. 233
Kulkarni, N. 81
Kumar, N. 81
Lai, J.G. 41
Lai, K. 186
Lee, J. 98
Lin, W. 274
Litchey, M. 1
Liu, K. 275
Liu, O. 90
Liu, X. 154
Lo, W.S. 154
Loo, R. 51
Loup, V. 147
Loyo-Prado, J. 186
Ma, Z. 275
Magnin, L. 147
Maharjan, K. 113
Mcintosh, B. 121
McLaughlin, K.J. 203
Medikonda, M. 154
Mehrabi, K. 1
Mezhenny, S. 233
Nakanishi, M. 233
Nakano, T. 41, 51, 186
Nogami, J. 162
Oberreit, D. 222
Pacco, A. 186
Park, H. 73
Park, J.G. 98, 127, 246
Passelande, A. 277
Pignie, M.C. 147
Powell, J. 272
Ra, Y.J. 98
Rad, Z.J. 213
Ramirez, A.V. 154
Reddy, N. 11
Reynoso, S. 154
Rimal, S. 154

Rotondaro, A. 162
Sampath, S. 113
Sanchez, E.A. 41, 51, 186
Sankarapandian, M. 154
Saucedo, I. 11
Schultz, A. 22, 113
Sebaai, F. 51, 62
Siegelman, R. 33
Simms, I. 162
Slaton, M. 233
Sun, P. 274
Taddei, J. 11
Tanaka, T. 41
Thomas, J. 222
Tripp, M.K. 1
Tyler, P. 11
Ueda, D. 41
Van Schooneveld, G. 1, 121
Villanueva, F.J.L. 62
Walker, M. 222
Wang, C.W. 33
Wang, J. 275
White, V. 273
Wiederin, D.R. 22
Williams, N. 113
Wu, Y. 275
Yang, C.J. 154
Yang, J. 33, 90, 98
Yeh, S.H. 33
Zazzera, L. 1, 121
Zhang, H. 154
Zhang, K. 273
Zhang, X. 275
Zhou, H. 273
Zou, X. 276